Svetlin G. Georgiev

Tensor Calculus on Time Scales

Also of Interest

Differential and Integral Calculus
Implicit Functions, Stieltjes Integrals and Curvilinear Integrals
Svetlin G. Georgiev, Khaled Zennir, 2025
ISBN 978-3-11-914462-9, e-ISBN (PDF) 978-3-11-221808-2

Multiple Integrals in Calculus
Improper Integrals, Line Integrals, Surface Integrals
Svetlin G. Georgiev, Khaled Zennir, 2025
ISBN 978-3-11-914345-5, e-ISBN (PDF) 978-3-11-221960-7

Partial Dynamic Equations
Wave, Parabolic and Elliptic Equations on Time Scales
Svetlin G. Georgiev, 2025
ISBN 978-3-11-163551-4, e-ISBN (PDF) 978-3-11-163614-6

Differential Equations
Projector Analysis on Time Scales
Svetlin G. Georgiev, Khaled Zennir, 2024
ISBN 978-3-11-137509-0, e-ISBN (PDF) 978-3-11-137715-5

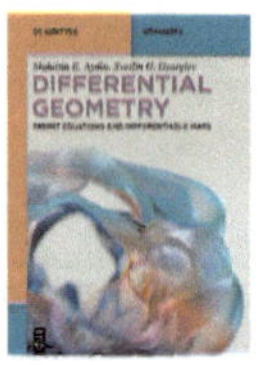

Differential Geometry
Frenet Equations and Differentiable Maps
Muhittin E. Aydin, Svetlin G. Georgiev, 2024
ISBN 978-3-11-150089-8, e-ISBN (PDF) 978-3-11-150185-7

Functional Analysis with Applications
Svetlin G. Georgiev, Khaled Zennir, 2019
ISBN 978-3-11-065769-2, e-ISBN (PDF) 978-3-11-065772-2

Svetlin G. Georgiev

Tensor Calculus on Time Scales

—

Dynamic Calculus and Riemannian Spaces

DE GRUYTER

Mathematics Subject Classification 2020
Primary: 14N07, 15A69; Secondary: 15A72, 53A45

Author
Dr. Svetlin G. Georgiev
Sorbonne University
1 Rue Victor Hugo
70005 Paris
France
svetlingeorgiev1@gmail.com

ISBN 978-3-11-222848-7
e-ISBN (PDF) 978-3-11-222849-4
e-ISBN (EPUB) 978-3-11-222850-0

Library of Congress Control Number: 2025947715

Bibliographic information published by the Deutsche Nationalbibliothek
The Deutsche Nationalbibliothek lists this publication in the Deutsche Nationalbibliografie;
detailed bibliographic data are available on the Internet at http://dnb.dnb.de.

© 2026 Walter de Gruyter GmbH, Berlin/Boston, Genthiner Straße 13, 10785 Berlin
Cover image: agsandrew / iStock / Getty Images Plus
Typesetting: VTeX UAB, Lithuania
Printing and binding: CPI books GmbH, Leck

www.degruyterbrill.com
Questions about General Product Safety Regulation:
productsafety@degruyterbrill.com

Preface

This book encompasses recent developments of tensor calculus on time scales. It is intended for use in the field of tensor calculus and dynamic calculus on time scales. It is also suitable for graduate courses in these fields. The book contains six chapters. The chapters in the book are pedagogically organized. This book is specially designed for those who wish to understand tensor calculus on time scales without having extensive mathematical background.

In Chapter 1, N-dimensional spaces, contravariant vectors, covariant vectors, and invariants are introduced. Some of their properties are deduced. Transformations of coordinates are investigated. Chapter 2 provides an informative introduction concerning the origin and nature of the tensor concept and the scope of the tensor calculus. The tensor algebra has been developed in an N-dimensional space. Contravariant, covariant, and mixed tensors of arbitrary order are defined. Some of their properties are deduced. The quotient law of the tensors is formulated and proved. Outer product and contractions are introduced. In Chapter 3, an N-dimensional Riemannian space has been chosen for the development of tensor calculus. Metric and associated tensors are defined, and some of their properties are explored. Affine and curvilinear coordinates are introduced. In Chapter 4, the Christoffel symbols of the first and second kinds are defined. Some of their basic properties are established. Covariant derivatives are introduced. The divergence, Laplace operator, and curl are defined and explored. Intrinsic differentiation is studied. Chapter 5 is devoted to the Riemann–Christoffel tensor, Ricci tensor, covariant curvature tensor, Riemann curvature, and Einsten tensor, and we deduct some of their properties. In Chapter 6, we present some applications of tensor calculus in relativistic dynamics and kinematics. Lorentz transformations are derived on arbitrary time scales. Velocity and acceleration vectors are defined and developed. Lagrange equations are deducted. Conservation laws for the energy momentum vector and angular momentum tensor are obtained.

The aim of this book is to present a clear and well-organized treatment of the concept behind the development of mathematics and solution techniques. The text material of this book is presented in highly readable, mathematically solid format. Many practical problems are illustrated displaying a wide variety of solution techniques.

The author welcomes any suggestions for the improvement of the text.

Paris, France, August 2025
Svetlin G. Georgiev

https://doi.org/10.1515/9783112228494-202

Contents

1 Introduction

In this chapter, N-dimensional spaces, contravariant vectors, covariant vectors, and invariants are introduced. Some of their properties are deduced. Transformations of coordinates are investigated.

1.1 N-Dimensional spaces

Suppose that $\mathbb{T}, \widetilde{\mathbb{T}}, \mathbb{T}_1, \ldots, \mathbb{T}_N, \mathbb{T}_{(1)}, \ldots, \mathbb{T}_{(M)}$ are time scales with forward jump operators and delta differentiation operators $\sigma, \widetilde{\sigma}, \sigma_1, \ldots, \sigma_N, \sigma_{(1)}, \ldots, \sigma_{(M)}$ and $\Delta, \widetilde{\Delta}, \Delta_1, \ldots, \Delta_N,$ $\Delta_{(1)}, \ldots, \Delta_{(M)}$, respectively.

Consider an ordered set of N variables $x^1 \in \mathbb{T}_1, \ldots, x^N \in \mathbb{T}_N$. These variables will be called the coordinates of a point. All the points corresponding to all values of the coordinates are said to form an N-dimensional space, denoted by V_N. Several or all of the coordinates may be restricted between points of V_N and the set of coordinates.

A curve in V_N is defined as the assemblage of points that satisfy N equations

$$x^j = x^j(u), \quad j \in \{1, \ldots, N\},$$

where $u \in \mathbb{T}$ is a parameter, and $x^j(u)$ are N functions of u, which obey certain continuity conditions. In general, it will be sufficient that derivatives exist up to any required order.

A subspace V_M of V_N is defined for $M < N$ as a collection of points that satisfy N equations

$$x^j = x^j(u^1, \ldots, u^M), \quad j \in \{1, \ldots, N\},$$

with M parameters $u^1 \in \mathbb{T}_{(1)}, \ldots, u^M \in \mathbb{T}_{(M)}$. Here $x^j(u^1, \ldots, u^M)$ are N functions of $u^1, \ldots, u^M$ satisfying certain conditions of continuity. In addition, the $M \times N$ matrix formed by the partial derivatives

$$\frac{\Delta_j x^j}{\Delta_{(i)} u^i} \quad \text{or} \quad \frac{\Delta_j x^j}{\Delta_{(i)} u^{i\sigma_{(0)}\sigma_{(1)}\cdots\sigma_{(i-1)}}}$$

is assumed to be of rank M. When $M = N - 1$, the subspace is called a hypersurface.

1.2 Transformations of coordinates

Now consider a space V_N with coordinate system $x^1, \ldots, x^N$. The equations

$$\overline{x}^j = \phi^j(x^1, x^2, \ldots, x^N), \quad j \in \{1, \ldots, N\}, \tag{1.1}$$

https://doi.org/10.1515/9783112228494-001

where ϕ_j are single-valued continuous differentiable functions of the coordinates, define a new coordinate system $\bar{x}^1, \bar{x}^2, \ldots, \bar{x}^N$.

Definition 1.1. Equations (1.1) are said to define a transformation of coordinates.

It is essential that the functions ϕ^j are independent. A necessary and sufficient condition is that the Jacobian determinant formed by the partial derivatives

$$\frac{\Delta_{(i)}\bar{x}^i}{\Delta_j x^j}$$

does not vanish. Under this condition, we can solve (1.1) for the functions x^j as functions of $\bar{x}^i$ and obtain

$$x^i = \psi^i(\bar{x}^1, \bar{x}^2, \ldots, \bar{x}^N), \quad i \in \{1, \ldots, N\}.$$

We will now introduce the following two conventions.
1. Latin indices, used as subscripts or superscripts, will take on all values from 1 to N unless otherwise specified. Thus, equations (1.1) are briefly written

$$\bar{x}^i = \phi^i(x^1, x^2, \ldots, x^N).$$

2. If a Latin index is repeated in a term, then it is understood that a summation with respect to that index over the range $1, 2, \ldots, N$ is implied. Thus, instead of the expression

$$\sum_{j=1}^{N} a_j x^j,$$

we merely write

$$a_j x^j.$$

Now a differentiation of (1.1) yields

$$\Delta_{(j)}\bar{x}^j = \sum_{r=1}^{N} \frac{\partial \phi^j}{\Delta_r x^r} \Delta_r x^r$$

$$= \sum_{r=1}^{N} \frac{\Delta_{(j)}\bar{x}^j}{\Delta_r x^r} \Delta_r x^r, \quad j \in \{1, \ldots, N\},$$

which, following the above conventions, simplify to

$$\Delta_{(j)}\bar{x}^j = \frac{\Delta_{(j)}\bar{x}^j}{\Delta_r x^r} \Delta_r x^r. \tag{1.2}$$

Definition 1.2. The repeated index r is called a dummy index.

Equation (1.2) can equivalently be written

$$\Delta_{(j)}\overline{x}^j = \frac{\Delta_{(j)}\overline{x}^j}{\Delta_m x^m}\Delta_m x^m$$

or

$$\Delta_{(l)}\overline{x}^l = \frac{\Delta_{(l)}\overline{x}^l}{\Delta_i x^i}\Delta_i x^i.$$

Example 1.1. In the expressions

$$a_i b^i, \quad a_i^i, \quad \text{or} \quad a_i^k x^i,$$

the index i is dummy.

Example 1.2. In the expression

$$a_{ij}x^i x^j$$

both the indices are dummy.

Definition 1.3. In an indexed expression, if an index is not dummy, then it is called a free index.

Example 1.3. In the expression

$$a_{ij}x^i,$$

the index i is dummy, and the index j is free.

To avoid a confusion, the same index must not be used more than twice in any single term. For example,

$$\left(\sum_{j=1}^{n} a_j x^j\right)^2$$

will not be written

$$a_j x^j a_j x^j,$$

but rather

$$a_l x^l a_j x^j.$$

It will always be clear from the context whether x^2 means x with superscript 2 or x squared. Usually, powers will be indicated by using brackets. Thus

$$\left(x^N\right)^2$$

means the square of x^N.

Example 1.4. Consider the time scale spherical transformation

$$\begin{aligned}
\overline{x}^1 &= x^1 \sin_\alpha(x^2, x_0^2) \cos_\beta(x^3, x_0^3), \\
\overline{x}^2 &= x^1 \sin_\alpha(x^2, x_0^2) \sin_\beta(x^3, x_0^3), \\
\overline{x}^3 &= x^1 \cos_\alpha(x^2, x_0^2),
\end{aligned}$$

where $x^1 \in \mathbb{T}_1$, $x^2, x_0^2 \in \mathbb{T}_2$, $x^3, x_0^3 \in \mathbb{T}_3$ and $\alpha, \beta \in \mathbb{R}$ are such that

$$1 + \mu_2(x^2)\alpha^2 \neq 0, \quad x^2 \in \mathbb{T}_2,$$

and

$$1 + \mu_3(x^3)\beta^2 \neq 0, \quad x^3 \in \mathbb{T}_3.$$

Figure 1.1 shows the time scale spherical transformation in the case where $\mathbb{T}_1 = \mathbb{Z}$, $\mathbb{T}_2 = \{1 - \frac{1}{n}\}_{n \in \mathbb{N}}$, and $\mathbb{T}_3 = \{1 + \frac{1}{n}\}_{n \in \mathbb{N}}$. We will find the Jacobian of this transformation. We have

$$\frac{\Delta_{(1)}\overline{x}^1}{\Delta_1 x^1} = \sin_\alpha(x^2, x_0^2) \cos_\beta(x^3, x_0^3),$$

$$\frac{\Delta_{(1)}\overline{x}^1}{\Delta_2 x^2} = \alpha x^1 \cos_\alpha(x^2, x_0^2) \cos_\beta(x^3, x_0^3),$$

$$\frac{\Delta_{(1)}\overline{x}^1}{\Delta_3 x^3} = -\beta x^1 \sin_\alpha(x^2, x_0^2) \sin_\beta(x^3, x_0^3),$$

$$\frac{\Delta_{(2)}\overline{x}^2}{\Delta_1 x^1} = \sin_\alpha(x^2, x_0^2) \sin_\beta(x^3, x_0^3),$$

$$\frac{\Delta_{(2)}\overline{x}^2}{\Delta_2 x^2} = \alpha x^1 \cos_\alpha(x^2, x_0^2) \sin_\beta(x^3, x_0^3),$$

$$\frac{\Delta_{(2)}\overline{x}^2}{\Delta_3 x^3} = \beta x^1 \sin_\alpha(x^2, x_0^2) \cos_\beta(x^3, x_0^3),$$

$$\frac{\Delta_{(3)}\overline{x}^3}{\Delta_1 x^1} = \cos_\alpha(x^2, x_0^2),$$

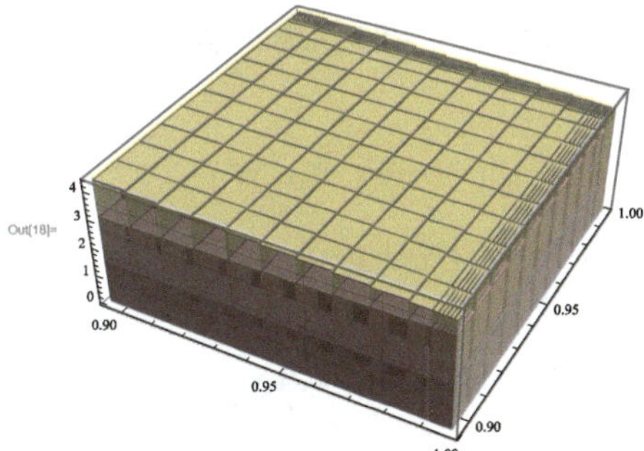

Figure 1.1: Coordinate surface of spherical coordinates.

$$\frac{\Delta_{(3)}\bar{x}^3}{\Delta_2 x^2} = -\alpha x^1 \sin_\alpha(x^2, x_0^2),$$

$$\frac{\Delta_{(3)}\bar{x}^3}{\Delta_3 x^3} = 0.$$

Then

$$J = \det \begin{pmatrix} \dfrac{\Delta_{(1)}\bar{x}^1}{\Delta_1 x^1} & \dfrac{\Delta_{(1)}\bar{x}^1}{\Delta_2 x^2} & \dfrac{\Delta_{(1)}\bar{x}^1}{\Delta_3 x^3} \\[2mm] \dfrac{\Delta_{(2)}\bar{x}^2}{\Delta_1 x^1} & \dfrac{\Delta_{(2)}\bar{x}^2}{\Delta_2 x^2} & \dfrac{\Delta_{(2)}\bar{x}^2}{\Delta_3 x^3} \\[2mm] \dfrac{\Delta_{(3)}\bar{x}^3}{\Delta_1 x^1} & \dfrac{\Delta_{(3)}\bar{x}^3}{\Delta_2 x^2} & \dfrac{\Delta_{(3)}\bar{x}^3}{\Delta_3 x^3} \end{pmatrix}$$

$$= \frac{\Delta_{(1)}\bar{x}^1}{\Delta_1 x^1}\frac{\Delta_{(2)}\bar{x}^2}{\Delta_2 x^2}\frac{\Delta_{(3)}\bar{x}^3}{\Delta_3 x^3} + \frac{\Delta_{(1)}\bar{x}^1}{\Delta_2 x^2}\frac{\Delta_{(2)}\bar{x}^2}{\Delta_3 x^3}\frac{\Delta_{(3)}\bar{x}^3}{\Delta_1 x^1}$$

$$+ \frac{\Delta_{(1)}\bar{x}^1}{\Delta_3 x^3}\frac{\Delta_{(2)}\bar{x}^2}{\Delta_1 x^1}\frac{\Delta_{(3)}\bar{x}^3}{\Delta_2 x^2} - \frac{\Delta_{(1)}\bar{x}^1}{\Delta_3 x^3}\frac{\Delta_{(2)}\bar{x}^2}{\Delta_2 x^2}\frac{\Delta_{(3)}\bar{x}^3}{\Delta_1 x^1}$$

$$- \frac{\Delta_{(1)}\bar{x}^1}{\Delta_1 x^1}\frac{\Delta_{(2)}\bar{x}^2}{\Delta_3 x^3}\frac{\Delta_{(3)}\bar{x}^3}{\Delta_2 x^2} - \frac{\Delta_{(1)}\bar{x}^1}{\Delta_2 x^2}\frac{\Delta_{(2)}\bar{x}^2}{\Delta_1 x^1}\frac{\Delta_{(3)}\bar{x}^3}{\Delta_3 x^3}$$

$$= (\sin_\alpha(x^2, x_0^2)\cos_\beta(x^3, x_0^3))(\alpha x^1 \cos_\alpha(x^2, x_0^2)\sin_\beta(x^3, x_0^3))0$$

$$+ (\alpha x^1 \cos_\alpha(x^2, x_0^2)\cos_\beta(x^3, x_0^3))(\beta x^1 \sin_\alpha(x^2, x_0^2)\cos_\beta(x^3, x_0^3))(\cos_\alpha(x^2, x_0^2))$$

$$+ (-\beta x^1 \sin_\alpha(x^2, x_0^2)\sin_\beta(x^3, x_0^3))(\sin_\alpha(x^2, x_0^2)\sin_\beta(x^3, x_0^3))(-\alpha x^1 \sin_\alpha(x^2, x_0^2))$$

$$- (-\beta x^1 \sin_\alpha(x^2, x_0^2)\sin_\beta(x^3, x_0^3))(\alpha x^1 \cos_\alpha(x^2, x_0^2)\sin_\beta(x^3, x_0^3))(\cos_\alpha(x^2, x_0^2))$$

$$- (\sin_\alpha(x^2, x_0^2)\cos_\beta(x^3, x_0^3))(\beta x^1 \sin_\alpha(x^2, x_0^2)\cos_\beta(x^3, x_0^3))(-\alpha x^1 \sin_\alpha(x^2, x_0^2))$$

$$- (\alpha x^1 \cos_\alpha(x^2, x_0^2)\cos_\beta(x^3, x_0^3))(\sin_\alpha(x^2, x_0^2)\sin_\beta(x^3, x_0^3))0$$

$$= \alpha\beta(x^1)^2 \sin_\alpha(x^2, x_0^2)(\cos_\alpha(x^2, x_0^2))^2(\cos_\beta(x^3, x_0^3))^2 + \alpha\beta(x^1)^2(\sin_\alpha(x^2, x_0^2))^2$$

$$\times (\sin_\beta(x^3, x_0^3))^2$$

$$
\begin{aligned}
&+ \alpha\beta(x^1)^2 \sin_\alpha(x^2, x_0^2)(\cos_\alpha(x^2, x_0^2))^2(\sin_\beta(x^3, x_0^3))^2 + \alpha\beta(x^1)^2(\sin_\alpha(x^2, x_0^2))^3 \\
&\quad \times (\cos_\beta(x^3, x_0^3))^2 \\
&= \alpha\beta(x^1)^2 \sin_\alpha(x^2, x_0^2)(\cos_\alpha(x^2, x_0^2))^2((\sin_\beta(x^3, x_0^3))^2 + (\cos_\beta(x^3, x_0^3))^2) \\
&\quad + \alpha\beta(x^1)^2(\sin_\alpha(x^2, x_0^2))^3((\sin_\beta(x^3, x_0^3))^2 + (\cos_\beta(x^3, x_0^3))^2) \\
&= \alpha\beta(x^1)^2 \sin_\alpha(x^2, x_0^2)((\sin_\alpha(x^2, x_0^2))^2 + (\cos_\alpha(x^2, x_0^2))^2)e_{\beta^2\mu_3}(x^3, x_0^3) \\
&= \alpha\beta(x^1)^2 \sin_\alpha(x^2, x_0^2)e_{\alpha^2\mu_2}(x^2, x_0^2)e_{\beta^2\mu_3}(x^3, x_0^3), \quad (x^1, x^2, x^3) \in \mathbb{T}_1 \times \mathbb{T}_2 \times \mathbb{T}_3,
\end{aligned}
$$

$x_0^2 \in \mathbb{T}_2, x_0^3 \in \mathbb{T}_3$.

Example 1.5. In $\mathbb{T}_{(1)} \times \mathbb{T}_{(2)}$, let a curvilinear coordinate system $\overline{x}^i$ be defined from rectangular coordinates x^j in $\mathbb{T}_1 \times \mathbb{T}_2$ by the equations

$$
\begin{aligned}
\overline{x}^1 &= x^1 x^2, \\
\overline{x}^2 &= (x^2)^2.
\end{aligned}
$$

Figure 1.2 shows the considered transformation in the case where $\mathbb{T}_1 = \mathbb{T}_2 = \mathbb{Z}$. We compute

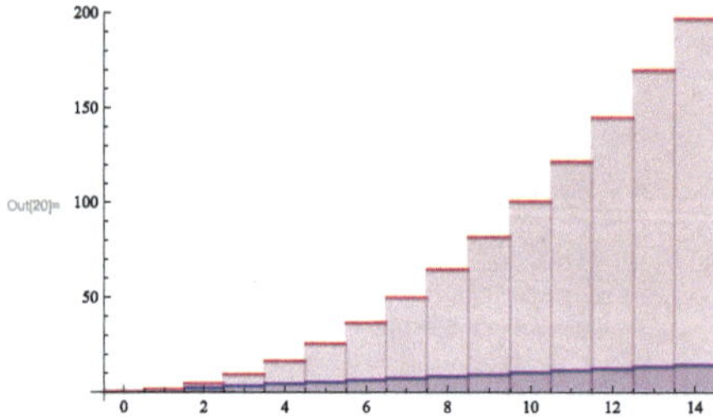

Figure 1.2: Coordinate plane of the coordinates $\overline{x}^1 = x^1 x^2, \overline{x}^2 = (x^2)^2$.

$$
\begin{aligned}
\frac{\Delta_{(1)}\overline{x}^1}{\Delta_1 x^1} &= x^2, \\[4pt]
\frac{\Delta_{(1)}\overline{x}^1}{\Delta_2 x^2} &= x^1, \\[4pt]
\frac{\Delta_{(2)}\overline{x}^2}{\Delta_1 x^1} &= 0, \\[4pt]
\frac{\Delta_{(2)}\overline{x}^2}{\Delta_2 x^2} &= \sigma_2(x^2) + x^2.
\end{aligned}
$$

Then for the Jacobian of the considered transformation, we have

$$
J = \det \begin{pmatrix} \frac{\Delta_{(1)}\overline{x}^1}{\Delta_1 x^1} & \frac{\Delta_{(1)}\overline{x}^1}{\Delta_2 x^2} \\ \frac{\Delta_{(2)}\overline{x}^2}{\Delta_1 x^1} & \frac{\Delta_{(2)}\overline{x}^2}{\Delta_2 x^2} \end{pmatrix}
$$

$$
= \frac{\Delta_{(1)}\overline{x}^1}{\Delta_1 x^1} \frac{\Delta_{(1)}\overline{x}^1}{\Delta_2 x^2} - \frac{\Delta_{(2)}\overline{x}^2}{\Delta_1 x^1} \frac{\Delta_{(2)}\overline{x}^2}{\Delta_2 x^2}
$$

$$
= x^2\big(\sigma_2(x^2) + x^2\big) + x^1 0
$$

$$
= x^2\big(\sigma_2(x^2) + x^2\big).
$$

Example 1.6. The relation between the cylindrical coordinates $\overline{x}^i$ and the rectangular coordinates x^i is given by

$$
\overline{x}^1 = x^1 \cos_1(x^2, x_0^2),
$$
$$
\overline{x}^2 = x^1 \sin_1(x^2, x_0^2),
$$
$$
\overline{x}^3 = x^3,
$$

where $x_0^2 \in \mathbb{T}_2$ is fixed. Figure 1.3 shows the time scale cylindrical transformation in the case where $\mathbb{T}_1 = \mathbb{Z}$, $\mathbb{T}_2 = \{1 - \frac{1}{n}\}_{n\in\mathbb{N}}$, and $\mathbb{T}_3 = \{1 + \frac{1}{n}\}_{n\in\mathbb{N}}$. We compute

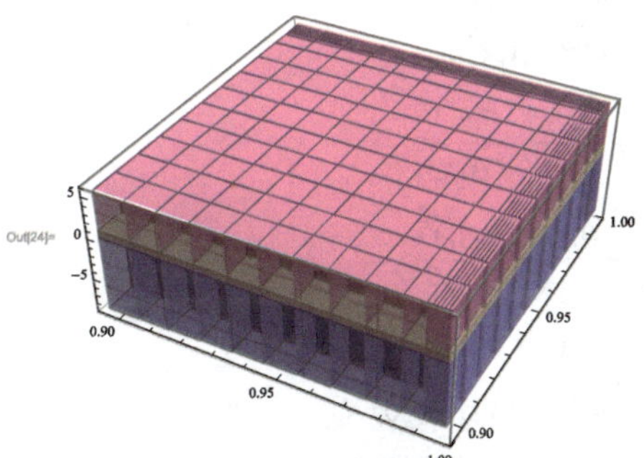

Figure 1.3: Coordinate surface of cylindrical coordinates.

$$
\frac{\Delta_{(1)}\overline{x}^1}{\Delta_1 x^1} = \cos_1(x^2, x_0^2),
$$

$$
\frac{\Delta_{(1)}\overline{x}^1}{\Delta_2 x^2} = -x^1 \sin_1(x^2, x_0^2),
$$

$$
\frac{\Delta_{(1)}\overline{x}^1}{\Delta_3 x^3} = 0,
$$

$$
\frac{\Delta_{(2)}\overline{x}^2}{\Delta_1 x^1} = \sin_1(x^2, x_0^2),
$$

$$\frac{\Delta_{(2)}\overline{x}^2}{\Delta_2 x^2} = x^1 \cos_1(x^2, x_0^2),$$

$$\frac{\Delta_{(2)}\overline{x}^2}{\Delta_3 x^3} = 0,$$

$$\frac{\Delta_{(3)}\overline{x}^3}{\Delta_1 x^1} = 0,$$

$$\frac{\Delta_{(3)}\overline{x}^3}{\Delta_2 x^2} = 0,$$

$$\frac{\Delta_{(3)}\overline{x}^3}{\Delta_3 x^3} = 1.$$

Then for the Jacobian of the considered transformation, we obtain

$$
J = \det \begin{pmatrix}
\frac{\Delta_{(1)}\overline{x}^1}{\Delta_1 x^1} & \frac{\Delta_{(1)}\overline{x}^1}{\Delta_2 x^2} & \frac{\Delta_{(1)}\overline{x}^1}{\Delta_3 x^3} \\[2mm]
\frac{\Delta_{(2)}\overline{x}^2}{\Delta_1 x^1} & \frac{\Delta_{(2)}\overline{x}^2}{\Delta_2 x^2} & \frac{\Delta_{(2)}\overline{x}^2}{\Delta_3 x^3} \\[2mm]
\frac{\Delta_{(3)}\overline{x}^3}{\Delta_1 x^1} & \frac{\Delta_{(3)}\overline{x}^3}{\Delta_2 x^2} & \frac{\Delta_{(3)}\overline{x}^3}{\Delta_3 x^3}
\end{pmatrix}
$$

$$
= \frac{\Delta_{(1)}\overline{x}^1}{\Delta_1 x^1}\frac{\Delta_{(2)}\overline{x}^2}{\Delta_2 x^2}\frac{\Delta_{(3)}\overline{x}^3}{\Delta_3 x^3} + \frac{\Delta_{(1)}\overline{x}^1}{\Delta_2 x^2}\frac{\Delta_{(2)}\overline{x}^2}{\Delta_3 x^3}\frac{\Delta_{(3)}\overline{x}^3}{\Delta_1 x^1}
$$

$$
+ \frac{\Delta_{(1)}\overline{x}^1}{\Delta_3 x^3}\frac{\Delta_{(2)}\overline{x}^2}{\Delta_1 x^1}\frac{\Delta_{(3)}\overline{x}^3}{\Delta_2 x^2} - \frac{\Delta_{(1)}\overline{x}^1}{\Delta_3 x^3}\frac{\Delta_{(2)}\overline{x}^2}{\Delta_2 x^2}\frac{\Delta_{(3)}\overline{x}^3}{\Delta_1 x^1}
$$

$$
- \frac{\Delta_{(1)}\overline{x}^1}{\Delta_1 x^1}\frac{\Delta_{(2)}\overline{x}^2}{\Delta_3 x^3}\frac{\Delta_{(3)}\overline{x}^3}{\Delta_2 x^2} - \frac{\Delta_{(1)}\overline{x}^1}{\Delta_2 x^2}\frac{\Delta_{(2)}\overline{x}^2}{\Delta_1 x^1}\frac{\Delta_{(3)}\overline{x}^3}{\Delta_3 x^3}
$$

$$
= (\cos_1(x^2, x_0^2))(x^1 \cos_1(x^2, x_0^2))1 + (-x^1 \sin_1(x^2, x_0^2))(0)(0)
$$

$$
+ (\sin_1(x^2, x_0^2))(0)(0) - (x^1 \cos_1(x^2, x_0^2))(0)(0)
$$

$$
- (\cos_1(x^2, x_0^2))(0)(0) - (-x^1 \sin_1(x^2, x_0^2))(\sin_1(x^2, x_0^2))1
$$

$$
= x^1(\cos_1(x^2, x_0^2))^2 + x^1(\sin_1(x^2, x_0^2))^2
$$

$$
= x^1((\cos_1(x^2, x_0^2))^2 + (\sin_1(x^2, x_0^2))^2)
$$

$$
= x^1 e_{\mu_2}(x^2, x_0^2).
$$

Example 1.7. The relation between the parabolic cylindrical coordinates $\overline{x}^i$ and the rectangular Cartesian coordinates x^i is given by

$$\overline{x}^1 = \frac{1}{2}((x^1)^2 - (x^2)^2),$$

$$\overline{x}^2 = x^1 x^2,$$

$$\overline{x}^3 = x^3.$$

Figure 1.4 shows the time scale parabolic cylindrical transformation in the case where $\mathbb{T}_1 = \mathbb{Z}$, $\mathbb{T}_2 = \{1 - \frac{1}{n}\}_{n\in\mathbb{N}}$, and $\mathbb{T}_3 = \{1 + \frac{1}{n}\}_{n\in\mathbb{N}}$. We compute

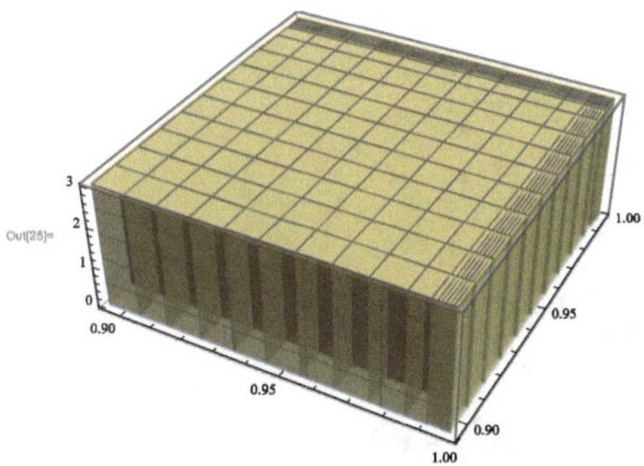

Figure 1.4: Coordinate surface of spherical coordinates.

$$\frac{\Delta_{(1)}\overline{x}^1}{\Delta_1 x^1} = \frac{1}{2}(\sigma_1(x^1) + x^1),$$

$$\frac{\Delta_{(1)}\overline{x}^1}{\Delta_2 x^2} = -\frac{1}{2}(\sigma_2(x^2) + x^2),$$

$$\frac{\Delta_{(1)}\overline{x}^1}{\Delta_3 x^3} = 0,$$

$$\frac{\Delta_{(2)}\overline{x}^2}{\Delta_1 x^1} = x^2,$$

$$\frac{\Delta_{(2)}\overline{x}^2}{\Delta_2 x^2} = x^1,$$

$$\frac{\Delta_{(2)}\overline{x}^2}{\Delta_3 x^3} = 0,$$

$$\frac{\Delta_{(3)}\overline{x}^3}{\Delta_1 x^1} = 0,$$

$$\frac{\Delta_{(3)}\overline{x}^3}{\Delta_2 x^2} = 0,$$

$$\frac{\Delta_{(3)}\overline{x}^3}{\Delta_3 x^3} = 1.$$

Then for the Jacobian of the considered transformation, we obtain

$$
J = \det \begin{pmatrix} \dfrac{\Delta_{(1)}\overline{x}^1}{\Delta_1 x^1} & \dfrac{\Delta_{(1)}\overline{x}^1}{\Delta_2 x^2} & \dfrac{\Delta_{(1)}\overline{x}^1}{\Delta_3 x^3} \\[2ex] \dfrac{\Delta_{(2)}\overline{x}^2}{\Delta_1 x^1} & \dfrac{\Delta_{(2)}\overline{x}^2}{\Delta_2 x^2} & \dfrac{\Delta_{(2)}\overline{x}^2}{\Delta_3 x^3} \\[2ex] \dfrac{\Delta_{(3)}\overline{x}^3}{\Delta_1 x^1} & \dfrac{\Delta_{(3)}\overline{x}^3}{\Delta_2 x^2} & \dfrac{\Delta_{(3)}\overline{x}^3}{\Delta_3 x^3} \end{pmatrix}
$$

$$
= \frac{\Delta_{(1)}\overline{x}^1}{\Delta_1 x^1}\frac{\Delta_{(2)}\overline{x}^2}{\Delta_2 x^2}\frac{\Delta_{(3)}\overline{x}^3}{\Delta_3 x^3} + \frac{\Delta_{(1)}\overline{x}^1}{\Delta_2 x^2}\frac{\Delta_{(2)}\overline{x}^2}{\Delta_3 x^3}\frac{\Delta_{(3)}\overline{x}^3}{\Delta_1 x^1}
$$

$$
+ \frac{\Delta_{(1)}\overline{x}^1}{\Delta_3 x^3}\frac{\Delta_{(2)}\overline{x}^2}{\Delta_1 x^1}\frac{\Delta_{(3)}\overline{x}^3}{\Delta_2 x^2} - \frac{\Delta_{(1)}\overline{x}^1}{\Delta_3 x^3}\frac{\Delta_{(2)}\overline{x}^2}{\Delta_2 x^2}\frac{\Delta_{(3)}\overline{x}^3}{\Delta_1 x^1}
$$

$$
- \frac{\Delta_{(1)}\overline{x}^1}{\Delta_1 x^1}\frac{\Delta_{(2)}\overline{x}^2}{\Delta_3 x^3}\frac{\Delta_{(3)}\overline{x}^3}{\Delta_2 x^2} - \frac{\Delta_{(1)}\overline{x}^1}{\Delta_2 x^2}\frac{\Delta_{(2)}\overline{x}^2}{\Delta_1 x^1}\frac{\Delta_{(3)}\overline{x}^3}{\Delta_3 x^3}
$$

$$
= \left(\frac{1}{2}(\sigma_1(x^1) + x^1)\right)(x^1)(1) + \left(-\frac{1}{2}(\sigma_2(x^2) + x^2)\right)(0)(0)
$$

$$
+ x^2(0)(0) - x^1(0)(0) - \left(\frac{1}{2}(\sigma_1(x^1) + x^1)\right)(0)(0)
$$

$$
- \left(-\frac{1}{2}(\sigma_2(x^2) + x^2)\right)(x^2)(1)
$$

$$
= \frac{1}{2}x^1(\sigma_1(x^1) + x^2) + \frac{1}{2}x^2(\sigma_2(x^2) + x^2).
$$

Example 1.8. The relation between the elliptic cylindrical coordinates $\overline{x}^i$ and the rectangular Cartesian coordinates x^i is given by

$$
\begin{aligned}
\overline{x}^1 &= a\,\cosh_1(x^1, x_0^1)\cos_1(x^2, x_0^2), \\
\overline{x}^2 &= a\,\sinh_1(x^1, x_0^1)\sin_1(x^2, x_0^2), \\
\overline{x}^3 &= x^3,
\end{aligned}
$$

where $x_0^1 \in \mathbb{T}_1$ and $x_0^2 \in \mathbb{T}_2$ are fixed, and a is a real constant. Figure 1.5 shows the time scale elliptic cylindrical transformation in the case where $\mathbb{T}_1 = \mathbb{Z}$, $\mathbb{T}_2 = \{1 - \frac{1}{n}\}_{n \in \mathbb{N}}$, and $\mathbb{T}_3 = \{1 + \frac{1}{n}\}_{n \in \mathbb{N}}$. We compute

$$
\frac{\Delta_{(1)}\overline{x}^1}{\Delta_1 x^1} = a\,\sinh_1(x^1, x_0^1)\cos_1(x^2, x_0^2),
$$

$$
\frac{\Delta_{(1)}\overline{x}^1}{\Delta_2 x^2} = -a\,\cosh_1(x^1, x_0^1)\sin_1(x^2, x_0^2),
$$

$$
\frac{\Delta_{(1)}\overline{x}^1}{\Delta_3 x^3} = 0,
$$

$$
\frac{\Delta_{(2)}\overline{x}^2}{\Delta_1 x^1} = a\,\cosh_1(x^1, x_0^1)\sin_1(x^2, x_0^2),
$$

$$
\frac{\Delta_{(2)}\overline{x}^2}{\Delta_2 x^2} = a\,\sinh_1(x^1, x_0^1)\cos_1(x^2, x_0^2),
$$

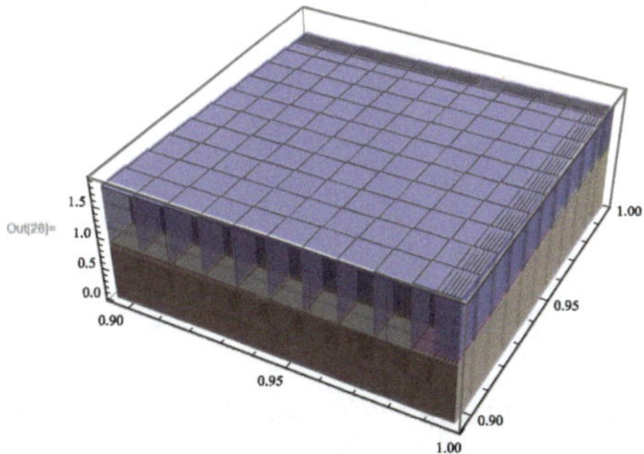

Figure 1.5: Coordinate surface of elliptic cylindrical coordinates.

$$\frac{\Delta_{(2)}\overline{x}^2}{\Delta_3 x^3} = 0,$$

$$\frac{\Delta_{(3)}\overline{x}^3}{\Delta_1 x^1} = 0,$$

$$\frac{\Delta_{(3)}\overline{x}^3}{\Delta_2 x^2} = 0,$$

$$\frac{\Delta_{(3)}\overline{x}^3}{\Delta_3 x^3} = 1.$$

Then for the Jacobian of the considered transformation, we get

$$
J = \det \begin{pmatrix}
\dfrac{\Delta_{(1)}\overline{x}^1}{\Delta_1 x^1} & \dfrac{\Delta_{(1)}\overline{x}^1}{\Delta_2 x^2} & \dfrac{\Delta_{(1)}\overline{x}^1}{\Delta_3 x^3} \\[2ex]
\dfrac{\Delta_{(2)}\overline{x}^2}{\Delta_1 x^1} & \dfrac{\Delta_{(2)}\overline{x}^2}{\Delta_2 x^2} & \dfrac{\Delta_{(2)}\overline{x}^2}{\Delta_3 x^3} \\[2ex]
\dfrac{\Delta_{(3)}\overline{x}^3}{\Delta_1 x^1} & \dfrac{\Delta_{(3)}\overline{x}^3}{\Delta_2 x^2} & \dfrac{\Delta_{(3)}\overline{x}^3}{\Delta_3 x^3}
\end{pmatrix}
$$

$$
= \frac{\Delta_{(1)}\overline{x}^1}{\Delta_1 x^1}\frac{\Delta_{(2)}\overline{x}^2}{\Delta_2 x^2}\frac{\Delta_{(3)}\overline{x}^3}{\Delta_3 x^3} + \frac{\Delta_{(1)}\overline{x}^1}{\Delta_2 x^2}\frac{\Delta_{(2)}\overline{x}^2}{\Delta_3 x^3}\frac{\Delta_{(3)}\overline{x}^3}{\Delta_1 x^1}
$$

$$
+ \frac{\Delta_{(1)}\overline{x}^1}{\Delta_3 x^3}\frac{\Delta_{(2)}\overline{x}^2}{\Delta_1 x^1}\frac{\Delta_{(3)}\overline{x}^3}{\Delta_2 x^2} - \frac{\Delta_{(1)}\overline{x}^1}{\Delta_3 x^3}\frac{\Delta_{(2)}\overline{x}^2}{\Delta_2 x^2}\frac{\Delta_{(3)}\overline{x}^3}{\Delta_1 x^1}
$$

$$
- \frac{\Delta_{(1)}\overline{x}^1}{\Delta_1 x^1}\frac{\Delta_{(2)}\overline{x}^2}{\Delta_3 x^3}\frac{\Delta_{(3)}\overline{x}^3}{\Delta_2 x^2} - \frac{\Delta_{(1)}\overline{x}^1}{\Delta_2 x^2}\frac{\Delta_{(2)}\overline{x}^2}{\Delta_1 x^1}\frac{\Delta_{(3)}\overline{x}^3}{\Delta_3 x^3}
$$

$$
= (a\sinh_1(x^1,x_0^1)\cos_1(x^2,x_0^2))(a\sinh_1(x^1,x_0^1)\cos_1(x^2,x_0^2))(1)
$$

$$
+ (-a\cosh_1(x^1,x_0^1)\sin_1(x^2,x_0^2))(0)(0) + (a\cosh_1(x^1,x_0^1)\sin_1(x^2,x_0^2))(0)(0)
$$

$$
- (a\sinh_1(x^1,x_0^1)\cos_1(x^2,x_0^2))(0)(0) - (a\sinh_1(x^1,x_0^1)\cos_1(x^2,x_0^2))(0)(0)
$$

$$
- (-a\cosh_1(x^1,x_0^1)\sin_1(x^2,x_0^2))(a\cosh_1(x^1,x_0^1)\sin_1(x^3,x_0^3))(1)
$$

$$
= a^2(\sinh_1(x^1,x_0^1))^2(\cos_1(x^2,x_0^2))^2 + a^2(\cosh_1(x^1,x_0^1))^2(\sin_1(x^2,x_0^2))^2
$$

$$
\begin{aligned}
&= a^2 \left(\sinh_1(x^1, x_0^1)\right)^2 \left(e_{\mu_2}(x^2, x_0^2) - \left(\sin_1(x^2, x_0^2)\right)^2\right) \\
&\quad + a^2 \left(\cosh_1(x^1, x_0^1)\right)^2 \left(\sin_1(x^2, x_0^2)\right)^2 \\
&= a^2 \left(\sinh_1(x^1, x_0^1)\right)^2 e_{\mu_2}(x^2, x_0^2) \\
&\quad + a^2 \left(\left(\cosh_1(x^1, x_0^1)\right)^2 - \left(\sinh_1(x^1, x_0^1)\right)^2\right)\left(\sin_1(x^2, x_0^2)\right)^2 \\
&= a^2 \left(\sinh_1(x^1, x_0^1)\right)^2 e_{\mu_2}(x^2, x_0^2) + a^2 e_{-\mu_1}(x^1, x_0^1)\left(\sin_1(x^2, x_0^2)\right)^2.
\end{aligned}
$$

Example 1.9. The relation between the paraboidal coordinates $\overline{x}^i$ and the rectangular Cartesian coordinates x^i is given by

$$
\begin{aligned}
\overline{x}^1 &= x^1 x^2 \cos_1(x^3, x_0^3), \\
\overline{x}^2 &= x^1 x^2 \sin_1(x^3, x_0^3), \\
\overline{x}^3 &= \frac{1}{2}\left((x^1)^2 - (x^2)^2\right),
\end{aligned}
$$

where $x_0^3 \in \mathbb{T}_3$ is fixed. Figure 1.6 shows the time scale paraboidal transformation in the case where $\mathbb{T}_1 = \mathbb{Z}$, $\mathbb{T}_2 = \{1 - \frac{1}{n}\}_{n \in \mathbb{N}}$, and $\mathbb{T}_3 = \{1 + \frac{1}{n}\}_{n \in \mathbb{N}}$. We compute

$$
\frac{\Delta_{(1)}\overline{x}^1}{\Delta_1 x^1} = x^2 \cos_1(x^3, x_0^3),
$$

$$
\frac{\Delta_{(1)}\overline{x}^1}{\Delta_2 x^2} = x^1 \cos_1(x^3, x_0^3),
$$

$$
\frac{\Delta_{(1)}\overline{x}^1}{\Delta_3 x^3} = -x^1 x^2 \sin_1(x^3, x_0^3),
$$

$$
\frac{\Delta_{(2)}\overline{x}^2}{\Delta_1 x^1} = x^2 \sin_1(x^3, x_0^3),
$$

$$
\frac{\Delta_{(2)}\overline{x}^2}{\Delta_2 x^2} = x^1 \sin_1(x^3, x_0^3),
$$

$$
\frac{\Delta_{(2)}\overline{x}^2}{\Delta_3 x^3} = x^1 x^2 \sin_1(x^3, x_0^3),
$$

$$
\frac{\Delta_{(3)}\overline{x}^3}{\Delta_1 x^1} = \frac{1}{2}\left(\sigma_1(x^1) + x^1\right),
$$

$$
\frac{\Delta_{(3)}\overline{x}^3}{\Delta_2 x^2} = -\frac{1}{2}\left(\sigma_2(x^2) + x^2\right),
$$

$$
\frac{\Delta_{(3)}\overline{x}^3}{\Delta_3 x^3} = 0.
$$

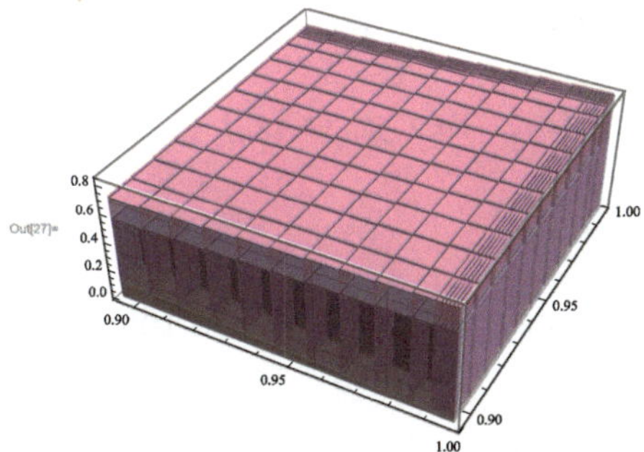

Figure 1.6: Coordinate surface of paraboidal coordinates.

Then for the Jacobian J of the considered transformation, we find

$$
J = \det \begin{pmatrix} \dfrac{\Delta_{(1)}\overline{x}^1}{\Delta_1 x^1} & \dfrac{\Delta_{(1)}\overline{x}^1}{\Delta_2 x^2} & \dfrac{\Delta_{(1)}\overline{x}^1}{\Delta_3 x^3} \\[2mm] \dfrac{\Delta_{(2)}\overline{x}^2}{\Delta_1 x^1} & \dfrac{\Delta_{(2)}\overline{x}^2}{\Delta_2 x^2} & \dfrac{\Delta_{(2)}\overline{x}^2}{\Delta_3 x^3} \\[2mm] \dfrac{\Delta_{(3)}\overline{x}^3}{\Delta_1 x^1} & \dfrac{\Delta_{(3)}\overline{x}^3}{\Delta_2 x^2} & \dfrac{\Delta_{(3)}\overline{x}^3}{\Delta_3 x^3} \end{pmatrix}
$$

$$
= \frac{\Delta_{(1)}\overline{x}^1}{\Delta_1 x^1}\frac{\Delta_{(2)}\overline{x}^2}{\Delta_2 x^2}\frac{\Delta_{(3)}\overline{x}^3}{\Delta_3 x^3} + \frac{\Delta_{(1)}\overline{x}^1}{\Delta_2 x^2}\frac{\Delta_{(2)}\overline{x}^2}{\Delta_3 x^3}\frac{\Delta_{(3)}\overline{x}^3}{\Delta_1 x^1}
$$

$$
+ \frac{\Delta_{(1)}\overline{x}^1}{\Delta_3 x^3}\frac{\Delta_{(2)}\overline{x}^2}{\Delta_1 x^1}\frac{\Delta_{(3)}\overline{x}^3}{\Delta_2 x^2} - \frac{\Delta_{(1)}\overline{x}^1}{\Delta_3 x^3}\frac{\Delta_{(2)}\overline{x}^2}{\Delta_2 x^2}\frac{\Delta_{(3)}\overline{x}^3}{\Delta_1 x^1}
$$

$$
- \frac{\Delta_{(1)}\overline{x}^1}{\Delta_1 x^1}\frac{\Delta_{(2)}\overline{x}^2}{\Delta_3 x^3}\frac{\Delta_{(3)}\overline{x}^3}{\Delta_2 x^2} - \frac{\Delta_{(1)}\overline{x}^1}{\Delta_2 x^2}\frac{\Delta_{(2)}\overline{x}^2}{\Delta_1 x^1}\frac{\Delta_{(3)}\overline{x}^3}{\Delta_3 x^3}
$$

$$
= (x^2 \cos_1(x^3, x_0^3))(x^1 \sin_1(x^3, x_0^3))(0)
$$

$$
+ (x^1 \cos_1(x^3, x_0^3))(x^1 x^2 \cos_1(x^3, x_0^3))\left(\frac{1}{2}(\sigma_1(x^1) + x^1)\right)
$$

$$
+ (x^2 \sin_1(x^3, x_0^3))\left(-\frac{1}{2}(\sigma_2(x^2) + x^2)\right)(-x^1 x^2 \sin_1(x^3, x_0^3))
$$

$$
- \left(\frac{1}{2}(\sigma_1(x^1) + x^1)\right)(x^1 \sin_1(x^3, x_0^3))(-x^1 x^2 \sin_1(x^3, x_0^3))
$$

$$
- \left(-\frac{1}{2}(\sigma_2(x^2) + x^2)\right)(x^1 x^2 \cos_1(x^3, x_0^3))(x^2 \cos_1(x^3, x_0^3))
$$

$$
- (x^2 \sin_1(x^3, x_0^3))(x^1 \cos_1(x^3, x_0^3))(0)
$$

$$
= \frac{1}{2}(x^1)^2 x^2 (\sigma_1(x^1) + x^1)(\cos_1(x^3, x_0^3))^2 + \frac{1}{2}x^1(x^2)^2(\sin_1(x^3, x_0^3))^2
$$

$$
+ \frac{1}{2}(x^1)^2 x^2 (\sigma_1(x^1) + x^1)(\sin_1(x^3, x_0^3))^2 + \frac{1}{2}x^1(x^2)^2(\sigma_2(x^2) + x^2)(\cos_1(x^3, x_0^3))^2
$$

$$
= \frac{1}{2}(x^1)^2 x^2 (\sigma_1(x^1) + x^1)\left((\cos_1(x^3, x_0^3))^2 + (\sin_1(x^3, x_0^3))^2\right)
$$

$$+ \frac{1}{2}x^1(x^2)^2(\sigma_2(x^2) + x^2)((\cos_1(x^3, x_0^3))^2 + (\sin_1(x^3, x_0^3))^2)$$

$$= \frac{1}{2}(x^1)^2 x^2(\sigma_1(x^1) + x^1)e_{\mu_3}(x^3, x_0^3) + \frac{1}{2}x^1(x^2)^2(\sigma_2(x^2) + x^2)e_{\mu_3}(x^3, x_0^3)$$

$$= \frac{1}{2}x^1 x^2(x^1(\sigma_1(x^1) + x^1) + x^2(\sigma_2(x^2) + x^2))e_{\mu_3}(x^3, x_0^3).$$

Exercise 1.1. Suppose that the following transformation connects x^i to $\bar{x}^i$ coordinate system:

$$\bar{x}^1 = e_2(x^1, x_0^1)e_{-2}(x^2, x_0^2),$$
$$\bar{x}^2 = e_1(x^1, x_0^1)e_{-3}(x^2, x_0^2),$$

where $x_0^1 \in \mathbb{T}_1$ and $x_0^2 \in \mathbb{T}_2$ are fixed, provided that

$$1 - 2\mu_2(x^2) \neq 0,$$
$$1 - 3\mu_2(x^2) \neq 0, \quad x^2 \in \mathbb{T}_2.$$

Find the Jacobian of this transformation.

Answer 1.1.

$$-6e_2(x^1, x_0^1)e_{-3}(x^2, x_0^2) - 2e_1(x^1, x_0^1)e_{-2}(x^2, x_0^2).$$

Definition 1.4. By a system of order zero we will mean a single quantity having no index.

Example 1.10. For example, A is a system of order zero.

Definition 1.5. The upper and lower indices of a system are called its indices of contravariance and covariance, respectively.

Example 1.11. For the system

$$A^i_{jk},$$

the index i is the index of contravariance, and the indices j, k are indices of covariance.

Definition 1.6. Accordingly, A^{ij} is called a contravariant system, the system A_{ij} is called a covariant system, and the system A^i_j is called a mixed system.

Similarly, according to the above convention, the expression

$$\sum_{i=1}^{N}\sum_{j=1}^{N} a_{ij}x^i x^j$$

will be written as

$$a_{ij}x^i x^j$$

in the N-dimensional space.

Remark 1.1. Note the following nonidentities:

$$a_{ij}(x^i + y^j) \neq a_{ij}x^i + a_{ij}y^j,$$
$$a_{ij}x^i y^j \neq a_{ij}y^i x^j,$$

and

$$(a_{ij} + a_{ji})x^i y^j \neq 2a_{ij}x^i y^j.$$

Remark 1.2. Note the following identities:

$$a_{ij}(x^j + y^j) = a_{ij}x^j + a_{ij}y^j,$$
$$a_{ij}x^i x^j = a_{ji}x^i x^j,$$
$$a_{ij}x^i y^j = a_{ij}y^j x^i,$$
$$(a_{ij} + a_{ji})x^i x^j = a_{ij}x^i x^j + a_{ji}x^i x^j$$
$$= a_{ij}x^i x^j + a_{ij}x^i x^j$$
$$= 2a_{ij}x^i x^j$$
$$= a_{ji}x^i x^j + a_{ji}x^i x^j$$
$$= 2a_{ji}x^i x^j,$$

and

$$(a_{ij} - a_{ji})x^i x^j = a_{ij}x^i x^j - a_{ji}x^i x^j$$
$$= a_{ij}x^i x^j - a_{ij}x^i x^j$$
$$= 0.$$

Example 1.12. We will express the sum

$$\sum_{i=1}^{N}\sum_{j=1}^{N} a_{ij}u^i v^j$$

by using the summation convention, and then we will find all the terms of the sum in which each of indices takes values from 1 to 3.

Since each of the indices i and j occurs twice, once as lower and again as an upper index, the required expression is

$$a_{ij}u^i v^j.$$

Since both indices are dummy, we have

$$
\begin{aligned}
a_{ij}u^i v^j &= \sum_{i=1}^{3}\sum_{j=1}^{3} a_{ij}u^i v^j \\
&= \sum_{i=1}^{3}\left(a_{i1}u^i v^1 + a_{i2}u^i v^2 + a_{i3}u^i v^3\right) \\
&= \sum_{i=1}^{3} a_{i1}u^i v^1 + \sum_{i=1}^{3} a_{i2}u^i v^2 + \sum_{i=1}^{3} a_{i3j}u^i v^3 \\
&= \left(a_{11}u^1 v^1 + a_{21}u^2 v^1 + a_{31}u^3 v^1\right) + \left(a_{12}u^1 v^2 + a_{22}u^2 v^2 + a_{32}u^3 v^2\right) \\
&\quad + \left(a_{13}u^1 v^3 + a_{23}u^2 v^3 + a_{33}u^3 v^3\right) \\
&= a_{11}u^1 v^1 + a_{22}u^2 v^2 + a_{33}u^3 v^3 + a_{12}u^1 v^2 + a_{21}u^2 v^1 \\
&\quad + a_{13}u^1 v^3 + a_{31}u^3 v^1 + a_{23}u^2 v^3 + a_{32}u^3 v^2.
\end{aligned}
$$

Example 1.13. If a_{ij} are constants, then we will compute

$$
\frac{\partial}{\Delta_k x^k}\left(a_{ij}x^i x^j\right).
$$

Returning to the summation convention, we have

$$
\begin{aligned}
a_{ij}x^i x^j &= \sum_{i=1}^{N}\sum_{j=1}^{N} a_{ij}x^i x^j \\
&= \sum_{i=1,\,i\neq k}^{N}\sum_{j=1,\,j\neq k}^{N} a_{ij}x^i x^j + \sum_{j=1}^{N} a_{kj}x^k x^j + \sum_{i=1}^{N} a_{ik}x^i x^k \\
&= \sum_{i=1,\,i\neq k}^{N}\sum_{j=1,\,j\neq k}^{N} a_{ij}x^i x^j + \left(\sum_{j=1}^{N} a_{kj}x^j\right)x^k + \left(\sum_{i=1}^{N} a_{ik}x^i\right)x^k.
\end{aligned}
$$

Hence

$$
\begin{aligned}
\frac{\partial}{\Delta_k x^k}\left(a_{ij}x^i x^j\right) &= \sum_{j=1}^{N} a_{kj}x^j + \sum_{i=1}^{N} a_{ik}x^i \\
&= a_{kj}x^j + a_{ik}x^i,
\end{aligned}
$$

where we return to the summation convention.

Now we introduce the Kronecker delta

$$
\delta_{jk} = \delta^{jk} = \delta_j^k =
\begin{cases}
1 & \text{if } j = k, \\
0 & \text{if } j \neq k.
\end{cases}
$$

Some of the properties of the Kronecker delta are as follows.

1. $\delta_j^k A^j = A^k$. Indeed, we have

$$
\delta_j^k A^j = \sum_{j=1}^{N} \delta_j^k A^j
$$
$$
= \delta_1^k A^1 + \cdots + \delta_{k-1}^k A^{k-1} + \delta_k^k A^k + \delta_{k+1}^k A^{k+1} + \cdots + \delta_N^k A^N
$$
$$
= 0 + \cdots + 0 + 1 + 0 + \cdots + 0
$$
$$
= 1.
$$

2. The next property of the Kronecker delta reads

$$
\frac{\Delta_k x^k}{\Delta_j x^j} = \delta_j^k
$$

because the coordinates x^j are independent.

3. Note that

$$
\delta_j^i \delta_k^j = \delta_k^i.
$$

Indeed, we have

$$
\delta_j^i \delta_k^j = \delta_1^i \delta_k^1 + \delta_2^i \delta_k^2 + \cdots + \delta_{i-1}^i \delta_k^{i-1} + \delta_i^i \delta_k^i + \delta_{i+1}^i \delta_k^{i+1} + \cdots + \delta_N^i \delta_k^N
$$
$$
= 0\delta_k^1 + 0\delta_k^2 + \cdots + 0\delta_k^{i-1} + 1\delta_k^i + 0\delta_k^{i+1} + \cdots + 0\delta_k^N
$$
$$
= \delta_k^i.
$$

In particular, when $k = i$, we get

$$
\delta_j^i \delta_i^j = \delta_i^i
$$
$$
= \delta_1^1 + \delta_2^2 + \cdots + \delta_N^N
$$
$$
= \underbrace{1 + 1 + \cdots + 1}_{N}
$$
$$
= N.
$$

4. Using the definition of δ_j^i, we obtain

$$
\frac{\Delta_k x^k}{\Delta_i x^i} \frac{\Delta_i x^i}{\Delta_j x^j} = \delta_i^k \delta_j^i
$$
$$
= \delta_j^k
$$
$$
= \frac{\Delta_k x^k}{\Delta_j x^j}.
$$

Example 1.14. Let

$$A^i = g^i_r a_{rs} y_s,$$
$$y_i = b_{ir} x_r,$$

and

$$a_{ir} b_{rj} = \delta^i_j.$$

We will find A^i in terms of x_r.

We have

$$A^i = g^i_r a_{rs} y_s$$
$$= g^i_r a_{rs} b_{st} x_t$$
$$= g^i_r \delta^r_t x_t$$
$$= g^i_r x_r.$$

Example 1.15. We will compute

$$\delta^i_j x_i x_j$$

when $N = 3$.

Using the summation convention, we obtain

$$\delta^i_j x_i x_j = \delta^i_1 x_i x_1 + \delta^i_2 x_i x_2 + \delta^i_3 x_i x_3$$
$$= (\delta^1_1 x_1 + \delta^2_1 x_2 + \delta^3_1 x_3) x_1 + (\delta^1_2 x_1 + \delta^2_2 x_2 + \delta^3_2 x_3) x_2$$
$$+ (\delta^1_3 x_1 + \delta^2_3 x_2 + \delta^3_3 x_3) x_3$$
$$= (x_1)^2 + (x_2)^2 + (x_3)^2.$$

In general, we have

$$\delta^i_j x_i x_j = x_i x_i$$

and

$$\delta^r_j a_{ir} x_i = a_{ij} x_i.$$

Exercise 1.2.

1. If

$$x^i = a^i_p y^p,$$
$$y^i = b^i_q z^q,$$

write x^i in terms of z^p.

2. If

$$b^{ij} c_{ik} = \delta^j_k$$

and

$$y_i = c_{ij} x_j,$$

write

$$b^{ij} y_i y_j$$

in terms of x_i.

Answer 1.2.
1. $x^i = z^i$.
2. $b^{ij} y_i y_j = c_{jl} x_j x_l$.

1.3 *e*-Systems

In this section, we define two completely skew-symmetric systems.

Definition 1.7. Define

$$e^{j_1 \cdots j_N} = \begin{cases} 1 & \text{if } j_1 \cdots j_N \text{ is an even permutation of } j_1 \cdots j_N, \\ -1 & \text{if } j_1 \cdots j_N \text{ is an odd permutation of } j_1 \cdots j_N, \\ 0 & \text{if any pair of indices equal to each other} \end{cases}$$

and

$$e_{j_1 \cdots j_N} = \begin{cases} 1 & \text{if } j_1 \cdots j_N \text{ is an even permutation of } j_1 \cdots j_N, \\ -1 & \text{if } j_1 \cdots j_N \text{ is an odd permutation of } j_1 \cdots j_N, \\ 0 & \text{if any pair of indices equal to each other.} \end{cases}$$

Example 1.16. For $N = 2$, we have

$$e_{11} = 0,$$
$$e_{12} = 1,$$
$$e_{21} = -1,$$
$$e_{22} = 0.$$

Example 1.17. Let us now write down all six permutations of 123:

$$
\begin{aligned}
(123) \quad & \text{zeroth permutation,} \\
(132) \quad & \text{first permutation,} \\
(312) \quad & \text{second permutation,} \\
(321) \quad & \text{third permutation,} \\
(231) \quad & \text{fourth permutation,} \\
(213) \quad & \text{fifth permutation.}
\end{aligned}
$$

Note that (123), (312), and (231) are even permutations, and (132), (321), and (231) are odd permutations. Therefore

$$
\begin{aligned}
e^{123} &= e^{312} \\
&= e^{231} \\
&= 1
\end{aligned}
$$

and

$$
\begin{aligned}
e^{132} &= e^{321} \\
&= e^{231} \\
&= -1.
\end{aligned}
$$

We will now establish some results using e-systems of second and third orders.

1. According to the definition of determinants, we have

$$
\begin{aligned}
\det(a_j^i) = |a_j^i| \\
= \begin{vmatrix} a_1^1 & a_2^1 \\ a_1^2 & a_2^2 \end{vmatrix} \\
= a_1^1 a_2^2 - a_1^2 a_2^1 \\
= e_{12} a_1^1 a_2^2 + e_{21} a_1^2 a_2^1.
\end{aligned}
$$

Now using the summation convention, we may write

$$
|a_j^i| = e_{ij} a_1^i a_2^j.
$$

Similarly, we have

$$
\begin{aligned}
|a_j^i| &= e^{12} a_1^1 a_2^2 + e^{21} a_2^1 a_1^2 \\
&= e^{ij} a_i^1 a_j^2.
\end{aligned}
$$

Let us now consider the expression

$$e_{ij}a_p^i a_q^j,$$

where the indices p and q are free and can be assigned the values 1 and 2 at will. We have

$$
\begin{aligned}
e_{ij}a_1^i a_2^j &= e_{i1}a_1^i a_2^1 + e_{i2}a_1^i a_2^2 \\
&= (e_{11}a_1^1 + e_{21}a_1^2)a_2^1 + (e_{12}a_1^1 + e_{22}a_1^2)a_2^2 \\
&= -a_1^2 a_2^1 + a_1^1 a_2^2 \\
&= e_{12}a_1^1 a_2^2 + e_{21}a_1^2 a_2^1 \\
&= e_{12}a_1^1 a_2^2 - e_{12}a_1^2 a_2^1 \\
&= e_{12}(a_1^1 a_2^2 - a_1^2 a_2^1) \\
&= e_{12}|a_j^i|.
\end{aligned}
$$

Next,

$$
\begin{aligned}
e_{ij}a_2^i a_1^j &= e_{i1}a_2^i a_1^1 + e_{i2}a_2^i a_1^2 \\
&= (e_{11}a_2^1 + e_{21}a_2^2)a_1^1 + (e_{12}a_2^1 + e_{22}a_2^2)a_1^2 \\
&= e_{21}a_2^2 a_1^1 + e_{12}a_2^1 a_1^2 \\
&= e_{21}a_1^1 a_2^2 - e_{21}a_2^1 a_1^2 \\
&= e_{21}(a_1^1 a_2^2 - a_2^1 a_1^2) \\
&= e_{21}|a_j^i|.
\end{aligned}
$$

From these two results we can write

$$e_{ij}a_p^i a_q^j = e_{pq}|a_j^i|.$$

Exercise 1.3. Prove that

$$e^{ij}a_i^p a_j^q = e^{pq}|a_j^i|.$$

2. Consider the case $N = 3$. We have

$$
\begin{aligned}
|a_j^i| &= \begin{vmatrix} a_1^1 & a_2^1 & a_3^1 \\ a_1^2 & a_2^2 & a_3^2 \\ a_1^3 & a_2^3 & a_3^3 \end{vmatrix} \\
&= a_1^1 a_2^2 a_3^3 + a_2^1 a_3^2 a_1^3 + a_3^1 a_1^2 a_2^3 \\
&\quad - a_3^1 a_2^2 a_1^3 - a_1^1 a_3^2 a_2^3 - a_2^1 a_1^2 a_3^3 \\
&= e^{123}a_1^1 a_2^2 a_3^3 + e^{231}a_2^1 a_3^2 a_1^3 + e^{312}a_3^1 a_1^2 a_2^3 \\
&\quad + e^{321}a_3^1 a_2^2 a_1^3 + e^{132}a_1^1 a_3^2 a_2^3 + e^{213}a_2^1 a_1^2 a_3^3.
\end{aligned}
$$

Now applying the summation convention, we arrive at

$$\left| a_j^i \right| = e^{ijk} a_i^1 a_j^2 a_k^3.$$

Moreover,

$$
\begin{aligned}
\left| a_j^i \right| &= a_1^1 a_2^2 a_3^3 + a_1^3 a_2^1 a_3^2 + a_1^2 a_2^3 a_3^1 \\
&\quad - a_1^3 a_2^2 a_3^1 - a_1^1 a_2^3 a_3^2 - a_1^2 a_2^1 a_3^3 \\
&= e_{123} a_1^1 a_2^2 a_3^3 + e_{312} a_1^3 a_2^1 a_3^2 + e_{231} a_1^2 a_2^3 a_3^1 \\
&\quad + e_{321} a_1^3 a_2^2 a_3^1 + e_{132} a_1^1 a_2^3 a_3^2 + e_{213} a_1^2 a_2^1 a_3^3 \\
&= e_{ijk} a_1^i a_2^j a_3^k.
\end{aligned}
$$

Now consider the expression

$$e_{ijk} a_p^i a_q^j a_r^k,$$

where p, q, and r are free indices and can be assigned the values 1, 2, and 3 at will. Then

$$
\begin{aligned}
e_{ijk} a_1^i a_2^j a_3^k &= a_{ij1} a_1^i a_2^j a_3^1 + e_{ij2} a_1^i a_2^j a_3^2 + e_{ij3} a_1^i a_2^j a_3^3 \\
&= e_{i11} a_1^i a_2^1 a_3^1 + e_{i21} a_1^i a_2^2 a_3^1 + e_{i31} a_1^i a_2^3 a_3^1 \\
&\quad + e_{i12} a_1^i a_2^1 a_3^2 + e_{i22} a_1^i a_2^2 a_3^2 + e_{i32} a_1^i a_2^3 a_3^2 \\
&\quad + e_{i13} a_1^i a_2^1 a_3^3 + e_{i23} a_1^i a_2^2 a_3^3 + e_{i33} a_1^i a_2^3 a_3^3 \\
&= e_{i21} a_1^i a_2^2 a_3^1 + e_{i31} a_1^i a_2^3 a_3^1 + e_{i12} a_1^i a_2^1 a_3^2 \\
&\quad + e_{i32} a_1^i a_2^3 a_3^2 + e_{i13} a_1^i a_2^1 a_3^3 + e_{i23} a_1^i a_2^2 a_3^3 \\
&= e_{121} a_1^1 a_2^2 a_3^1 + e_{221} a_1^2 a_2^2 a_3^3 + e_{321} a_1^3 a_2^2 a_3^3 \\
&\quad + e_{131} a_1^1 a_2^3 a_3^1 + e_{231} a_1^2 a_2^3 a_3^1 + e_{331} a_1^3 a_2^3 a_3^1 \\
&\quad + e_{112} a_1^1 a_2^1 a_3^2 + e_{212} a_1^2 a_2^1 a_3^2 + e_{312} a_1^3 a_2^1 a_3^2 \\
&\quad + e_{132} a_1^1 a_2^3 a_3^2 + e_{232} a_1^2 a_2^3 a_3^2 + e_{332} a_1^3 a_2^3 a_3^2 \\
&\quad + e_{113} a_1^1 a_2^1 a_3^2 + e_{213} a_1^2 a_2^1 a_3^2 + e_{313} a_1^3 a_2^1 a_3^2 \\
&\quad + e_{123} a_1^1 a_2^2 a_3^3 + e_{223} a_1^2 a_2^2 a_3^3 + e_{323} a_1^3 a_2^2 a_3^3 \\
&= e_{321} a_1^3 a_2^2 a_3^3 + e_{231} a_1^2 a_2^3 a_3^1 + e_{312} a_1^3 a_2^1 a_3^2 \\
&\quad + e_{132} a_1^1 a_2^3 a_3^2 + e_{213} a_1^2 a_2^1 a_3^2 + e_{123} a_1^1 a_2^2 a_3^3 \\
&= e_{123} \left| a_j^i \right|.
\end{aligned}
$$

Exercise 1.4. Prove that:

a.

$$e_{ijk} a_1^i a_3^j a_2^k = e_{132} \left| a_j^i \right|.$$

b.
$$e_{ijk}a_2^i a_1^j a_3^k = e_{213}\left|a_j^i\right|.$$

c.
$$e_{ijk}a_2^i a_3^j a_1^k = e_{231}\left|a_j^i\right|.$$

d.
$$e_{ijk}a_3^i a_1^j a_2^k = e_{312}\left|a_j^i\right|.$$

e.
$$e_{ijk}a_3^i a_2^j a_1^k = e_{321}\left|a_j^i\right|.$$

f.
$$e_{ijk}a_p^i a_q^j a_r^k = e_{pqr}\left|a_j^i\right|.$$

g.
$$e^{ijk}a_i^1 a_j^2 a_k^3 = e^{123}\left|a_j^i\right|.$$

h.
$$e^{ijk}a_i^1 a_j^3 a_k^2 = e^{132}\left|a_j^i\right|.$$

i.
$$e^{ijk}a_i^2 a_j^1 a_k^3 = e^{213}\left|a_j^i\right|.$$

j.
$$e^{ijk}a_i^2 a_j^3 a_k^1 = e^{231}\left|a_j^i\right|.$$

k.
$$e^{ijk}a_i^3 a_j^1 a_k^2 = e^{312}\left|a_j^i\right|.$$

l.
$$e^{ijk}a_i^3 a_j^2 a_k^1 = e^{321}\left|a_j^i\right|.$$

m.

$$e^{ijk} a_i^p a_j^q a_k^r = e^{pqr} |a_j^i|.$$

Exercise 1.5. Prove that

a.

$$e_{i_1 i_2 \cdots i_N} a_1^{i_1} a_2^{i_2} \cdots a_N^{i_N} = |a_j^i|.$$

b.

$$e^{i_1 i_2 \cdots i_N} a_{i_1}^1 a_{i_2}^2 \cdots a_{i_N}^N = |a_j^i|.$$

Exercise 1.6. Let $p_1, p_2, \ldots, p_N$ be free indices that can be assigned the values $1, 2, \ldots, N$. Prove that

a.

$$e_{p_1 p_2 \cdots p_N} |a_j^i| = e_{i_1 i_2 \cdots i_N} a_{p_1}^{i_1} a_{p_2}^{i_2} \cdots a_{p_N}^{i_N}.$$

b.

$$e^{p_1 p_2 \cdots p_N} |a_j^i| = e^{i_1 i_2 \cdots i_N} a_{i_1}^{p_1} a_{i_2}^{p_2} \cdots a_{i_N}^{p_N}.$$

Definition 1.8. The product of e^{ij} and e_{pq} is called the generalized Kronecker delta and is denoted by δ_{pq}^{ij}, i. e.,

$$\delta_{pq}^{ij} = e^{ij} e_{pq}.$$

Similarly,

$$\delta_{pqr}^{ijk} = e^{ijk} e_{pqr}$$

and

$$\delta_{p_1 p_2 \cdots p_N}^{i_1 i_2 \cdots i_N} = e^{i_1 i_2 \cdots i_N} e_{p_1 p_2 \cdots p_N}.$$

Example 1.18. We will evaluate

$$e_{ij} e^{ik}, \quad i, j, k \in \{1, 2, 3\}.$$

Using the summation convention, we obtain

$$e_{ij} e^{ik} = e_{1j} e^{1k} + e_{2j} e^{2k}.$$

We have the following cases.

1. $j = k = 1$. Then

$$\begin{aligned}
e_{ij}e^{ik} &= e_{11}e^{11} + e_{21}e^{21}\\
&= 0(0) + (-1)(-1)\\
&= 1.
\end{aligned}$$

2. $j = 1, k = 2$. Then

$$\begin{aligned}
e_{ij}e^{ik} &= e_{11}e^{12} + e_{21}e^{22}\\
&= 0(1) + 1(0)\\
&= 0.
\end{aligned}$$

3. $j = 2, k = 1$. Then

$$\begin{aligned}
e_{ij}e^{ik} &= e_{12}e^{11} + e_{22}e^{21}\\
&= 1(0) + 0(-1)\\
&= 0.
\end{aligned}$$

4. $j = k = 2$. Then

$$\begin{aligned}
e_{ij}e^{ik} &= e_{12}e^{12} + e_{21}e^{22}\\
&= 1(1) + 0(0)\\
&= 1.
\end{aligned}$$

Therefore

$$e_{ij}e^{ik} = \begin{cases} 1 & \text{if } j = k,\\ 0 & \text{if } j \neq k. \end{cases}$$

Hence

$$e_{ij}e^{ik} = \delta_j^k.$$

In particular, when $j = k$, we get

$$\begin{aligned}
e_{ij}e^{ij} &= \delta_j^j\\
&= 2.
\end{aligned}$$

Example 1.19. We will establish the following identity:

$$e_{rij}e_{rkl} = \delta_{ik}\delta_{jl} - \delta_{il}\delta_{jk}, \quad i, j, k, r, l \in \{1, 2, 3\}. \tag{1.3}$$

We will consider the following cases.

1. $i = j$ or $k = l$. Then

$$e_{rij} = 0 \quad \text{and} \quad e_{rkl} = 0,$$

whereupon

$$e_{rij}e_{rkl} = 0.$$

Moreover, if $i = j$, then

$$\delta_{ik}\delta_{jl} - \delta_{il}\delta_{jk} = \delta_{jk}\delta_{jl} - \delta_{jl}\delta_{jk}$$
$$= 0.$$

If $k = l$, then

$$\delta_{ik}\delta_{jl} - \delta_{il}\delta_{jk} = \delta_{il}\delta_{jl} - \delta_{il}\delta_{jl}$$
$$= 0.$$

Thus (1.3) holds.

2. $i \neq j$ and $k \neq l$. Then

$$e_{rij}e_{rkl} = e_{1ij}e_{1kl} + e_{2ij}e_{2kl} + e_{3ij}e_{3kl}$$
$$= e_{pqs}e_{pkl}, \quad i = q, \, j = s,$$

where (pqr) denotes some permutation of (123). Thus we arrive at the following two cases.

a. $e_{pqs}e_{pkl} \neq 0$.
   Then $k = q$ and $l = s$, or $k = s$ and $l = q$. If $k = q$ and $l = s$, then we get

$$e_{pqs}e_{pkl} = e_{pqs}e_{pqs}$$
$$= 1$$

and

$$\delta_{ik}\delta_{jl} - \delta_{il}\delta_{jk} = \delta_{qk}\delta_{sl} - \delta_{ql}\delta_{sk}$$
$$= \delta_{kk}\delta_{ss} - \delta_{ks}\delta_{sk}$$
$$= 1(1) - 0(0)$$
$$= 1.$$

If $k = s$ and $l = q$, then

$$e_{pqs}e_{pkl} = e_{pqs}e_{psq}$$
$$= -1$$

and

$$\delta_{ik}\delta_{jl} - \delta_{il}\delta_{jk} = \delta_{qk}\delta_{sl} - \delta_{ql}\delta_{sk}$$
$$= \delta_{lk}\delta_{kl} - \delta_{ll}\delta_{kk}$$
$$= 0(0) - 1(1)$$
$$= -1.$$

Consequently, (1.3) holds.

b. $e_{pqs}e_{pkl} = 0$

Since $k \neq l$, either $k = p$ or $l = p$. If $k = p$, then

$$\delta_{ik}\delta_{jl} - \delta_{il}\delta_{jk} = \delta_{qk}\delta_{sl} - \delta_{ql}\delta_{sk}$$
$$= \delta_{qp}\delta_{sk} - \delta_{ql}\delta_{sp}$$
$$= 0.$$

If $l = p$, then

$$\delta_{ik}\delta_{jl} - \delta_{il}\delta_{jk} = \delta_{qk}\delta_{sl} - \delta_{ql}\delta_{sk}$$
$$= \delta_{qk}\delta_{sp} - \delta_{qp}\delta_{sk}$$
$$= 0.$$

Consequently, (1.3) holds.

Exercise 1.7. Let $N = 3$. Prove that

1.

$$e^{ijk}e_{pqk} = \delta^{ij}_{pq}.$$

2.

$$e^{ijk}e_{pjk} = 2\delta^i_p.$$

1.4 Tensor notation on matrices

It is known that if the range of the indices in a system of second order is from 1 to N, then the number of its components is N^2. A system of second order can be of three types

$$a^i_j, \quad a_{ij}, \quad \text{or} \quad a^{ij}.$$

By matrices of systems of second order we mean the matrices

$$(a^i_j), \quad (a_{ij}), \quad \text{or} \quad (a^{ij}),$$

i. e.,

$$(a_j^i) = \begin{pmatrix} a_1^1 & a_2^1 & \cdots & a_N^1 \\ a_1^2 & a_2^2 & \cdots & a_N^2 \\ \vdots & \vdots & \vdots & \vdots \\ a_1^N & a_2^N & \cdot & a_N^N \end{pmatrix},$$

$$(a_{ij}) = \begin{pmatrix} a_{11} & a_{12} & \cdots & a_{1N} \\ a_{21} & a_{22} & \cdots & a_{2N} \\ \vdots & \vdots & \vdots & \vdots \\ a_{N1} & a_{N2} & \cdots & a_{NN} \end{pmatrix},$$

and

$$(a^{ij}) = \begin{pmatrix} a^{11} & a^{12} & \cdots & a^{1N} \\ a^{21} & a^{22} & \cdots & a^{2N} \\ \vdots & \vdots & \vdots & \vdots \\ a^{N1} & a^{N2} & \cdots & a^{NN} \end{pmatrix},$$

each of which is an $N \times N$ matrix. The determinants of the matrices are

$$|a_j^i|, \quad |a_{ij}|, \quad \text{and} \quad |a^{ij}|.$$

We will establish the following results on matrices and determinants of systems of second order.

In terms of the Kronecker deltas, the identity matrix of order N is

$$I = (\delta_{ij}), \quad I = (\delta_j^i), \quad \text{or} \quad I = (\delta^{ij}),$$

which has the property

$$IA = AI$$
$$= A$$

for any square matrix of order N. A square matrix $A = (a_{ij})$ is invertible if there exists a unique matrix $B = (b_{ij})$, called the inverse of A, such that

$$AB = BA$$
$$= I.$$

In terms of components, the criterion reads

$$a_{ir}b_{rj} = b_{ir}a - rj$$
$$= \delta_{ij},$$

$$a_r^i b_j^r = b_i^r a_r^j$$
$$= \delta_j^i,$$

and

$$a^{ir} b^{rj} = b^{ir} a^{rj}$$
$$= \delta^{ij}.$$

If $A = (a_j^i)$ and $B = (b_j^i)$ are two matrices conformable for multiplication, then

$$AB = a_j^i b_p^j$$

is the product of both matrices, where i and p are not summed.

1. If

$$a_j^i b_p^j = c_p^i,$$

then

$$(a_j^i)(b_p^j) = (c_p^i)$$

and

$$|a_j^i||b_p^j| = |c_p^i|.$$

Without loss of generality, consider the case $N = 2$. Applying the summation convention, we get

$$c_p^i = a_j^i b_p^j$$
$$= a_1^i b_p^1 + a_2^i b_p^2, \quad i, p \in \{1, 2\}.$$

Hence

$$\begin{pmatrix} a_1^1 & a_2^1 \\ a_1^2 & a_2^2 \end{pmatrix} \begin{pmatrix} b_1^1 & b_2^1 \\ b_1^2 & b_2^2 \end{pmatrix} = \begin{pmatrix} a_1^1 b_1^1 + a_2^1 b_1^2 & a_1^1 b_2^1 + a_2^1 b_2^2 \\ a_1^2 b_1^1 + a_2^2 b_1^2 & a_1^2 b_2^1 + a_2^2 b_2^2 \end{pmatrix}$$
$$= \begin{pmatrix} c_1^1 & c_2^1 \\ c_1^2 & c_2^2 \end{pmatrix},$$

or

$$(a_j^i)(b_p^j) = (c_p^i).$$

Taking the determinants of both sides, we have

$$|a_j^i||b_p^i| = |c_p^i|.$$

Generalizing this to a finite number, we find

$$|a_j^i b_k^j c_l^k \cdots p_t^s| = |a_j^i||b_k^j||c_l^k| \cdots |p_t^s|.$$

2. If

$$a_{ij}b^{ik} = c_j^k,$$

then

$$(b^{ik})^T (a_{ij}) = (c_j^k)$$

and

$$|a_{ij}||b^{ik}| = |c_j^k|,$$

where $(b^{ik})^T$ is the transpose of (b^{ik}).
Indeed, applying the summation convention, we find

$$c_j^k = a_{ij}b^{ik}$$
$$= a_{1j}b^{1k} + a_{2j}b^{2k},$$

or

$$\begin{pmatrix} c_1^1 & c_2^1 \\ c_1^2 & c_2^2 \end{pmatrix} = \begin{pmatrix} a_{11}b^{11} + a_{21}b^{21} & a_{12}b^{11} + a_{22}b^{21} \\ a_{11}b^{12} + a_{21}b^{22} & a_{12}b^{12} + a_{22}b^{22} \end{pmatrix}$$
$$= \begin{pmatrix} b^{11} & b^{21} \\ b^{12} & b^{22} \end{pmatrix} \begin{pmatrix} a_{11} & a_{12} \\ a_{21} & a_{22} \end{pmatrix},$$

or

$$(c_j^k) = (b^{ik})^T (a_{ij}).$$

Taking the determinants of both sides, we obtain

$$|c_j^k| = |b^{ik}|^T |a_{ij}|$$
$$= |b^{ik}||a_{ij}|,$$

because

$$|B^T| = |B|.$$

3. Let the cofactor of the element a_j^i in the determinant $a = |a_j^i|$ be denoted by A_i^j. Then applying the summation convention, we get

$$a_j^i A_k^j = a_1^i A_k^1 + a_2^i A_k^2 + \cdots + a_N^i A_k^N$$
$$= \delta_k^i a$$

and

$$a_j^i A_i^k = a_j^1 A_1^k + a_j^2 A_2^k + \cdots + a_j^N A_N^k$$
$$= \delta_j^k a.$$

If the elements of the determinant a are denoted by a_{ij}, then we will write the cofactor of a_{ij} with A_{ij}. Then the above results can be written as

$$a_{ij} A_{jk} = \delta_{ik} a$$

and

$$a_{ij} A_{ki} = \delta_{jk} a.$$

4. Now consider a system of n linear equations

$$a_j^i x^j = b^i, \quad i, j \in \{1, \ldots, n\}, \tag{1.4}$$

in n unknown x^i, where

$$a = |a_j^i| \neq 0.$$

Multiplying both sides of (1.4) by A_i^k, we get

$$a_j^i A_i^k x^j = b^i A_i^k,$$

or

$$\delta_j^k a x^j = A_i^k b^i,$$

or

$$a x^k = A_i^k b^i,$$

whereupon

$$x^k = \frac{1}{a} A_i^k x^i.$$

This is the Crammer rule for the solution of the system of n linear equations (1.4).

5. Consider the determinant $a = |a_j^i|$ and suppose that a_j^i and A_i^k are functions of the independent variables $x^1, x^2, \ldots, x^N$. Then using

$$\delta_j^k a = a_j^i A_i^k,$$

we find

$$\delta_j^k \frac{\partial a}{\Delta_p x^p} = \frac{\partial}{\Delta_p x^p}(a_j^i A_i^k)$$

$$= \frac{\partial a_j^i}{\Delta_p x^p} A_i^k + (a_j^i)^{\sigma_p} \frac{\partial A_i^k}{\Delta_p x^p}$$

$$= \frac{\partial a_j^i}{\Delta_p x^p}(A_i^k)^{\sigma_p} + a_j^i \frac{\partial A_i^k}{\Delta_p x^p}.$$

Remark 1.3. Note that on general time scales, we have

$$\delta_j^k \frac{\partial a}{\Delta_p x^p} \neq \frac{\partial a_j^i}{\Delta_p x^p} A_i^k.$$

Indeed, let

$$a(t) = \det \begin{pmatrix} a_1^1(t) & a_2^1(t) \\ a_1^2(t) & a_2^2(t) \end{pmatrix}$$

$$= \det \begin{pmatrix} 2t^2 & t \\ t & t^2 \end{pmatrix}, \quad t \in \mathbb{T}.$$

Then

$$a(t) = 2t^4 - t^2, \quad t \in \mathbb{T}.$$

Hence

$$a^\Delta(t) = 2((\sigma(t))^3 + (\sigma(t))^2 t + \sigma(t)t^2 + t^3) - (\sigma(t) + t)$$
$$= 2(\sigma(t))^3 + 2(\sigma(t))^2 t + 2\sigma(t)t^2 + 2t^3 - \sigma(t) - t$$
$$= 2(\sigma(t))^3 + 2(\sigma(t))^2 t + \sigma(t)(2t^2 - 1) + 2t^3 - t, \quad t \in \mathbb{T}.$$

On the other hand, we have

$$\det \begin{pmatrix} (a_1^1)^\Delta(t) & (a_2^1)^\Delta(t) \\ a_1^2(t) & a_2^2(t) \end{pmatrix} + \det \begin{pmatrix} a_1^1(t) & a_2^1(t) \\ (a_1^2)^\Delta(t) & (a_2^2)^\Delta(t) \end{pmatrix}$$

$$= \det \begin{pmatrix} 2(\sigma(t) + t) & 1 \\ t & t^2 \end{pmatrix} + \det \begin{pmatrix} 2t^2 & t \\ 1 & \sigma(t) + t \end{pmatrix}$$

$$
= 2t^2(\sigma(t) + t) - t + 2t^2(\sigma(t) + t) - t
$$
$$
= 4t^2(\sigma(t) + t) - t
$$
$$
\neq a^A(t), \quad t \in \mathbb{T}.
$$

6. Consider the transformations

$$
z^i = z^i(\overline{x}^k),
$$
$$
\overline{x}^i = \overline{x}^i(x^j).
$$

Then using the chain rule, we find

$$
\frac{\partial z^i}{\Delta_k x^k} = \frac{\partial z^{i\sigma_{(0)}\sigma_{(1)}\cdots\sigma_{(j-1)}}}{\Delta_{(j)}\overline{x}^j}\frac{\Delta_{(j)}\overline{x}^j}{\Delta_k x^k}.
$$

Hence

$$
\left|\frac{\partial z^i}{\Delta_k x^k}\right| = \left|\frac{\partial z^{i\sigma_{(0)}\sigma_{(1)}\cdots\sigma_{(j-1)}}}{\Delta_{(j)}\overline{x}^j}\right|\left|\frac{\Delta_{(j)}\overline{x}^j}{\Delta_k x^k}\right|.
$$

In particular, when $z^i = x^i$, we obtain

$$
\left|\frac{\Delta_i x^i}{\Delta_k x^k}\right| = \left|\frac{\Delta_i x^{i\sigma_{(0)}\sigma_{(1)}\cdots\sigma_{(j-1)}}}{\Delta_{(j)}\overline{x}^j}\right|\left|\frac{\Delta_{(j)}\overline{x}^j}{\Delta_k x^k}\right|,
$$

or

$$
\left|\delta_i^k\right| = \left|\frac{\Delta_i x^{i\sigma_{(0)}\sigma_{(1)}\cdots\sigma_{(j-1)}}}{\Delta_{(j)}\overline{x}^j}\right|\left|\frac{\Delta_{(j)}\overline{x}^j}{\Delta_k x^k}\right|,
$$

or

$$
1 = \left|\frac{\Delta_i x^{i\sigma_{(0)}\sigma_{(1)}\cdots\sigma_{(j-1)}}}{\Delta_{(j)}\overline{x}^j}\right|\left|\frac{\Delta_{(j)}\overline{x}^j}{\Delta_k x^k}\right|.
$$

Therefore

$$
\left|\frac{\Delta_i x^{i\sigma_{(0)}\sigma_{(1)}\cdots\sigma_{(j-1)}}}{\Delta_{(j)}\overline{x}^j}\right| = \frac{1}{\left|\frac{\Delta_{(j)}\overline{x}^j}{\Delta_k x^k}\right|}.
$$

Example 1.20. Consider the third-order systems a_j^i, b_m^k, c_j^i. Let

$$
c_m^i = a_p^i b_m^j.
$$

We will prove that

$$|a_j^i||b_m^k| = |c_j^i|.$$

We have

$$
\begin{aligned}
|a_j^i||b_m^k| &= |a_j^i|(e_{kms}b_1^k b_2^m b_3^s)\\
&= (e_{kms}|a_j^i|)b_1^k b_2^m b_3^s\\
&= (e_{ijt}a_k^i a_m^j a_s^t)(b_1^k b_2^m b_3^s)\\
&= e_{ijt}(a_k^i b_1^k)(a_m^j b_2^m)(a_s^t b_3^s)\\
&= e_{ijt}c_1^i c_2^j c_3^t\\
&= |c_j^i|.
\end{aligned}
$$

1.5 Contravariant vectors

Suppose that

$$\bar{x}^j(x^1,\ldots,x^{l-1},\sigma_l(x^l),x^{l+1},\ldots,x^n) = \sigma_{(j)}(\bar{x}^j(x^1,\ldots,x^N)). \tag{1.5}$$

In V_N, consider a curve C defined by the set of parametric equations

$$C: x^i = x^i(t), \quad i \in \{1,2,\ldots,N\},$$

where $t \in \mathbb{T}$ is a parameter. The tangent vector to the curve C is given by

$$\vec{T} = \left(\frac{\Delta_1 x^1}{\Delta t}, \frac{\Delta_2 x^2}{\Delta t},\ldots,\frac{\Delta_N x^N}{\Delta t}\right).$$

In index notation, which focuses attention on the components, the tangent vector is denoted by

$$T^i = \frac{\Delta_i x^i}{\Delta t}.$$

For a coordinate transformation

$$x^i = x^i(\bar{x}^1,\bar{x}^2,\ldots,\bar{x}^N)$$

with its transformation

$$\bar{x}^j = \bar{x}^j(x^1,x^2,\ldots,x^N),$$

the curve C is represented in the barred space by

$$\overline{x}^i = \overline{x}^i(x^1(t), x^2(t), \ldots, x^N(t))$$
$$= \overline{x}^i(t),$$

with t unchanged. The tangent to the curve in the barred system of coordinates is represented by

$$\frac{\Delta_{(i)}\overline{x}^i}{\Delta t} = \frac{\Delta_{(i)}\overline{x}^{i\sigma_0\sigma_1\cdots\sigma_{j-1}}}{\Delta_j x^j}\frac{\Delta_j x^j}{\Delta t}.$$

Letting $\overline{T}^i$ denote the components of this tangent vector in the barred system of coordinates, the last equation can be expressed in the form

$$\overline{T}^i = \frac{\Delta_{(i)}\overline{x}^{i\sigma_0\sigma_1\cdots\sigma_{j-1}}}{\Delta_j x^j}T^j.$$

This equation is said to define the transformation law associated with a contravariant vector.

Definition 1.9. A set of N functions A^j of the N coordinates x^k are said to be the components of a contravariant vector if they transform according to the equation

$$\overline{A}^j = \frac{\Delta_{(j)}\overline{x}^{j\sigma_0\sigma_1\cdots\sigma_{l-1}}}{\Delta_l x^l}A^l \tag{1.6}$$

on change of coordinates x^l to $\overline{x}^l$. The contravariant vector is also called a contravariant tensor of order 1.

This means that any N functions can be chosen as the components of a contravariant vector in the coordinate system x^l and equations (1.6) define N components in the new coordinate system $\overline{x}^l$.

Multiplying equations (1.6) by $\frac{\Delta_l x^l}{\Delta_{(k)}\overline{x}^k}$ and using that

$$\delta^j_k = \frac{\Delta_{(j)}\overline{x}^j}{\Delta_{(k)}\overline{x}^k}$$
$$= \frac{\Delta_{(j)}\overline{x}^{j\sigma_0\sigma_1\cdots\sigma_{l-1}}}{\Delta_l x^l}\frac{\Delta_l x^l}{\Delta_{(k)}\overline{x}^k},$$

we get

$$\frac{\Delta_l x^l}{\Delta_{(k)}\overline{x}^k}\overline{A}^j = \frac{\Delta_{(j)}\overline{x}^{j\sigma_0\sigma_1\cdots\sigma_{l-1}}}{\Delta_l x^l}\frac{\Delta_l x^l}{\Delta_{(k)}\overline{x}^k}A^l$$

$$= \frac{\Delta_{(j)}\overline{x}^j}{\Delta_{(k)}\overline{x}^k}A^l$$

$$= \delta_k^j A^l,$$

whereupon

$$A^l = \frac{\Delta_l x^l}{\Delta_{(j)}\overline{x}^j}\overline{A}^j.$$

Hence the solution of equation (1.6) is

$$A^k = \frac{\Delta_k x^k}{\Delta_{(j)}\overline{x}^j}\overline{A}^j.$$

Examining equations (1.2), we see that the differentials $\Delta_r x^r$ form the components of a contravariant vector, whose components in any other system are the differentials $\Delta_{(j)}\overline{x}^j$.

Example 1.21. Let the components of a contravariant vector in the x^i coordinate system be $(2,1)$, and let $\mathbb{T}_1 = \mathbb{Z}$ and $\mathbb{T}_2 = 2^{\mathbb{N}_0}$. We will find its components in the $\overline{x}^i$ coordinate system, where

$$\overline{x}^1 = x^1 + e_2(x^2, x_0^2),$$
$$\overline{x}^2 = 3x^1 - e_2(x^2, x_0^2),$$

where $x_0^2 \in \mathbb{T}_2$ is fixed. Here

$$A^1 = 2,$$
$$A^2 = 1$$

and

$$\sigma_1(x^1) = x^1 + 1, \quad x^1 \in \mathbb{T}_1,$$
$$\sigma_2(x^2) = 2x^2, \quad x^2 \in \mathbb{T}_2.$$

Now we will express x^1 and x^2 in terms of $\overline{x}^1$ and $\overline{x}^2$. We have

$$x^1 = \frac{\overline{x}^1 + \overline{x}^2}{4}$$

and

$$e_2(x^2, x_0^2) = \overline{x}^1 - x^1$$
$$= \overline{x}^1 - \frac{\overline{x}^1 + \overline{x}^2}{4}$$
$$= \frac{3\overline{x}^1 - \overline{x}^2}{4}.$$

Now we will compute the partial derivatives

$$\frac{\Delta_{(j)}\overline{x}^{j\sigma_0\sigma_1\cdots\sigma_{k-1}}}{\Delta_k x^k}, \quad j, k \in \{1, 2\}.$$

We have

$$\frac{\Delta_{(1)}\overline{x}^1}{\Delta_1 x^1} = 3,$$

$$\frac{\Delta_{(1)}\overline{x}^1}{\Delta_2 x^2} = 2e_2(x^2, x_0^2),$$

$$\frac{\Delta_{(1)}\overline{x}^{1\sigma_1}}{\Delta_2 x^2} = 2e_2(x^2, x_0^2),$$

$$\frac{\Delta_{(2)}\overline{x}^2}{\Delta_1 x^1} = 3,$$

$$\frac{\Delta_{(2)}\overline{x}^2}{\Delta_2 x^2} = -2e_2(x^2, x_0^2),$$

$$\frac{\Delta_{(2)}\overline{x}^{2\sigma_1}}{\Delta_2 x^2} = -2e_2(x^2, x_0^2).$$

Then using the summation convention, we find

$$\overline{A}^1 = \frac{\Delta_{(1)}\overline{x}^{1\sigma_0\sigma_1\cdots\sigma_{i-1}}}{\Delta_i x^i} A^i$$
$$= \frac{\Delta_{(1)}\overline{x}^1}{\Delta_1 x^1} A^1 + \frac{\Delta_{(1)}\overline{x}^{1\sigma_1}}{\Delta_2 x^2} A^2$$
$$= 1(2) + (2e_2(x^2, x_0^2))1$$
$$= 2(1 + e_2(x^2, x_0^2))$$

and

$$\overline{A}^2 = \frac{\Delta_{(2)}\overline{x}^{2\sigma_0\sigma_1\cdots\sigma_{i-1}}}{\Delta_i x^i} A^i$$
$$= \frac{\Delta_{(2)}\overline{x}^2}{\Delta_1 x^1} A^1 + \frac{\Delta_{(2)}\overline{x}^{2\sigma_1}}{\Delta_2 x^2} A^2$$

$$\begin{aligned}
&= 3(2) + (-2e_2(x^2, x_0^2))1 \\
&= 6 - 2e_2(x^2, x_0^2) \\
&= 2(3 - e_2(x^2, x_0^2)) \\
&= 2\left(3 - \frac{3\overline{x}^1 - \overline{x}^2}{4}\right) \\
&= \frac{12 - 3\overline{x}^1 + \overline{x}^2}{2}.
\end{aligned}$$

Example 1.22. Let the components of a contravariant vector in the x^i coordinate system be (x^2, x^1). We will find its components in the $\overline{x}^i$ coordinate system, where

$$\begin{aligned}
\overline{x}^1 &= (x^2)^2, \\
\overline{x}^2 &= x^1 x^2, \quad x^1 > 0, \ x^2 > 0.
\end{aligned}$$

Here

$$\begin{aligned}
A^1 &= x^2, \\
A^2 &= x^1.
\end{aligned}$$

We will express x^1 and x^2 in terms of $\overline{x}^1$ and $\overline{x}^2$. We have

$$\begin{aligned}
x^2 &= \sqrt{\overline{x}^1}, \\
x^1 &= \frac{\overline{x}^2}{x^2} \\
&= \frac{\overline{x}^2}{\sqrt{\overline{x}^1}}.
\end{aligned}$$

Now we will compute the partial derivatives

$$\frac{\Delta_{(j)}\overline{x}^{j\sigma_0\sigma_1\cdots\sigma_{k-1}}}{\Delta_k x^k}, \quad j, k \in \{1, 2\}.$$

We have

$$\frac{\Delta_{(1)}\overline{x}^1}{\Delta_1 x^1} = 0,$$

$$\frac{\Delta_{(1)}\overline{x}^1}{\Delta_2 x^2} = \sigma_2(x^2) + x^2,$$

$$\frac{\Delta_{(1)}\overline{x}^{1\sigma_1}}{\Delta_2 x^2} = \sigma_2(x^2) + x^2,$$

$$\frac{\Delta_{(2)}\overline{x}^2}{\Delta_1 x^1} = x^2,$$

$$\frac{\Delta_{(2)}\overline{x}^2}{\Delta_2 x^2} = x^1,$$

$$\frac{\Delta_{(2)}\overline{x}^{2\sigma_1}}{\Delta_2 x^2} = \sigma_1(x^1).$$

Then applying the summation convention, we find

$$\overline{A}^1 = \frac{\Delta_{(1)}\overline{x}^{1\sigma_0\sigma_1\cdots\sigma_{i-1}}}{\Delta_i x^i} A^i$$

$$= \frac{\Delta_{(1)}\overline{x}^1}{\Delta_1 x^1} A^1 + \frac{\Delta_{(1)}\overline{x}^{1\sigma_1}}{\Delta_2 x^2} A^2$$

$$= 0(x^2) + (\sigma_2(x^2) + x^2)x^1$$

$$= x^1(\sigma_2(x^2) + x^2)$$

$$= \frac{\overline{x}^2}{\sqrt{\overline{x}^1}}\left(\sqrt{\sigma_{(1)}(\overline{x}^1)} + \sqrt{\overline{x}^1}\right)$$

and

$$\overline{A}^2 = \frac{\Delta_{(2)}\overline{x}^2}{\Delta_i x^i} A^i$$

$$= \frac{\Delta_{(2)}\overline{x}^2}{\Delta_1 x^1} A^1 + \frac{\Delta_{(2)}\overline{x}^{2\sigma_1}}{\Delta_2 x^2} A^2$$

$$= x^2(x^2) + \sigma_1(x^1)(x^1)$$

$$= \sigma_1(x^1)x^1 + (x^2)^2$$

$$= \frac{(\overline{x}^2)^2}{\sqrt{\sigma_{(1)}(\overline{x}^1)}\sqrt{\overline{x}^1}} + \overline{x}^1.$$

In the particular case where $\mathbb{T}_{(1)} = 3\mathbb{Z}$ and $\mathbb{T}_{(2)} = 4^{\mathbb{N}_0}$, we have

$$\sigma_{(1)}(\overline{x}^1) = \overline{x}^1 + 3,$$
$$\sigma_{(2)}(\overline{x}^2) = 4\overline{x}^2,$$

and then

$$\overline{A}^1 = \frac{\overline{x}^2}{\sqrt{\overline{x}^1}}\left(\sqrt{\overline{x}^1 + 3} + \sqrt{\overline{x}^1}\right)$$

and

$$\overline{A}^2 = \frac{(\overline{x^2})^2}{\sqrt{\overline{x^1}} + 3\sqrt{\overline{x^1}}} + \overline{x}^1.$$

Example 1.23. Consider

$$x = r\cos_1(\theta, \theta_0),$$
$$y = r\sin_1(\theta, \theta_0),$$

where $r \in \mathbb{T}_1$, $\theta, \theta_0 \in \mathbb{T}_2$, and

$$r = r(t),$$
$$\theta = \theta(t), \quad t \in \mathbb{T},$$
$$r(\sigma(t)) = \sigma_1(r(t)),$$
$$\theta(\sigma(t)) = \sigma_2(\theta_2(t)), \quad t \in \mathbb{T}.$$

Suppose that σ_1 is Δ-differentiable. Then

$$\frac{\Delta x}{\Delta t} = \frac{\Delta r}{\Delta t}\cos_1(\theta, \theta_0) - r^{\sigma_1}\sin_1(\theta, \theta_0)\frac{\Delta\theta}{\Delta t}$$

and

$$\frac{\Delta y}{\Delta t} = \frac{\Delta r}{\Delta t}\sin_1(\theta, \theta_0) + r^{\sigma_1}\cos_1(\theta, \theta_0)\frac{\Delta\theta}{\Delta t}.$$

Moreover,

$$\frac{\Delta^2 x}{\Delta t^2} = \frac{\Delta^2 r}{\Delta t^2}\cos_1(\theta, \theta_0) - \left(\frac{\Delta r}{\Delta t}\right)^{\sigma_1}\sin_1(\theta, \theta_0)\frac{\Delta\theta}{\Delta t} - \frac{\Delta r^{\sigma_1}}{\Delta t}\sin_1(\theta, \theta_0)\frac{\Delta\theta}{\Delta t}$$
$$- r^{\sigma_1\sigma_1}\sin_1(\theta, \theta_0)\frac{\Delta\theta}{\Delta t} - r^{\sigma_1\sigma_1}\cos_1(\theta, \theta_0)\left(\frac{\Delta\theta}{\Delta t}\right)^2 - r^{\sigma_1}\sin_1^{\sigma_2}(\theta, \theta_0)\frac{\Delta^2\theta}{\Delta t^2}$$
$$= \frac{\Delta^2 r}{\Delta t^2}\cos_1(\theta, \theta_0) - \left(\frac{\Delta r}{\Delta t}\right)^{\sigma_1}\frac{\Delta\theta}{\Delta t}\sin_1(\theta, \theta_0) - \frac{\Delta r^{\sigma_1}}{\Delta t}\sin_1(\theta, \theta_0)\frac{\Delta\theta}{\Delta t}$$
$$- r^{\sigma_1\sigma_1}\cos_1(\theta, \theta_0)\left(\frac{\Delta\theta}{\Delta t}\right)^2 - r^{\sigma_1}(\sin_1(\theta, \theta_0) + \mu_2(\theta)\cos_1(\theta, \theta_0))\frac{\Delta^2\theta}{\Delta t^2}$$
$$= \left(\frac{\Delta^2 r}{\Delta t^2} - r^{\sigma_1\sigma_1}\left(\frac{\Delta\theta}{\Delta t}\right)^2 - r^{\sigma_1}\mu_2(\theta)\frac{\Delta^2\theta}{\Delta t^2}\right)\cos_1(\theta, \theta_0)$$
$$- \left(r^{\sigma_1}\frac{\Delta^2\theta}{\Delta t^2} + \frac{\Delta r^{\sigma_1}}{\Delta t}\frac{\Delta\theta}{\Delta t} + \left(\frac{\Delta r}{\Delta t}\right)^{\sigma_1}\frac{\Delta\theta}{\Delta t}\right)\sin_1(\theta, \theta_0)$$
$$= \left(\frac{\Delta^2 r}{\Delta t^2} - r^{\sigma_1\sigma_1}\left(\frac{\Delta\theta}{\Delta t}\right)^2 - r^{\sigma_1}\mu_2(\theta)\frac{\Delta^2\theta}{\Delta t^2}\right)\cos_1(\theta, \theta_0)$$
$$- \left(\frac{\Delta^2\theta}{\Delta t^2} + \frac{1}{r^{\sigma_1}}\left(\frac{\Delta r^{\sigma_1}}{\Delta t}\frac{\Delta\theta}{\Delta t} + \left(\frac{\Delta r}{\Delta t}\right)^{\sigma_1}\frac{\Delta\theta}{\Delta t}\right)\right)r^{\sigma_1}\sin_1(\theta, \theta_0),$$

and

$$\frac{\Delta^2 y}{\Delta t^2} = \frac{\Delta^2 r}{\Delta t^2}\sin_1(\theta,\theta_0) + \left(\frac{\Delta r}{\Delta t}\right)^{\sigma_1}\cos_1(\theta,\theta_0)\frac{\Delta\theta}{\Delta t} + \frac{\Delta r^{\sigma_1}}{\Delta t}\cos_1(\theta,\theta_0)\frac{\Delta\theta}{\Delta t}$$

$$- r^{\sigma_1\sigma_1}\sin_1(\theta,\theta_0)\left(\frac{\Delta\theta}{\Delta t}\right)^2 + r^{\sigma_1}\cos_1^{\sigma_2}(\theta,\theta_0)\frac{\Delta^2\theta}{\Delta t^2}$$

$$= \frac{\Delta^2 r}{\Delta t^2}\sin_1(\theta,\theta_0) + \left(\frac{\Delta r}{\Delta t}\right)^{\sigma_1}\cos_1(\theta,\theta_0)\frac{\Delta\theta}{\Delta t} + \frac{\Delta r^{\sigma_1}}{\Delta t}\frac{\Delta\theta}{\Delta t}\cos_1(\theta,\theta_0)$$

$$- r^{\sigma_1\sigma_1}\sin_1(\theta,\theta_0)\left(\frac{\Delta\theta}{\Delta t}\right)^2 + r^{\sigma_1}(\cos_1(\theta,\theta_0) - \mu_2(\theta)\sin_1(\theta,\theta_0))\frac{\Delta^2\theta}{\Delta t^2}$$

$$= \left(\frac{\Delta^2 r}{\Delta t^2} - r^{\sigma_1\sigma_1}\left(\frac{\Delta\theta}{\Delta t}\right)^2 - \mu_2(\theta)r^{\sigma_1}\frac{\Delta^2\theta}{\Delta t^2}\right)\sin_1(\theta,\theta_0)$$

$$+ \left(r^{\sigma_1}\frac{\Delta^2\theta}{\Delta t^2} + \frac{\Delta r^{\sigma_1}}{\Delta t}\frac{\Delta\theta}{\Delta t} + \left(\frac{\Delta r}{\Delta t}\right)^{\sigma_1}\frac{\Delta\theta}{\Delta t}\right)\cos_1(\theta,\theta_0)$$

$$= \left(\frac{\Delta^2 r}{\Delta t^2} - r^{\sigma_1\sigma_1}\left(\frac{\Delta\theta}{\Delta t}\right)^2 - \mu_2(\theta)r^{\sigma_1}\frac{\Delta^2\theta}{\Delta t^2}\right)\sin_1(\theta,\theta_0)$$

$$+ \left(\frac{\Delta^2\theta}{\Delta t^2} + \frac{1}{r^{\sigma_1}}\left(\frac{\Delta r^{\sigma_1}}{\Delta t}\frac{\Delta\theta}{\Delta t} + \left(\frac{\Delta r}{\Delta t}\right)^{\sigma_1}\frac{\Delta\theta}{\Delta t}\right)\right)\cos_1(\theta,\theta_0).$$

Thus, if a vector has components

$$\frac{\Delta^2 x}{\Delta t^2} \quad \text{and} \quad \frac{\Delta^2 y}{\Delta t^2}$$

in rectangular Cartesian coordinates, then they are

$$\frac{\Delta^2 r}{\Delta t^2} - r^{\sigma_1\sigma_1}\left(\frac{\Delta\theta}{\Delta t}\right)^2 - \mu_2(\theta)r^{\sigma_1}\frac{\Delta^2\theta}{\Delta t^2}$$

and

$$\frac{\Delta^2\theta}{\Delta t^2} + \frac{1}{r^{\sigma_1}}\left(\frac{\Delta r^{\sigma_1}}{\Delta t}\frac{\Delta\theta}{\Delta t} + \left(\frac{\Delta r}{\Delta t}\right)^{\sigma_1}\frac{\Delta\theta}{\Delta t}\right)$$

in coordinates (r,θ).

Example 1.24. If x^i are the coordinates of a point in the N-dimensional space and

$$\bar{x}^i = \bar{x}^i(x^1, x^2, \ldots, x^N),$$

then

$$\Delta_{(i)}\bar{x}^i = \frac{\Delta_{(i)}\bar{x}^i}{\Delta_1 x^1}\Delta_1 x^1 + \frac{\Delta_{(i)}\bar{x}^{i\sigma_1}}{\Delta_2 x^2}\Delta_2 x^2 + \cdots + \frac{\Delta_{(i)}\bar{x}^{i\sigma_1\sigma_2\cdots\sigma_{N-1}}}{\Delta_N x^N}\Delta_N x^N$$

$$= \frac{\Delta_{(i)}\bar{x}^{i\sigma_0\sigma_1\cdots\sigma_{j-1}}}{\Delta_j x^j}\Delta_j x^j.$$

Therefore the coordinate differentials $\Delta_{(i)}\bar{x}^i$ are the components of a contravariant vector.

Example 1.25. Let the coordinates of a point in the fluid be $x^i(t)$ at any time t. Then the velocity v^i at any point in the coordinate system x^i is given by

$$v^i = \frac{\Delta_i x^i}{\Delta t}.$$

Let the coordinates x^i be transformed to new coordinates $\bar{x}^i$. Then

$$\bar{v}^i = \frac{\Delta_{(i)}\bar{x}^i}{\Delta t}$$

$$= \frac{\Delta_{(i)}\bar{x}^{i\sigma_0\sigma_1\cdots\sigma_{k-1}}}{\Delta_k x^k}\frac{\Delta_k x^k}{\Delta t}$$

$$= \frac{\Delta_{(i)}\bar{x}^{i\sigma_0\sigma_1\cdots\sigma_{k-1}}}{\Delta_k x^k}v^k.$$

Thus the velocities v^i are components of a contravariant vector.

The coordinates x^i in $\frac{\Delta_i x^i}{\Delta t}$ are the coordinates of, say, a particle in motion, whereas the coefficients

$$\frac{\Delta_{(i)}\bar{x}^{i\sigma_0\sigma_1\cdots\sigma_{k-1}}}{\Delta_k x^k}$$

only denote a relation between two coordinate systems, which is independent of t. Thus for the acceleration a^i, we find

$$\bar{a}^i = \frac{\Delta v^i}{\Delta t}$$

$$= \frac{\Delta_{(i)}\bar{x}^{i\sigma_0\sigma_1\cdots\sigma_{k-1}}}{\Delta_k x^k}\frac{\Delta v^k}{\Delta t}$$

$$= \frac{\Delta_{(i)}\bar{x}^{i\sigma_0\sigma_1\cdots\sigma_{k-1}}}{\Delta_k x^k}a^k.$$

This shows that the acceleration a^i is also a contravariant vector.

Exercise 1.8. If the components of a contravariant vector in the x^i coordinate system are $(1, 4)$, then show that its components in the $\bar{x}^i$ coordinate system are $(24, 52)$, where

$$\overline{x}^1 = 3x^1,$$
$$\overline{x}^2 = 5x^1 + 3x^2.$$

Theorem 1.3. *The transformations of contravariant vectors form a group.*

Proof. Let A^i be a contravariant vector. Let also, S be the set of transformations of contravariant vectors, and let T_1 and T_2 be two such transformations from the system x^i to the system $\overline{x}^i$ and from the system $\overline{x}^i$ to the system x'^i, respectively, where $x'^j \in \widetilde{\mathbb{T}}_j$, $j \in \{1,\dots,N\}$, and $\widetilde{\mathbb{T}}_j, j \in \{1,\dots,N\}$, are time scales with forward jump operators and delta differentiation operators $\widetilde{\sigma}_j$ and $\widetilde{\Delta}_j$, respectively. Thus

$$T_1(A^i) = \frac{\Delta_{(i)}\overline{x}^i}{\Delta_p x^p}A^p,$$

and

$$T_2(A^i) = \frac{\widetilde{\Delta}_i x'^i}{\Delta_{(r)}\overline{x}^r}\overline{A}^r.$$

Then

$$T_2 T_1(A^j) = \frac{\widetilde{\Delta}_j x'^j}{\Delta_{(j)}\overline{x}^j}\overline{A}^j$$

$$= \frac{\widetilde{\Delta}_j x'^j}{\Delta_{(j)}\overline{x}^j}\frac{\Delta_{(j)}\overline{x}^j}{\Delta_k x^k}A^k$$

$$= \frac{\widetilde{\Delta}_j x'^j}{\Delta_k x^k}A^k.$$

This equation is of the same form as (1.6), and this shows that $T_2 T_1 \in S$. Let I be the transformation given by

$$I : A^i = \frac{\Delta_p x^p}{\Delta_p^p}\overline{A}^i.$$

Then

$$IT_1(A^i) = \frac{\Delta_p x^p}{\Delta_p x^p}\overline{A}^i$$

$$= \frac{\Delta_p x^p}{\Delta_p x^p}\frac{\Delta_{(i)}\overline{x}^i}{\Delta_j x^j}A^j$$

$$= \frac{\Delta_{(i)}\overline{x}^i}{\Delta_j x^j}A^j$$

$$= T_1(A^i),$$

and

$$T_1 I(A^i) = \frac{\Delta_{(i)}\overline{x}^i}{\Delta_k x^k} \frac{\Delta_p x^p}{\Delta_p x^p} A^k$$

$$= \frac{\Delta_{(i)}\overline{x}^i}{\Delta_k x^k} A^k$$

$$= T_1(A^i).$$

Therefore I is the identity transformation, and $I \in S$. Next, consider the transformation

$$T_1^*(\overline{A}^i) = \frac{\Delta_p x^p}{\Delta_{(i)}\overline{x}^i} \overline{A}^i.$$

Then

$$T_1^* T_1(A^p) = \frac{\Delta_p x^p}{\Delta_{(i)}\overline{x}^i} \frac{\Delta_{(i)}\overline{x}^i}{\Delta_j x^j} A^j$$

$$= \frac{\Delta_{(i)}\overline{x}^i}{\Delta_{(i)}\overline{x}^i} A^p$$

$$= I(A^p),$$

and

$$T_1 T_1^*(\overline{A}^i) = \frac{\Delta_{(i)}\overline{x}^i}{\Delta_p x^p} \frac{\Delta_p x^p}{\Delta_{(i)}\overline{x}^i} \overline{A}^i$$

$$= \frac{\Delta_p x^p}{\Delta_p x^p} \overline{A}^i$$

$$= I(\overline{A}^i).$$

Consequently,

$$T_1^* T_1 = T_1 T_1^* = I.$$

In other words, T_1^* is the inverse of T_1, and $T_1^* \in S$.

Let now T_3 represent a transformation from the system x'^i to $\overline{\overline{x}}^i$, where $\overline{\overline{x}}^i \in \tilde{\tilde{T}}_i$, $\tilde{\tilde{T}}_i$ are time scales with forward jump operators and delta differentiation operators $\tilde{\tilde{\sigma}}_i$ and $\tilde{\tilde{\Delta}}_i$, respectively. Then

$$(T_3(T_2 T_1))(A^i) = \frac{\widetilde{\widetilde{\Delta}}_i \overline{\overline{x}}^i}{\widetilde{\Delta}_p x'^p}\left(\frac{\widetilde{\Delta}_p x'^p}{\Delta_{(j)} \overline{x}^j}\frac{\Delta_{(j)} \overline{x}^j}{\Delta_k x^k}A^k\right)$$

$$= \frac{\widetilde{\widetilde{\Delta}}_i \overline{\overline{x}}^i}{\widetilde{\Delta}_p x'^p}\frac{\widetilde{\Delta}_p x'^p}{\Delta_{(j)} \overline{x}^j}\frac{\Delta_{(j)} \overline{x}^j}{\Delta_k x^k}A^k$$

$$= \frac{\widetilde{\widetilde{\Delta}}_i \overline{\overline{x}}^i}{\Delta_k x^k}A^k,$$

and

$$((T_3 T_2)T_1)(A^i) = \left(\frac{\widetilde{\widetilde{\Delta}}_i \overline{\overline{x}}^i}{\widetilde{\Delta}_p x'^p}\frac{\widetilde{\Delta}_p x'^p}{\Delta_{(j)} \overline{x}^j}\right)\frac{\Delta_{(j)} \overline{x}^j}{\Delta_k x^k}A^k$$

$$= \frac{\widetilde{\widetilde{\Delta}}_i \overline{\overline{x}}^i}{\widetilde{\Delta}_p x'^p}\frac{\widetilde{\Delta}_p x'^p}{\Delta_{(j)} \overline{x}^j}\frac{\Delta_{(j)} \overline{x}^j}{\Delta_k x^k}A^k$$

$$= \frac{\widetilde{\widetilde{\Delta}}_i \overline{\overline{x}}^i}{\Delta_k x^k}A^k.$$

Consequently,

$$T_3(T_2 T_1) = (T_3 T_2)T_1.$$

Thus S is a group. This completes the proof. $\square$

With the exception of the coordinates x^j themselves, a single superscript will always denote a contravariant vector unless the contrary is explicitly stated. The coordinates x^j will only behave like the components of a contravariant vector with respect to linear transformations of the type

$$\overline{x}^j = a^j_k x^k,$$

where a^j_k are a set of N^2 constants. In this case, we have

$$\frac{\Delta_{(j)} \overline{x}^j}{\Delta_k x^k} = a^j_k,$$

and the transformation is given by

$$\overline{x}^j = \frac{\Delta_{(j)} \overline{x}^j}{\Delta_k x^k} x^k.$$

With respect to general transformations of coordinates, the x^j do not form the components of a contravariant vector. Essentially, this means that if we select

$$A^i = x^i,$$

then the components $\overline{A}^i$ with respect to the coordinate system $\overline{x}^i$ do not satisfy the equation

$$\overline{A}^i = \overline{x}^i.$$

1.6 Covariant vectors

In this section, assume that

$$x^j(\overline{x}^1, \ldots, \overline{x}^{l-1}, \sigma_{(l)}(\overline{x}^l), \overline{x}^{l+1}, \ldots, \overline{x}^n) = \sigma_j(x^j(\overline{x}^1, \ldots, \overline{x}^n)).$$

Consider a scalar

$$A(x) = \overline{A}(\overline{x}).$$

Differentiating and using the chain rule, we find the components of the gradient

$$\frac{\partial \overline{A}}{\Delta_{(i)}\overline{x}^i} = \frac{\partial A^{\sigma_0 \sigma_1 \cdots \sigma_{j-1}}}{\Delta_j x^j} \frac{\Delta_j x^j}{\Delta_{(i)}\overline{x}^i}.$$

Let

$$A_j = \frac{\partial A^{\sigma_0 \sigma_1 \cdots \sigma_{j-1}}}{\Delta_j x^j},$$

$$\overline{A}_i = \frac{\partial \overline{A}}{\Delta_{(i)}\overline{x}^i}.$$

Then

$$\overline{A}_i = A_j \frac{\Delta_j x^j}{\Delta_{(i)}\overline{x}^i}.$$

This is the transformation law for a covariant vector.

Definition 1.10. N functions A_i of the N coordinates are said to be the components of a covariant vector if they transform according to the equation

$$\overline{A}_i = \frac{\Delta_j x^j}{\Delta_{(i)}\overline{x}^i} A_j \tag{1.7}$$

on change of coordinates x^i to $\overline{x}^i$.

Any N functions can be chosen as the components of a covariant vector in the coordinate system x^i, and equations (1.7) define the N components in the new coordinate system $\bar{x}^i$.

Multiplying (1.7) by

$$\frac{\Delta_{(k)}\bar{x}^{k\sigma_0\sigma_1\cdots\sigma_{j-1}}}{\Delta_j x^j}$$

and using that

$$\frac{\Delta_{(k)}\bar{x}^k}{\Delta_{(i)}\bar{x}^i} = \frac{\Delta_{(k)}\bar{x}^{k\sigma_0\sigma_1\cdots\sigma_{j-1}}}{\Delta_j x^j}\frac{\Delta_j x^j}{\Delta_{(i)}\bar{x}^i},$$

we find

$$\frac{\Delta_{(k)}\bar{x}^{k\sigma_0\sigma_1\cdots\sigma_{j-1}}}{\Delta_j x^j}\bar{A}_i = A_j\frac{\Delta_{(k)}\bar{x}^{k\sigma_0\sigma_1\cdots\sigma_{j-1}}}{\Delta_j x^j}\frac{\Delta_j x^j}{\Delta_{(i)}\bar{x}^i}$$

$$= A_j\frac{\Delta_{(k)}\bar{x}^k}{\Delta_{(i)}\bar{x}^i},$$

whereupon

$$\frac{\Delta_{(k)}\bar{x}^{k\sigma_0\sigma_1\cdots\sigma_{j-1}}}{\Delta_j x^j}\bar{A}_k = A_j.$$

Example 1.26. Since

$$\frac{\partial f}{\Delta_{(i)}\bar{x}^i} = \left(\frac{\partial f}{\Delta_j x^j}\right)^{\sigma_0\cdots\sigma_{j-1}}\frac{\Delta_j x^j}{\Delta_{(i)}\bar{x}^i}$$

$$= \frac{\partial f^{\sigma_0\cdots\sigma_{j-1}}}{\Delta_j x^j}\frac{\Delta_j x^j}{\Delta_{(i)}\bar{x}^i},$$

it follows immediately from (1.7) that the components

$$\frac{\partial f}{\Delta_j x^j}$$

are the components of a covariant vector, whose components in any other system are the corresponding partial derivatives

$$\frac{\partial f}{\Delta_{(k)}x^k}.$$

Such a covariant vector is called the time scale gradient of f. A single subscript will always denote a covariant vector unless the contrary is explicitly stated. In conformity with this convention, we will regard the index k in the covariant vector

$$\frac{\Delta_k f^{\sigma_0 \cdots \sigma_{k-1}}}{\Delta_k x^k}$$

as a subscript.

Example 1.27. Let the components of a covariant vector in the x^i coordinate system be $(x^2, x^1 + 2x^2)$, $x^2 > 0$. We will find its coordinates in the $\overline{x}^i$ coordinate system, where

$$\overline{x}^1 = (x^2)^2,$$
$$\overline{x}^2 = x^1 x^2.$$

Here

$$A_1 = x^2,$$
$$A_2 = x^1 + 2x^2$$

and

$$x^1 = \frac{\overline{x}^2}{\sqrt{\overline{x}^1}},$$
$$x^2 = \sqrt{\overline{x}^1}.$$

We will find

$$\frac{\Delta_{(i)} \overline{x}^{i\sigma_0 \sigma_1 \cdots \sigma_{j-1}}}{\Delta_j x^j}.$$

We have

$$\frac{\Delta_{(1)} \overline{x}^1}{\Delta_1 x^1} = 0,$$

$$\frac{\Delta_{(1)} \overline{x}^1}{\Delta_2 x^2} = \sigma_2(x^2) + x^2,$$

$$\frac{\Delta_{(1)} \overline{x}^{1\sigma_1}}{\Delta_2 x^2} = \sigma_2(x^2) + x^2,$$

$$\frac{\Delta_{(2)} \overline{x}^2}{\Delta_1 x^1} = x^2,$$

$$\frac{\Delta_{(2)}\bar{x}^2}{\Delta_2 x^2} = x^1,$$

$$\frac{\Delta_{(2)}\bar{x}^{2\sigma_1}}{\Delta_2 x^2} = \sigma_1(x^1).$$

Then

$$\begin{aligned}
A_1 &= \frac{\Delta_{(k)}\bar{x}^k}{\Delta_1 x^1}\bar{A}_k \\
&= \frac{\Delta_{(1)}\bar{x}^1}{\Delta_1 x^1}\bar{A}_1 + \frac{\Delta_{(2)}\bar{x}^2}{\Delta_1 x^1}\bar{A}_2 \\
&= 0\bar{A}_1 + x^2\bar{A}_2 \\
&= x^2\bar{A}_2,
\end{aligned}$$

and

$$\begin{aligned}
A_2 &= \frac{\Delta_{(k)}\bar{x}^{k\sigma_1}}{\Delta_2 x^2}\bar{A}_k \\
&= \frac{\Delta_{(1)}\bar{x}^{1\sigma_1}}{\Delta_2 x^2}\bar{A}_1 + \frac{\Delta_{(2)}\bar{x}^{2\sigma_1}}{\Delta_2 x^2}\bar{A}_2 \\
&= \left(\sigma_2(x^2) + x^2\right)\bar{A}_1 + \sigma_1(x^1)\bar{A}_2.
\end{aligned}$$

Thus

$$\left(x^2\right)^2 = x^2\bar{A}_2,$$
$$x^1 + 2x^2 = \left(\sigma_2(x^2) + x^2\right)\bar{A}_1 + \sigma_1(x^1)\bar{A}_2.$$

Hence

$$\bar{A}_2 = x^2,$$
$$\left(\sigma_2(x^2) + x^2\right)\bar{A}_1 = x^1 + 2x^2 - \sigma_1(x^1)x^2,$$

that is,

$$\bar{A}_1 = \frac{x^1 + 2x^2 - \sigma_1(x^1)x^2}{x^2 + \sigma_2(x^2)}$$
$$\bar{A}_2 = x^2,$$

or

$$\bar{A}_1 = \frac{\dfrac{\bar{x}^2}{\sqrt{\bar{x}^1}} + 2\sqrt{\bar{x}^1} - \dfrac{\bar{x}^2}{\sqrt{\sigma_{(1)}(\bar{x}^1)}}\sqrt{\bar{x}^1}}{\sqrt{\sigma_{(1)}(\bar{x}^1)} + \sqrt{\bar{x}^1}},$$

$$\bar{A}_2 = \sqrt{\bar{x}^1}.$$

We now show that there is no distinction between contravariant and covariant vectors when we restrict ourselves to transformations of the type

$$\bar{x}^j = a^j_m x^m + b^j, \tag{1.8}$$

where b^j are N constants, which do not necessarily form the components of a contravariant vector, and a^j_m are constants such that

$$a^j_r a^j_m = \delta^r_m.$$

Multiplying equations (1.8) by a^j_r and summing over the index j from 1 to N, we obtain

$$a^j_r \bar{x}^j = a^j_r a^j_m x^m + a^j_r b^j,$$

or

$$a^j_r \bar{x}^j = \delta^m_r x^m + a^j_r b^j,$$

or

$$a^j_r \bar{x}^j = x^r + a^j_r b^j,$$

or

$$x^r = a^j_r \bar{x}^j - a^j_r b^j.$$

Thus

$$\frac{\Delta_{(j)}\bar{x}^j}{\Delta_k x^k} = a^j_k$$

and

$$\frac{\Delta_k x^k}{\Delta_{(j)}\bar{x}^j} = a^j_k.$$

This shows that equations (1.6) and (1.7) define the same type of entity.

1.7 Invariants

Definition 1.11. A function F of the N coordinates is called an invariant or a scalar with respect to the coordinate transformations if

$$F = \overline{F},$$

where $\overline{F}$ is the value of F in the new coordinate system $\overline{x}^i$.

Example 1.28. From the components A^i and B_i of a contravariant vector and a covariant vector, respectively, we can form the sum

$$A^j B_j. \tag{1.9}$$

When we change the new coordinates $\overline{x}^j$, the sum transforms to

$$\overline{A}^j \overline{B}_j.$$

We have

$$
\overline{A}^j \overline{B}_j = \frac{\Delta_{(j)} \overline{x}^{j\sigma_0 \sigma_1 \cdots \sigma_{k-1}}}{\Delta_k x^k} A^k \frac{\Delta_k x^k}{\Delta_{(l)} \overline{x}^j} B_j
$$

$$
= \frac{\Delta_{(j)} \overline{x}^j}{\Delta_{(l)} \overline{x}^l} A^k B_j
$$

$$
= A^j B_j,
$$

where we have used that

$$
\frac{\Delta_{(j)} \overline{x}^j}{\Delta_{(l)} \overline{x}^l} = \frac{\Delta_{(j)} \overline{x}^{j\sigma_0 \sigma_1 \cdots \sigma_{k-1}}}{\Delta_k x^k} \frac{\Delta_k x^k}{\Delta_{(l)} \overline{x}^l}.
$$

Thus the sum (1.9) is an invariant.

1.8 Advanced practical problems

Problem 1.1. Using the relation

$$
\frac{\Delta_p x^p}{\Delta_q x^q} = \delta^p_q,
$$

prove that

$$
\frac{\partial}{\Delta_k x^k}(a_{ij} x^i x^j) = (a_{ik} + a_{ki})x^k + a_{kk}(\sigma_k(x^k) + x^k), \quad i \neq k.
$$

Problem 1.2. If

$$A = a_p x^p, \quad a_p = \text{constant},$$

then prove that

$$\frac{\partial}{\Delta_j x^j}(a_p x^p) = a_j.$$

Problem 1.3. Compute

$$\frac{\partial}{\Delta_k x^k}(a_{ij} x^j), \quad a_{ij} = \text{constant}.$$

Answer 1.4. a_{ik}.

Problem 1.4. Compute

$$\frac{\partial}{\Delta_k x^k}(a_{ij} x^i (x^j)^2), \quad a_{ij} = \text{constant}.$$

Answer 1.5.

$$a_{ik} x^i (x^k + \sigma_k(x^k)) + a_{kj} x^j x^j + a_{kk}((\sigma_k(x^k))^2 + x^k \sigma_k(x^k) + x^k x^k), \quad i, j \neq k.$$

Problem 1.5. Compute

$$\frac{\partial}{\Delta_l x^l}(a_{ijk} x^i x^j x^k), \quad a_{ijk} = \text{constant}.$$

Answer 1.6.

$$a_{ijl} x^i x^j + a_{llk} x^k (x^l + \sigma_l(x^l)) + a_{ijl} x^j (x^l + \sigma_l(x^l))$$
$$+ a_{ill} x^i (x^l + \sigma_l(x^l)) + a_{lll}((\sigma_l(x^l))^2 + \sigma_l(x^l) x^l + x^l x^l), \quad i, j, k \neq l.$$

Problem 1.6. Find the following partial derivatives if a_{ij} are constants:

$$\frac{\partial}{\Delta_k x^k}(a_{11} x^1 + a_{12} x^2 + a_{13} x^3), \quad k \in \{1, 2, 3\}.$$

Answer 1.7. a_{1k}.

Problem 1.7. Show that for the independent functions

$$\bar{x}^i = \bar{x}^i(x^1, x^2, \ldots, x^N),$$

$$\frac{\Delta_{(i)} \bar{x}^{i\sigma_0 \sigma_1 \cdots \sigma_{r-1}}}{\Delta_r x^r} \frac{\Delta_r x^r}{\Delta_{(j)} \bar{x}^j} = \delta^i_j.$$

Take the partial derivative withe respect to x^k to get the following formulas:

$$\frac{\Delta^2_{(i)}\overline{x}^{i\sigma_0\sigma_1\cdots\sigma_{r-1}}}{\Delta_k x^k \Delta_r x^r}\frac{\Delta_r x^r}{\Delta_{(j)}\overline{x}^j} = -\left(\frac{\Delta_{(i)}\overline{x}^{i\sigma_0\sigma_1\cdots\sigma_{r-1}}}{\Delta_r x^r}\right)^{\sigma_k}\left(\frac{\Delta^2_r}{\Delta_{(m)}\overline{x}^m \Delta_{(j)}\overline{x}^j}\right)^{\sigma_0\sigma_1\cdots\sigma_{m-1}}\frac{\Delta_{(m)}\overline{x}^m}{\Delta_k x^k}$$

and

$$\frac{\Delta^2_{(i)}\overline{x}^{i\sigma_0\sigma_1\cdots\sigma_{r-1}}}{\Delta_k x^k \Delta_r x^r}\left(\frac{\Delta_r x^r}{\Delta_{(j)}\overline{x}^j}\right)^{\sigma_k} = -\frac{\Delta_{(i)}\overline{x}^{i\sigma_0\sigma_1\cdots\sigma_{r-1}}}{\Delta_r x^r}\left(\frac{\Delta^2_r}{\Delta_{(m)}\overline{x}^m \Delta_{(j)}\overline{x}^j}\right)^{\sigma_0\sigma_1\cdots\sigma_{m-1}}\frac{\Delta_{(m)}\overline{x}^m}{\Delta_k x^k}.$$

Problem 1.8. Suppose that the following transformation connects x^i to $\overline{x}^i$ coordinate system:

$$\overline{x}^1 = x^1 \sin_2(x^2, x_0^2)\cos_3(x^3, x_0^3),$$
$$\overline{x}^2 = x^1 \sin_2(x^2, x_0^2)\sin_3(x^3, x_0^3),$$
$$\overline{x}^3 = x^1 \cos_2(x^2, x_0^2),$$

where $x_0^2 \in \mathbb{T}_2$ and $x_0^3 \in \mathbb{T}_3$ are fixed. Find the Jacobian of the transformation.

Answer 1.8.

$$6(x^1)^2 \sin_2(x^2, x_0^2)e_{4\mu_2}(x^2, x_0^2)e_{9\mu_3}(x^3, x_0^3).$$

Problem 1.9. For $N = 3$, prove that

$$e_{ijk}e^{ijk} = 6.$$

Problem 1.10. For $N = 2$, expand
1.

$$e^{ij}a_i^1 aa_j^2.$$

2.

$$e^{ij}a_i^2 a_j^1.$$

Problem 1.11. Prove that
1.

$$\delta^{ij}_{\alpha\beta}a^{\alpha\beta} = a^{ij} - a^{ji}.$$

2.

$$\delta^{ijk}_{\alpha\beta\gamma}a^{\alpha\beta\gamma} = a^{ijk} - a^{ikj} + a^{jki} - a^{jik} + a^{kij} - a^{kji}.$$

Problem 1.12. Prove

1.

$$\delta^{ij}_{\alpha\beta} = \det \begin{pmatrix} \delta^i_\alpha & \delta^i_\beta \\ \delta^j_\alpha & \delta^j_\beta \end{pmatrix}.$$

2.

$$\delta^{ijk}_{\alpha\beta\gamma} = \det \begin{pmatrix} \delta^i_\alpha & \delta^i_\beta & \delta^i_\gamma \\ \delta^j_\alpha & \delta^j_\beta & \delta^j_\gamma \\ \delta^k_\alpha & \delta^k_\beta & \delta^k_\gamma \end{pmatrix}.$$

Problem 1.13. If the components of a contravariant vector in the x^i coordinate system are $(2, 1, 2)$, then show that its components in the $\overline{x}^i$ coordinate system are $(5, 5, 5)$, where

$$\overline{x}^1 = 3x^1 - 3x^2 + 2x^3,$$
$$\overline{x}^2 = 2x^2 + 3x^3,$$
$$\overline{x}^3 = x^1 + x^2 + 2x^3.$$

Problem 1.14. The components of a contravariant vector in the x^i coordinate system are $(2, 3)$. Find its components in the $\overline{x}^i$ coordinate system, where

$$\overline{x}^1 = 3\left(x^1\right)^2,$$
$$\overline{x}^2 = 5\left(x^1\right)^2 + 3\left(x^2\right)^2, \quad x^1 > 0, \ x^2 > 0.$$

Answer 1.9.

$$\left(6\left(\sqrt{\frac{\sigma_{(1)}(\overline{x}^1)}{3}} + \sqrt{\frac{\overline{x}^1}{3}} \right), 10\left(\sqrt{\frac{\sigma_{(1)}(\overline{x}^1)}{3}} + \sqrt{\frac{\overline{x}^1}{3}} \right) + 9\left(\frac{\sqrt{3\sigma_{(2)}(\overline{x}^2) - 5\overline{x}^1}}{3} + \sqrt{\frac{3\overline{x}^2 - 5\overline{x}^1}{3}} \right) \right).$$

2 Tensors

This chapter provides an informative introduction concerning the origin and nature of the tensor concept on time scales and the scope of the tensor calculus on time scales. The tensor algebra has been developed in an N-dimensional space. Contravariant, covariant, and mixed tensors of arbitrary order are defined. Some of their properties are deduced. The quotient law of the tensors is formulated and proved. Outer product and contractions are introduced.

Throughout this book, we suppose

$$\overline{x}^i(x^1,\ldots,x^{l-1},\sigma_l(x^l),x^{l+1},\ldots,x^n) = \sigma_{(i)}(\overline{x}^i(x^1,x^2,\ldots,x^N)), \quad i,l \in \{1,\ldots,N\},$$

and

$$x^i(\overline{x},\ldots,\overline{x}^{l-1},\sigma_{(l)}(\overline{x}^l),\overline{x}^{l+1},\ldots,\overline{x}^N) = \sigma_i(x^1(\overline{x}^1,\overline{x}^2,\ldots,\overline{x}^N)), \quad i,l \in \{1,\ldots,N\}.$$

2.1 Second-order tensors

Form the N^2 quantities

$$A^{ij} = B^i C^j,$$

where B^i and C^j are components of contravariant vectors. Using the transformation law of contravariant vectors, we get

$$\begin{aligned}
\overline{A}^{ij} &= \overline{B}^i \overline{C}^j \\
&= \frac{\Delta_{(i)}\overline{x}^{i\sigma_0\sigma_1\cdots\sigma_{k-1}}}{\Delta_k x^k} \frac{\Delta_{(j)}\overline{x}^{j\sigma_0\sigma_1\cdots\sigma_{l-1}}}{\Delta_l x^l} B^k C^l \\
&= \frac{\Delta_{(i)}\overline{x}^{i\sigma_0\sigma_1\cdots\sigma_{k-1}}}{\Delta_k x^k} \frac{\Delta_{(j)}\overline{x}^{j\sigma_0\sigma_1\cdots\sigma_{l-1}}}{\Delta_l x^l} A^{kl},
\end{aligned}$$

i. e.,

$$\overline{A}^{ij} = \frac{\Delta_{(i)}\overline{x}^{i\sigma_0\sigma_1\cdots\sigma_{k-1}}}{\Delta_k x^k} \frac{\Delta_{(j)}\overline{x}^{j\sigma_0\sigma_1\cdots\sigma_{l-1}}}{\Delta_l x^l} A^{kl}. \tag{2.1}$$

Moreover, we have

$$A^{ij} = B^i C^j$$

$$= \frac{\Delta_i x^i}{\Delta_{(k)}\overline{x}^k}\overline{B}^k \frac{\Delta_j x^j}{\Delta_{(l)}\overline{x}^l}\overline{C}^l \tag{2.2}$$

https://doi.org/10.1515/9783112228494-002

$$= \frac{\Delta_i x^i}{\Delta_{(k)}\bar{x}^k} \frac{\Delta_j x^j}{\Delta_{(l)}\bar{x}^l} \bar{B}^k \bar{C}^l$$

$$= \frac{\Delta_i x^i}{\Delta_{(k)}\bar{x}^k} \frac{\Delta_j x^j}{\Delta_{(l)}\bar{x}^l} \bar{A}^{kl}.$$

Definition 2.1. More generally, if we have N^2 functions A^{ij} whose transformation law is that of (2.1) or (2.2), then we call A^{ij} the components of a contravariant tensor of the second order.

This tensor is not necessarily the product of the components of two contravariant vectors. Any set of N^2 functions can be chosen as the components of a contravariant tensor of the second order, and then (2.1) defines the components in any other coordinate system $\bar{x}^i$.

A contravariant tensor of the second order can be written as a square matrix of order N, just as a contravariant tensor of the first order can be treated as an N-component vector. Thus a contravariant vector of the second order can be written as

$$A^{ij} = \begin{pmatrix} A^{11} & A^{12} & \cdots & A^{1N} \\ A^{21} & A^{22} & \cdots & A^{2N} \\ \vdots & \vdots & \vdots & \vdots \\ A^{N1} & A^{N2} & \cdots & A^{NN} \end{pmatrix}.$$

However, the converse is not true. The elements of an arbitrary square matrix do not form the components of a contravariant tensor of the second order.

Example 2.1. Let the components of a contravariant tensor A^{ij} of the second order be as follows:

$$A^{11} = 1, \quad A^{12} = 1, \quad A^{21} = -1, \quad A^{22} = 2.$$

We will find its components in the $\bar{x}^i$ coordinate system, where

$$\bar{x}^1 = (x^1)^2,$$
$$\bar{x}^2 = x^1 x^2, \quad x^1, x^2 > 0.$$

We have

$$x^1 = \sqrt{\bar{x}^1},$$
$$x^2 = \frac{\bar{x}^2}{\sqrt{\bar{x}^1}}.$$

Now we will compute the partial derivatives:

$$\frac{\Delta_{(j)}\overline{x}^{j\sigma_0\sigma_1\cdots\sigma_{k-1}}}{\Delta_k x^k}, \quad j,k \in \{1,2\}.$$

We have

$$\frac{\Delta_{(1)}\overline{x}^1}{\Delta_1 x^1} = \sigma_1(x^1) + x^1$$

$$= \sigma_1(\sqrt{\overline{x}^1}) + \sqrt{\overline{x}^1}$$

$$= \sqrt{\sigma_{(1)}(\overline{x}^1)} + \sqrt{\overline{x}^1},$$

$$\frac{\Delta_{(1)}\overline{x}^1}{\Delta_2 x^2} = 0,$$

$$\frac{\Delta_{(1)}\overline{x}^{1\sigma_1}}{\Delta_2 x^2} = 0,$$

$$\frac{\Delta_{(2)}\overline{x}^2}{\Delta_1 x^1} = x^2$$

$$= \frac{\overline{x}^2}{\sqrt{\overline{x}^1}},$$

$$\frac{\Delta_{(2)}\overline{x}^2}{\Delta_2 x^2} = x^1$$

$$= \sqrt{\overline{x}^1},$$

$$\frac{\Delta_{(2)}\overline{x}^{2\sigma_1}}{\Delta_2 x^2} = \sigma_1(x^1)$$

$$= \sqrt{\sigma_{(1)}(\overline{x}^1)}.$$

Then

$$\overline{A}^{11} = \frac{\Delta_{(1)}\overline{x}^{1\sigma_0\sigma_1\cdots\sigma_{k-1}}}{\Delta_k x^k}\frac{\Delta_{(1)}\overline{x}^{1\sigma_0\sigma_1\cdots\sigma_{l-1}}}{\Delta_1 x^l}A^{kl}$$

$$= \left(\frac{\Delta_{(1)}\overline{x}^1}{\Delta_1 x^1}\right)^2 A^{11} + \frac{\Delta_{(1)}\overline{x}^1}{\Delta_1 x^1}\frac{\Delta_{(1)}\overline{x}^{1\sigma_1}}{\Delta_2 x^2}A^{12}$$

$$+ \frac{\Delta_{(1)}\overline{x}^{1\sigma_1}}{\Delta_2 x^2}\frac{\Delta_{(1)}\overline{x}^1}{\Delta_1 x^1}A^{21} + \left(\frac{\Delta_{(1)}\overline{x}^{1\sigma_1}}{\Delta_2 x^2}\right)^2 A^{22}$$

$$= \left(\sqrt{\sigma_{(1)}(\overline{x}^1)} + \sqrt{\overline{x}^1}\right)^2(1) + \left(\sqrt{\sigma_{(1)}(\overline{x}^1)} + \sqrt{\overline{x}^1}\right)(0)(1)$$

$$+ 0\left(\sqrt{\sigma_{(1)}(\overline{x}^1)} + \sqrt{\overline{x}^1}\right)(-1) + 0(\sigma_1(x^1))(2)$$

$$= \left(\sqrt{\sigma_{(1)}(\overline{x}^1)} + \sqrt{\overline{x}^1}\right)^2,$$

$$\overline{A}^{12} = \frac{\Delta_{(1)}\overline{x}^{1\sigma_0\sigma_1\cdots\sigma_{k-1}}}{\Delta_k x^k}\frac{\Delta_{(2)}\overline{x}^{2\sigma_0\sigma_1\cdots\sigma_{l-1}}}{\Delta_l x^l}A^{kl}$$

$$= \frac{\Delta_{(1)}\overline{x}^1}{\Delta_1 x^1}\frac{\Delta_{(2)}\overline{x}^2}{\Delta_1 x^1}A^{11} + \frac{\Delta_{(1)}\overline{x}^1}{\Delta_1 x^1}\frac{\Delta_{(2)}\overline{x}^{2\sigma_1}}{\Delta_2 x^2}A^{12}$$

$$+ \frac{\Delta_{(1)}\overline{x}^{1\sigma_1}}{\Delta_2 x^2}\frac{\Delta_{(2)}\overline{x}^2}{\Delta_1 x^1}A^{21} + \frac{\Delta_{(1)}\overline{x}^{1\sigma_1}}{\Delta_2 x^2}\frac{\Delta_{(2)}\overline{x}^{2\sigma_1}}{\Delta_2 x^2}A^{22}$$

$$= \left(\sqrt{\sigma_{(1)}(\overline{x}^1)} + \sqrt{\overline{x}^1}\right)\left(\frac{\overline{x}^2}{\sqrt{\overline{x}^1}}\right)(1) + \left(\sqrt{\sigma_{(1)}(\overline{x}^1)} + \sqrt{\overline{x}^1}\right)\left(\sqrt{\overline{x}^1}\right)(1)$$

$$+ 0(x^2)(-1) + 0(\sigma_1(x^1))(2)$$

$$= \left(\sqrt{\sigma_{(1)}(\overline{x}^1)} + \sqrt{\overline{x}^1}\right)\left(\frac{\overline{x}^2}{\sqrt{\overline{x}^1}} + \sqrt{\overline{x}^1}\right)$$

$$= \left(\sqrt{\sigma_{(1)}(\overline{x}^1)} + \sqrt{\overline{x}^1}\right)\frac{\overline{x}^1 + \overline{x}^2}{\sqrt{\overline{x}^1}},$$

$$\overline{A}^{21} = \frac{\Delta_{(2)}\overline{x}^{2\sigma_0\sigma_1\cdots\sigma_{k-1}}}{\Delta_k x^k}\frac{\Delta_{(1)}\overline{x}^{1\sigma_0\sigma_1\cdots\sigma_{l-1}}}{\Delta_l x^l}A^{kl}$$

$$= \frac{\Delta_{(2)}\overline{x}^2}{\Delta_1 x^1}\frac{\Delta_{(1)}\overline{x}^1}{\Delta_1 x^1}A^{11} + \frac{\Delta_{(2)}\overline{x}^2}{\Delta_1 x^1}\frac{\Delta_{(1)}\overline{x}^{1\sigma_1}}{\Delta_2 x^2}A^{12}$$

$$+ \frac{\Delta_{(2)}\overline{x}^{2\sigma_1}}{\Delta_2 x^2}\frac{\Delta_{(1)}\overline{x}^1}{\Delta_1 x^1}A^{21} + \frac{\Delta_{(2)}\overline{x}^{2\sigma_1}}{\Delta_2 x^2}\frac{\Delta_{(1)}\overline{x}^{1\sigma_1}}{\Delta_2 x^2}A^{22}$$

$$= \frac{\overline{x}^2}{\sqrt{\overline{x}^1}}\left(\sqrt{\sigma_{(1)}(\overline{x}^1)} + \sqrt{\overline{x}^1}\right)(1) + \frac{\overline{x}^2}{\sqrt{\overline{x}^1}}(0)(1)$$

$$+ \sqrt{\sigma_{(1)}(\overline{x}^1)}\left(\sqrt{\sigma_{(1)}(\overline{x}^1)} + \sqrt{\overline{x}^1}\right)(-1) + \sqrt{\overline{x}^1}(0)(2)$$

$$= \left(\sqrt{\sigma_{(1)}(\overline{x}^1)} + \sqrt{\overline{x}^1}\right)\left(\frac{\overline{x}^2}{\sqrt{\overline{x}^1}} - \sqrt{\sigma_{(1)}(\overline{x}^1)}\right),$$

$$\overline{A}^{22} = \frac{\Delta_{(2)}\overline{x}^{2\sigma_0\sigma_1\cdots\sigma_{k-1}}}{\Delta_k x^k}\frac{\Delta_{(2)}\overline{x}^{2\sigma_0\sigma_1\cdots\sigma_{l-1}}}{\Delta_l x^l}A^{kl}$$

$$= \left(\frac{\Delta_{(2)}\overline{x}^2}{\Delta_1 x^1}\right)^2 A^{11} + \frac{\Delta_{(2)}\overline{x}^2}{\Delta_1 x^1}\frac{\Delta_{(2)}\overline{x}^{2\sigma_1}}{\Delta_2 x^2}A^{12}$$

$$+ \frac{\Delta_{(2)}\overline{x}^2}{\Delta_2 x^{2\sigma_1}}\frac{\Delta_{(2)}\overline{x}^2}{\Delta_1 x^1}A^{21} + \left(\frac{\Delta_{(2)}\overline{x}^{2\sigma_1}}{\Delta_2 x^2}\right)^2 A^{22}$$

$$= \left(\frac{\overline{x}^2}{\sqrt{\overline{x}^1}}\right)^2 (1) + \left(\frac{\overline{x}^2}{\sqrt{\overline{x}^1}}\right)\left(\sqrt{\sigma_{(1)}(\overline{x}^1)}\right)(1)$$

$$+ \left(\sqrt{\sigma_{(1)}(\bar{x}^1)} \right) \left(\frac{\bar{x}^2}{\sqrt{\bar{x}^1}} \right) (-1) + \left(\sqrt{\sigma_{(1)}(\bar{x}^1)} \right)^2 (2)$$

$$= \frac{(\bar{x}^2)^2}{\bar{x}^1} + 2\sigma_{(1)}(\bar{x}^1).$$

Exercise 2.1. Let the components of a contravariant tensor A^{ij} of the second order be as follows:

$$A^{11} = -1, \quad A^{12} = 2, \quad A^{21} = 3, \quad A^{22} = -1.$$

Find its components in the $\bar{x}^i$ coordinate system, where

$$\bar{x}^1 = 3x^1 - 2x^2,$$
$$\bar{x}^2 = (x^1)^2, \quad x^1 > 0, \ x^2 > 0.$$

Answer 2.1.

$$\bar{A}^{11} = -37,$$
$$\bar{A}^{12} = -9\left(\sqrt{\sigma_{(2)}(\bar{x}^2)} + \sqrt{\bar{x}^2} \right),$$
$$\bar{A}^{21} = -3\left(\sqrt{\sigma_{(2)}(\bar{x}^2)} + \sqrt{\bar{x}^2} \right),$$
$$\bar{A}^{22} = -\left(\sqrt{\sigma_{(2)}(\bar{x}^2)} + \sqrt{\bar{x}^2} \right)^2.$$

Similarly, if we have N^2 functions

$$A_{ij} = B_i C_j,$$

where B_i and C_j are the components of two covariant vectors, applying (1.7), we find

$$\bar{A}_{ij} = \bar{B}_i \bar{C}_j$$
$$= \frac{\Delta_k x^k}{\Delta_{(i)} \bar{x}^i} B_k \frac{\Delta_l x^l}{\Delta_{(j)} \bar{x}^j} C_l$$
$$= \frac{\Delta_k x^k}{\Delta_{(i)} \bar{x}^i} \frac{\Delta_l x^l}{\Delta_{(j)} \bar{x}^j} B_k C_l$$
$$= \frac{\Delta_k x^k}{\Delta_{(i)} \bar{x}^i} \frac{\Delta_l x^l}{\Delta_{(j)} \bar{x}^j} A_{kl},$$

i. e.,

$$\bar{A}_{ij} - \frac{\Delta_k x^k}{\Delta_{(i)} \bar{x}^i} \frac{\Delta_l x^l}{\Delta_{(j)} \bar{x}^j} A_{kl}. \tag{2.3}$$

Moreover, we get

$$A_{ij} = B_i C_j$$
$$= \frac{\Delta_{(l)} \overline{x}^{l\sigma_{(0)}\sigma_{(1)}\cdots\sigma_{(i-1)}}}{\Delta_i x^i} \overline{B}_l \frac{\Delta_{(m)} \overline{x}^{m\sigma_{(0)}\sigma_{(1)}\cdots\sigma_{(j-1)}}}{\Delta_j x^j} \overline{C}_m$$
$$= \frac{\Delta_{(l)} \overline{x}^{l\sigma_{(0)}\sigma_{(1)}\cdots\sigma_{(i-1)}}}{\Delta_i x^i} \frac{\Delta_{(m)} \overline{x}^{m\sigma_{(0)}\sigma_{(1)}\cdots\sigma_{(j-1)}}}{\Delta_j x^j} \overline{B}_l \overline{C}_m$$
$$= \frac{\Delta_{(l)} \overline{x}^{l\sigma_{(0)}\sigma_{(1)}\cdots\sigma_{(i-1)}}}{\Delta_i x^i} \frac{\Delta_{(m)} \overline{x}^{m\sigma_{(0)}\sigma_{(1)}\cdots\sigma_{(j-1)}}}{\Delta_j x^j} \overline{A}_{lm}.$$

Definition 2.2. We call A_{ij} the components of a covariant tensor of the second order.

Example 2.2. Let the components of a contravariant tensor A_{ij} of the second order in the x^i coordinate system be as follows:

$$A_{11} = x^2, \quad A_{12} = 0, \quad A_{21} = 0, \quad A_{22} = x^1.$$

We will find its components $\overline{A}_{ij}$ in the $\overline{x}^i$ coordinate system, where

$$\overline{x}^1 = \left(x^1\right)^2,$$
$$\overline{x}^2 = x^1 x^2, \quad x^1 > 0, \; x^2 > 0.$$

We have

$$x^1 = \sqrt{\overline{x}^1},$$
$$x^2 = \frac{\overline{x}^2}{\sqrt{\overline{x}^1}}.$$

Now we will compute the partial derivatives

$$\frac{\Delta_j x^j}{\Delta_{(k)} \overline{x}^k}, \quad j, k \in \{1, 2\}.$$

We have

$$\frac{\Delta_1 x^1}{\Delta_{(1)} \overline{x}^1} = \frac{1}{\sqrt{\sigma_{(1)}(\overline{x}^1)} + \sqrt{\overline{x}^1}},$$
$$\frac{\Delta_1 x^1}{\Delta_{(2)} \overline{x}^2} = 0$$

and

$$\frac{\Delta_2 x^2}{\Delta_{(1)}\overline{x}^1} = -\frac{\overline{x}^2}{(\sqrt{\sigma_{(1)}(\overline{x}^1)} + \sqrt{\overline{x}^1})\sqrt{\overline{x}^1}\sqrt{\sigma_{(1)}(\overline{x}^1)}},$$

$$\frac{\Delta_2 x^2}{\Delta_{(2)}\overline{x}^2} = \frac{1}{\sqrt{\overline{x}^1}}.$$

Then

$$\overline{A}_{11} = \frac{\Delta_i x^i}{\Delta_{(1)}\overline{x}^1}\frac{\Delta_j x^j}{\Delta_{(1)}\overline{x}^1}A_{ij}$$

$$= \left(\frac{\Delta_1 x^1}{\Delta_{(1)}\overline{x}^1}\right)^2 A_{11} + \frac{\Delta_1 x^1}{\Delta_{(1)}\overline{x}^1}\frac{\Delta_2 x^2}{\Delta_{(1)}\overline{x}^1}A_{12}$$

$$+ \frac{\Delta_2 x^2}{\Delta_{(1)}\overline{x}^1}\frac{\Delta_1 x^1}{\Delta_{(1)}\overline{x}^1}A_{21} + \left(\frac{\Delta_2 x^2}{\Delta_{(1)}\overline{x}^1}\right)^2 A_{22}$$

$$= \left(\frac{1}{\sqrt{\sigma_{(1)}(\overline{x}^1)} + \sqrt{\overline{x}^1}}\right)^2 \frac{\overline{x}^2}{\sqrt{\overline{x}^1}} + \left(-\frac{\overline{x}^2}{(\sqrt{\sigma_{(1)}(\overline{x}^1)} + \sqrt{\overline{x}^1})\sqrt{\overline{x}^1}\sqrt{\sigma_{(1)}(\overline{x}^1)}}\right)^2 \sqrt{\overline{x}^1}$$

$$= \left(\frac{1}{\sqrt{\sigma_{(1)}(\overline{x}^1)} + \sqrt{\overline{x}^1}}\right)^2 \left(\frac{\overline{x}^2}{\sqrt{\overline{x}^1}} + \frac{(\overline{x}^2)^2}{\overline{x}^1 \sigma_{(1)}(\overline{x}^1)}\sqrt{\overline{x}^1}\right)$$

$$= \frac{\overline{x}^2}{\sqrt{\overline{x}^1}}\left(\frac{1}{\sqrt{\sigma_{(1)}(\overline{x}^1)} + \sqrt{\overline{x}^1}}\right)^2 \left(\frac{\sigma_{(1)}(\overline{x}^1) + \overline{x}^2}{\sigma_{(1)}(\overline{x}^1)}\right),$$

$$\overline{A}_{12} = \frac{\Delta_i x^1}{\Delta_{(1)}\overline{x}^1}\frac{\Delta_j x^j}{\Delta_{(2)}\overline{x}^2}A_{ij}$$

$$= \frac{\Delta_1 x^1}{\Delta_{(1)}\overline{x}^1}\frac{\Delta_1 x^1}{\Delta_{(2)}\overline{x}^2}A_{11} + \frac{\Delta_1 x^1}{\Delta_{(1)}\overline{x}^1}\frac{\Delta_2 x^2}{\Delta_{(2)}\overline{x}^2}A_{12}$$

$$+ \frac{\Delta_2 x^2}{\Delta_{(1)}\overline{x}^1}\frac{\Delta_1 x^1}{\Delta_{(2)}\overline{x}^2}A_{21} + \frac{\Delta_2 x^2}{\Delta_{(1)}\overline{x}^1}\frac{\Delta_2 x^2}{\Delta_{(2)}\overline{x}^2}A_{22}$$

$$= -\frac{\overline{x}^2}{\sqrt{\overline{x}^1}(\sqrt{\sigma_{(1)}(\overline{x}^1)} + \sqrt{\overline{x}^1})\sqrt{\overline{x}^1}\sqrt{\sigma_{(1)}(\overline{x}^1)}}\sqrt{\overline{x}^1}$$

$$= -\frac{\overline{x}^2}{\sqrt{\overline{x}^1}(\sqrt{\sigma_{(1)}(\overline{x}^1)} + \sqrt{\overline{x}^1})\sqrt{\sigma_{(1)}(\overline{x}^1)}},$$

$$\overline{A}_{21} = \frac{\Delta_i x^i}{\Delta_{(2)}\overline{x}^2}\frac{\Delta_j x^j}{\Delta_{(1)}\overline{x}^1}A_{ij}$$

$$= \frac{\Delta_1 x^1}{\Delta_{(2)}\overline{x}^2}\frac{\Delta_1 x^1}{\Delta_{(1)}\overline{x}^1}A_{11} + \frac{\Delta_1 x^1}{\Delta_{(2)}\overline{x}^2}\frac{\Delta_2 x^2}{\Delta_{(1)}\overline{x}^1}A_{12}$$

$$+ \frac{\Delta_2 x^2}{\Delta_{(2)}\bar{x}^2} \frac{\Delta_1 x^1}{\Delta_{(1)}\bar{x}^1} A_{21} + \frac{\Delta_2 x^2}{\Delta_{(2)}\bar{x}^2} \frac{\Delta_2 x^2}{\Delta_{(1)}\bar{x}^1} A_{22}$$

$$= -\frac{\bar{x}^2}{\sqrt{\bar{x}^1}\left(\sqrt{\sigma_{(1)}(\bar{x}^1)} + \sqrt{\bar{x}^1}\right)\sqrt{\sigma_{(1)}(\bar{x}^1)}},$$

$$A_{22} = \frac{\Delta_i x^i}{\Delta_{(2)}\bar{x}^2} \frac{\Delta_j x^j}{\Delta_{(2)}\bar{x}^2} A_{ij}$$

$$= \left(\frac{\Delta_1 x^1}{\Delta_{(2)}\bar{x}^2}\right)^2 A_{11} + \frac{\Delta_1 x^1}{\Delta_{(2)}\bar{x}^2} \frac{\Delta_2 x^2}{\Delta_{(2)}\bar{x}^2} A_{12}$$

$$+ \frac{\Delta_2 x^2}{\Delta_{(2)}\bar{x}^2} \frac{\Delta_1 x^1}{\Delta_{(2)}\bar{x}^2} A_{21} + \left(\frac{\Delta_2 x^2}{\Delta_{(2)}\bar{x}^2}\right) A_{22}$$

$$= \frac{1}{\bar{x}^1}\sqrt{\bar{x}^1}$$

$$= \frac{1}{\sqrt{\bar{x}^1}}.$$

Exercise 2.2. Let the components of a covariant tensor A_{ij} in the x^i coordinate system be as follows:

$$A_{11} = -1, \quad A_{12} = 1, \quad A_{21} = -2, \quad A_{22} = 3.$$

Find its components in the $\bar{x}^i$ coordinate system, where

$$\bar{x}^1 = x^2,$$
$$\bar{x}^2 = (x^1)^2, \quad x^1 > 0, \ x^2 > 0.$$

Answer 2.2.

$$\bar{A}_{11} = 1,$$

$$\bar{A}_{12} = -\frac{1}{\sqrt{\sigma_{(2)}(\bar{x}^2)} + \sqrt{\bar{x}^2}},$$

$$\bar{A}_{21} = \frac{1}{\sqrt{\sigma_{(2)}(\bar{x}^2)} + \sqrt{\bar{x}^2}},$$

$$\bar{A}_{22} = \left(\frac{1}{\sqrt{\sigma_{(2)}(\bar{x}^2)} + \sqrt{\bar{x}^2}}\right)^2.$$

Example 2.3. Let A_{ij} be a covariant tensor of the second order, and let

$$B_{ij} = A_{ji}.$$

We will prove that B_{ij} is a covariant tensor of the second order.

We have

$$\overline{B}_{ij} = \overline{A}_{ji}$$
$$= \frac{\Delta_k x^k}{\Delta_{(j)}\overline{x}^j} \frac{\Delta_l x^l}{\Delta_{(i)}\overline{x}^i} A_{kl}$$
$$= \frac{\Delta_k x^k}{\Delta_{(j)}\overline{x}^j} \frac{\Delta_l x^l}{\Delta_{(i)}\overline{x}^i} B_{lk}$$
$$= \frac{\Delta_l x^l}{\Delta_{(i)}\overline{x}^i} \frac{\Delta_k x^k}{\Delta_{(j)}\overline{x}^j} B_{lk}.$$

From this relation it follows that B_{ij} is a covariant tensor of the second order.

Example 2.4. Let u^i be an arbitrary contravariant vector, and $a_{ij} u^i u^j$ let be an invariant. We will prove that

$$a_{ij} + a_{ji} \tag{2.4}$$

is a covariant tensor of the second order.

Since $a_{ij} u^i u^j$ is an invariant, we have

$$\overline{a}_{ij}\overline{u}^i\overline{u}^j = a_{ij} u^i u^j.$$

On the other hand, applying the transformation law for a contravariant vector, we obtain

$$\overline{a}_{ij}\overline{u}^i\overline{u}^j = a_{ij}\frac{\Delta_i x^i}{\Delta_{(p)}\overline{x}^p}\overline{u}^p \frac{\Delta_j x^j}{\Delta_{(q)}\overline{x}^q}\overline{u}^q$$
$$= a_{ij}\frac{\Delta_i x^i}{\Delta_{(p)}\overline{x}^p} \frac{\Delta_j x^j}{\Delta_{(q)}\overline{x}^q}\overline{u}^p\overline{u}^q.$$

Replacing the dummy indices i, j, p, q by p, q, i, j, respectively, we get

$$\overline{a}_{ij}\overline{u}^i\overline{u}^j = a_{pq}\frac{\Delta_p x^p}{\Delta_{(i)}\overline{x}^i} \frac{\Delta_q x^q}{\Delta_{(j)}\overline{x}^j}\overline{u}^i\overline{u}^j.$$

Interchanging the suffices i and j, we obtain

$$\overline{a}_{ji}\overline{u}^j\overline{u}^i = a_{pq}\frac{\Delta_p x^p}{\Delta_{(j)}\overline{x}^j} \frac{\Delta_q x^q}{\Delta_{(i)}\overline{x}^i}\overline{u}^j\overline{u}^i.$$

Now interchanging the dummy indices p and q, we find

$$\overline{a}_{ji}\overline{u}^{j}\overline{u}^{i} = a_{qp}\frac{\Delta_q x^q \, \Delta_p x^p}{\Delta_{(j)}\overline{x}^{j} \, \Delta_{(i)}\overline{x}^{i}}\overline{u}^{j}\overline{u}^{i}$$

$$= a_{qp}\frac{\Delta_p x^p \, \Delta_q x^q}{\Delta_{(i)}\overline{x}^{i} \, \Delta_{(j)}\overline{x}^{j}}\overline{u}^{j}\overline{u}^{i}.$$

Since

$$\overline{u}^{i}\overline{u}^{j} = \overline{u}^{j}\overline{u}^{i},$$

we arrive at

$$\overline{a}_{ji}\overline{u}^{j}\overline{u}^{i} = a_{qp}\frac{\Delta_p x^p \, \Delta_q x^q}{\Delta_{(i)}\overline{x}^{i} \, \Delta_{(j)}\overline{x}^{j}}\overline{u}^{i}\overline{u}^{j}.$$

Hence

$$(\overline{a}_{ij} + \overline{a}_{ji})\overline{u}^{j}\overline{u}^{i} = (a_{pq} + a_{qp})\frac{\Delta_p x^p \, \Delta_q x^q}{\Delta_{(i)}\overline{x}^{i} \, \Delta_{(j)}\overline{x}^{j}}\overline{u}^{i}\overline{u}^{j},$$

whereupon

$$\left((\overline{a}_{ij} + \overline{a}_{ji}) - (a_{pq} + a_{qp})\frac{\Delta_p x^p \, \Delta_q x^q}{\Delta_{(i)}\overline{x}^{i} \, \Delta_{(j)}\overline{x}^{j}}\right)\overline{u}^{i}\overline{u}^{j} = 0.$$

Because u^{i} and u^{j} are arbitrary, we get

$$(\overline{a}_{ij} + \overline{a}_{ji}) = (a_{pq} + a_{qp})\frac{\Delta_p x^p \, \Delta_q x^q}{\Delta_{(i)}\overline{x}^{i} \, \Delta_{(j)}\overline{x}^{j}}.$$

This confirms the covariant second-order tensor law of transformation. Thus (2.4) is a covariant tensor of the second order.

Example 2.5. Suppose that a_{ij} is a covariant tensor of the second order and $|a_{ij}| \neq 0$. We will check if $|a_{ij}|$ is an invariant.

By the transformation law for a covariant tensor of the second order we find

$$\overline{a}_{ij} = \frac{\Delta_i x^{i} \, \Delta_j x^{j}}{\Delta_{(p)}\overline{x}^{p} \, \Delta_{(q)}x^{q}}a_{pq}.$$

Hence

$$|\overline{a}_{ij}| = \left|\frac{\Delta_i x^i}{\Delta_{(p)}\overline{x}^p}\right|\left|\frac{\Delta_j x^j}{\Delta_{(q)}\overline{x}^q}\right||a_{pq}|$$

$$= \left|\frac{\Delta_i x^i}{\Delta_{(p)}\overline{x}^p}\right|^2 |a_{pq}|.$$

Thus, in general, $|a_{ij}|$ is not an invariant.

Example 2.6. Let a_{ij} be a covariant tensor of the second order, and let λ^i and β^j be contravariant vectors. We will show that

$$a_{ij}\lambda^i\beta^j$$

is an invariant. Indeed, we have

$$a_{ij}\lambda^i\beta^j = \frac{\Delta_{(p)}\overline{x}^{p\sigma_0\sigma_1\cdots\sigma_{i-1}}}{\Delta_i x^i}\frac{\Delta_{(q)}\overline{x}^{q\sigma_0\sigma_1\cdots\sigma_{j-1}}}{\Delta_j x^j}\overline{a}_{pq}\frac{\Delta_i x^i}{\Delta_{(r)}\overline{x}^r}\overline{\lambda}^r\frac{\Delta_j x^j}{\Delta_{(m)}\overline{x}^m}\overline{\beta}^m$$

$$= \left(\frac{\Delta_{(p)}\overline{x}^{p\sigma_0\sigma_1\cdots\sigma_{i-1}}}{\Delta_i x^i}\frac{\Delta_i x^i}{\Delta_{(r)}\overline{x}^r}\right)\left(\frac{\Delta_{(q)}\overline{x}^{q\sigma_0\sigma_1\cdots\sigma_{j-1}}}{\Delta_j x^j}\frac{\Delta_j x^j}{\Delta_{(m)}\overline{x}^m}\right)\overline{a}_{pq}\overline{\lambda}^r\overline{\beta}^m$$

$$= \delta^p_r\delta^q_m\overline{a}_{pq}\overline{\lambda}^r\overline{\beta}^m$$

$$= \overline{a}_{pq}\overline{\lambda}^p\overline{\beta}^q.$$

Replacing the dummy indices p and q with i and j, we obtain

$$a_{ij}\lambda^i\beta^j = \overline{a}_{ij}\overline{\lambda}^i\overline{\beta}^j.$$

Example 2.7. Let f be an invariant. Then

$$\overline{f} = f,$$

and

$$\frac{\partial\overline{f}}{\Delta_i x^i} = \frac{\partial f^{\sigma_{(0)}\sigma_{(1)}\cdots\sigma_{(p-1)}}}{\Delta_{(p)}\overline{x}^p}\frac{\Delta_{(p)}\overline{x}^p}{\Delta_i x^i}.$$

Hence

$$\frac{\partial^2\overline{f}}{\Delta_j x^j\Delta_i x^i} = \frac{\partial}{\Delta_j x^j}\left(\frac{\partial f^{\sigma_{(0)}\sigma_{(1)}\cdots\sigma_{(p-1)}}}{\Delta_{(p)}\overline{x}^p}\frac{\Delta_{(p)}\overline{x}^p}{\Delta_i x^i}\right)$$

$$= \frac{\partial}{\Delta_j x^j}\left(\frac{\partial f^{\sigma_{(0)}\sigma_{(1)}\cdots\sigma_{(p-1)}}}{\Delta_{(p)}\overline{x}^p}\right)\frac{\Delta_{(p)}\overline{x}^p}{\Delta_i x^i}$$

$$+ \left(\frac{\partial f^{\sigma_{(0)}\sigma_{(1)}\cdots\sigma_{(p-1)}}}{\Delta_{(p)}\overline{x}^p}\right)^{\sigma_j}\frac{\Delta_j(\Delta_{(p)}\overline{x}^p)}{\Delta_j x^j\Delta_i x^i}.$$

Thus

$$\frac{\partial^2 f}{\Delta_j x^j \Delta_i x^i}$$

is not a tensor.

Example 2.8. Let A_i be the components of a covariant vector. Applying the transformation law, we obtain

$$\overline{A}_i = \frac{\Delta_k x^k}{\Delta_{(i)} \overline{x}^i} A_k.$$

Now differentiating both sides, we find

$$\frac{\partial \overline{A}_i}{\Delta_{(j)} \overline{x}^j} = \frac{\partial}{\Delta_{(j)} \overline{x}^j}\left(\frac{\Delta_k x^k}{\Delta_{(i)} \overline{x}^i} A_k\right)$$

$$= \frac{\partial}{\Delta_{(j)} \overline{x}^j}\left(\frac{\Delta_k x^k}{\Delta_{(i)} \overline{x}^i}\right) A_k^{\sigma_{(j)}} + \frac{\Delta_k x^k}{\Delta_{(i)} \overline{x}^i} \frac{\partial A_k}{\Delta_{(j)} \overline{x}^j}$$

$$= \frac{\partial}{\Delta_{(j)} \overline{x}^j}\left(\frac{\Delta_k x^k}{\Delta_{(i)} \overline{x}^i}\right) A_k^{\sigma_{(j)}} + \frac{\Delta_k x^k}{\Delta_{(i)} \overline{x}^i} \frac{\Delta_l x^l}{\Delta_{(j)} \overline{x}^j} \frac{\partial A_k}{\Delta_l x^l}.$$

Interchanging i and j, we find

$$\frac{\partial \overline{A}_j}{\Delta_{(i)} \overline{x}^i} = \frac{\partial}{\Delta_{(i)} \overline{x}^i}\left(\frac{\Delta_k x^k}{\Delta_{(j)} \overline{x}^j}\right) A_k^{\sigma_{(i)}} + \frac{\Delta_k x^k}{\Delta_{(j)} \overline{x}^j} \frac{\Delta_l x^l}{\Delta_{(i)} \overline{x}^i} \frac{\partial A_k}{\Delta_l x^l}.$$

Interchanging the dummy indices k and l in the last representation, we have

$$\frac{\partial \overline{A}_j}{\Delta_{(i)} \overline{x}^i} = \frac{\partial}{\Delta_{(i)} \overline{x}^i}\left(\frac{\Delta_k x^k}{\Delta_{(j)} \overline{x}^j}\right) A_k^{\sigma_{(i)}} + \frac{\Delta_l x^l}{\Delta_{(j)} \overline{x}^j} \frac{\Delta_k x^k}{\Delta_{(i)} \overline{x}^i} \frac{\partial A_l}{\Delta_k x^k}.$$

Hence

$$\frac{\partial \overline{A}_i}{\Delta_{(j)} \overline{x}^j} - \frac{\partial \overline{A}_j}{\Delta_{(i)} \overline{x}^i} = \left(\frac{\partial}{\Delta_{(j)} \overline{x}^j}\left(\frac{\Delta_k x^k}{\Delta_{(i)} \overline{x}^i}\right) A_k^{\sigma_{(j)}} - \frac{\partial}{\Delta_{(i)} \overline{x}^i}\left(\frac{\Delta_k x^k}{\Delta_{(j)} \overline{x}^j}\right) A_k^{\sigma_{(i)}}\right)$$

$$+ \frac{\Delta_k x^k}{\Delta_{(i)} \overline{x}^i} \frac{\Delta_l x^l}{\Delta_{(j)} \overline{x}^j}\left(\frac{\partial A_k}{\Delta_l x^l} - \frac{\partial A_l}{\Delta_k x^k}\right).$$

The last expression shows that in the general case, we have that

$$\frac{\partial A_i}{\Delta_j x^j} - \frac{\partial A_j}{\Delta_i x^i}$$

are not the components of a covariant tensor of the second order.

Further, if we have N^2 functions

$$A^i_j = B^i C_j,$$

where B^i and C_j are the components of a contravariant vector and a covariant vector, respectively, then following (1.6) and (1.7), we find

$$\overline{A}^i_j = \overline{B}^i \overline{C}_j$$

$$= \frac{\Delta_{(i)} \overline{x}^{i\sigma_0 \sigma_1 \cdots \sigma_{k-1}}}{\Delta_k x^k} B^k \frac{\Delta_l x^l}{\Delta_{(j)} \overline{x}^j} C_l$$

$$= \frac{\Delta_{(i)} \overline{x}^{i\sigma_0 \sigma_1 \cdots \sigma_{k-1}}}{\Delta_k x^k} \frac{\Delta_l x^l}{\Delta_{(j)} \overline{x}^j} B^k C_l$$

$$= \frac{\Delta_{(i)} \overline{x}^{i\sigma_0 \sigma_1 \cdots \sigma_{k-1}}}{\Delta_k x^k} \frac{\Delta_l x^l}{\Delta_{(j)} \overline{x}^j} A^k_l,$$

i. e.,

$$\overline{A}^i_j = \frac{\Delta_{(i)} \overline{x}^{i\sigma_0 \sigma_1 \cdots \sigma_{k-1}}}{\Delta_k x^k} \frac{\Delta_l x^l}{\Delta_{(j)} \overline{x}^j} A^k_l. \tag{2.5}$$

Moreover, we have

$$A^i_j = B^i C_j$$

$$= \frac{\Delta_i x^i}{\Delta_{(k)} \overline{x}^k} \overline{B}^k \frac{\Delta_{(l)} \overline{x}^{l\sigma_0 \sigma_1 \cdots \sigma_{j-1}}}{\Delta_j x^j} \overline{C}_j$$

$$= \frac{\Delta_i x^i}{\Delta_{(k)} \overline{x}^k} \frac{\Delta_{(l)} \overline{x}^{l\sigma_0 \sigma_1 \cdots \sigma_{j-1}}}{\Delta_j x^j} \overline{B}^k \overline{C}_j$$

$$= \frac{\Delta_i x^i}{\Delta_{(k)} \overline{x}^k} \frac{\Delta_{(l)} \overline{x}^{l\sigma_0 \sigma_1 \cdots \sigma_{j-1}}}{\Delta_j x^j} \overline{A}^k_l.$$

Definition 2.3. We call A^i_j the components of a mixed tensor of the second order.

Note that the indices are placed on the tensors as superscripts when they denote contravariance and as subscripts when they denote covariance. In particular, the mixed tensor A^i_j transforms like a contravariant vector with respect to the index i and like a covariant vector with respect to the index j. Consequently, the i is placed as a superscript, and the j is placed as a subscript.

Example 2.9. Let the components of a mixed tensor A^i_j of the second order in the x^i coordinate system be as follows:

$$A_1^1 = -1, \quad A_2^1 = 2, \quad A_1^2 = 1, \quad A_2^2 = 3.$$

Find its components in the $\overline{x}^i$ coordinate system, where

$$\overline{x}^1 = (x^2)^3,$$
$$\overline{x}^2 = (x^1)^2, \quad x^1 > 0, \ x^2 > 0.$$

We have

$$x^2 = \sqrt[3]{\overline{x}^1},$$
$$x^1 = \sqrt{\overline{x}^2}.$$

Now we will find the partial derivatives

$$\frac{\Delta_{(j)}\overline{x}^{j\sigma_0\sigma_1\cdots\sigma_{k-1}}}{\Delta_k x^k}, \quad j, k \in \{1, 2\}.$$

We have

$$\frac{\Delta_{(1)}\overline{x}^1}{\Delta_1 x^1} = 0,$$

$$\frac{\Delta_{(1)}\overline{x}^1}{\Delta_2 x^2} = \left(\sigma_2(x^2)\right)^2 + x^2\sigma_2(x^2) + (x^2)^2$$

$$= \sqrt[3]{\left(\sigma_{(1)}(\overline{x}^1)\right)^2} + \sqrt[3]{\overline{x}^1}\sqrt[3]{\sigma_{(1)}(\overline{x}^1)} + \sqrt[3]{(\overline{x}^1)^2},$$

$$\frac{\Delta_{(1)}\overline{x}^{1\sigma_1}}{\Delta_2 x^2} = \sqrt[3]{\left(\sigma_{(1)}(\overline{x}^1)\right)^2} + \sqrt[3]{\overline{x}^1}\sqrt[3]{\sigma_{(1)}(\overline{x}^1)} + \sqrt[3]{(\overline{x}^1)^2},$$

$$\frac{\Delta_{(2)}\overline{x}^2}{\Delta_1 x^1} = \sigma_1(x^1) + x^1$$

$$= \sqrt{\sigma_{(2)}(\overline{x}^2)} + \sqrt{\overline{x}^2},$$

$$\frac{\Delta_{(2)}\overline{x}^2}{\Delta_2 x^2} = 0,$$

$$\frac{\Delta_{(2)}\overline{x}^{2\sigma_1}}{\Delta_2 x^2} = 0.$$

Next, we will compute the partial derivatives

$$\frac{\Delta_j x^j}{\Delta_{(k)}\overline{x}^k}, \quad j, k \in \{1, 2\}.$$

We find

$$\frac{\Delta_1 x^1}{\Delta_{(1)}\overline{x}^1} = 0,$$

$$\frac{\Delta_1 x^1}{\Delta_{(2)}\overline{x}^2} = \frac{1}{\sqrt{\sigma_{(2)}(\overline{x}^2)} + \sqrt{\overline{x}^2}},$$

$$\frac{\Delta_2 x^2}{\Delta_{(1)}\overline{x}^1} = \frac{1}{\sqrt[3]{(\sigma_{(1)}(\overline{x}^1))^2} + \sqrt[3]{\overline{x}^1 \sigma_{(1)}(\overline{x}^1)} + \sqrt[3]{(\overline{x}^1)^2}},$$

$$\frac{\Delta_2 x^2}{\Delta_{(2)}\overline{x}^2} = 0.$$

Hence

$$\overline{A}_1^1 = \frac{\Delta_{(1)}\overline{x}^{1\sigma_0\sigma_1\cdots\sigma_{k-1}}}{\Delta_k x^k} \frac{\Delta_l x^l}{\Delta_{(1)}\overline{x}^1} A_l^k$$

$$= \frac{\Delta_{(1)}\overline{x}^1}{\Delta_1 x^1} \frac{\Delta_1 x^1}{\Delta_{(1)}\overline{x}^1} A_1^1 + \frac{\Delta_{(1)}\overline{x}^1}{\Delta_1 x^1} \frac{\Delta_2 x^2}{\Delta_{(1)}\overline{x}^1} A_2^1$$

$$+ \frac{\Delta_{(1)}\overline{x}^{1\sigma_1}}{\Delta_2 x^2} \frac{\Delta_1 x^1}{\Delta_{(1)}\overline{x}^1} A_1^2 + \frac{\Delta_{(1)}\overline{x}^{1\sigma_1}}{\Delta_2 x^2} \frac{\Delta_2 x^2}{\Delta_{(1)}\overline{x}^1} A_2^2$$

$$= 3 \frac{(\sigma_2(x^2))^2 + x^2 \sigma_2(x^2) + (x^2)^2}{\sqrt[3]{(\sigma_{(1)}(\overline{x}^1))^2} + \sqrt[3]{\overline{x}^1 \sigma_{(1)}(\overline{x}^1)} + \sqrt[3]{(\overline{x}^1)^2}},$$

$$\overline{A}_2^1 = \frac{\Delta_{(1)}\overline{x}^{1\sigma_0\sigma_1\cdots\sigma_{k-1}}}{\Delta_k x^k} \frac{\Delta_l x^l}{\Delta_{(2)}\overline{x}^2} A_l^k$$

$$= \frac{\Delta_{(1)}\overline{x}^1}{\Delta_1 x^1} \frac{\Delta_1 x^1}{\Delta_{(2)}\overline{x}^2} A_1^1 + \frac{\Delta_{(1)}\overline{x}^1}{\Delta_1 x^1} \frac{\Delta_2 x^2}{\Delta_{(2)}\overline{x}^2} A_2^1$$

$$+ \frac{\Delta_{(1)}\overline{x}^{1\sigma_1}}{\Delta_2 x^2} \frac{\Delta_1 x^1}{\Delta_{(2)}\overline{x}^2} A_1^2 + \frac{\Delta_{(1)}\overline{x}^{1\sigma_1}}{\Delta_2 x^2} \frac{\Delta_2 x^2}{\Delta_{(2)}\overline{x}^2} A_2^2$$

$$= \frac{(\sigma_2(x^2))^2 + x^2 \sigma_2(x^2) + (x^2)^2}{\sqrt[3]{(\sigma_{(1)}(\overline{x}^1))^2} + \sqrt[3]{\overline{x}^1 \sigma_{(1)}(\overline{x}^1)} + \sqrt[3]{(\overline{x}^1)^2}},$$

$$\overline{A}_1^2 = \frac{\Delta_{(2)}\overline{x}^{2\sigma_0\sigma_1\cdots\sigma_{k-1}}}{\Delta_k x^k} \frac{\Delta_l x^l}{\Delta_{(1)}\overline{x}^1} A_l^k$$

$$= \frac{\Delta_{(2)}\overline{x}^2}{\Delta_1 x^1} \frac{\Delta_1 x^1}{\Delta_{(1)}\overline{x}^1} A_1^1 + \frac{\Delta_{(2)}\overline{x}^2}{\Delta_1 x^1} \frac{\Delta_2 x^2}{\Delta_{(1)}\overline{x}^1} A_2^1$$

$$+ \frac{\Delta_{(2)}\overline{x}^{2\sigma_1}}{\Delta_2 x^2} \frac{\Delta_1 x^1}{\Delta_{(1)}\overline{x}^1} A_1^2 + \frac{\Delta_{(2)}\overline{x}^{2\sigma_1}}{\Delta_2 x^2} \frac{\Delta_2 x^2}{\Delta_{(1)}\overline{x}^1} A_2^2$$

$$= 2\frac{\sigma_1(x^1) + x^1}{\sqrt[3]{(\sigma_{(1)}(\overline{x}^1))^2} + \sqrt[3]{\overline{x}^1 \sigma_{(1)}(\overline{x}^1)} + \sqrt[3]{(\overline{x}^1)^2}},$$

$$\overline{A}_2^2 = \frac{\Delta_{(2)}\overline{x}^{2\sigma_0\sigma_1\cdots\sigma_{k-1}}}{\Delta_k x^k}\frac{\Delta_l x^l}{\Delta_{(2)}\overline{x}^2}A_l^k$$

$$= \frac{\Delta_{(2)}\overline{x}^2}{\Delta_1 x^1}\frac{\Delta_1 x^1}{\Delta_{(2)}\overline{x}^2}A_1^1 + \frac{\Delta_{(2)}\overline{x}^2}{\Delta_1 x^1}\frac{\Delta_2 x^2}{\Delta_{(2)}\overline{x}^2}A_2^1$$

$$+ \frac{\Delta_{(2)}\overline{x}^{2\sigma_1}}{\Delta_2 x^2}\frac{\Delta_1 x^1}{\Delta_{(2)}\overline{x}^2}A_1^2 + \frac{\Delta_{(2)}\overline{x}^{2\sigma_1}}{\Delta_2 x^2}\frac{\Delta_2 x^2}{\Delta_{(2)}\overline{x}^2}A_2^2$$

$$= -\frac{\sigma_1(x^1) + x^1}{\sqrt[3]{(\sigma_{(1)}(\overline{x}^1))^2} + \sqrt[3]{\overline{x}^1 \sigma_{(1)}(\overline{x}^1)} + \sqrt[3]{(\overline{x}^1)^2}}.$$

Exercise 2.3. Let the components of a mixed tensor A_j^i of the second order in the x^i coordinate system be as follows:

$$A_1^1 = 1, \quad A_2^1 = 0, \quad A_1^2 = 1, \quad A_2^2 = -1.$$

Find its components in the $\overline{x}^i$ coordinate system, where

$$\overline{x}^1 = x^2,$$
$$\overline{x}^2 = x^1 x^2.$$

Answer 2.3.

$$\overline{A}_1^1 = 0,$$

$$\overline{A}_2^1 = \frac{1}{\overline{x}^1},$$

$$\overline{A}_1^2 = -\frac{\overline{x}^2}{\sigma_{(1)}(\overline{x}^1)} - \frac{(\overline{x}^2)^2}{(\overline{x}^1)^2 \sigma_{(1)}(\overline{x}^1)} - \frac{\overline{x}^2}{\overline{x}^1},$$

$$\overline{A}_2^2 = 1 + \frac{\overline{x}^2}{(\overline{x}^1)^2}.$$

Example 2.10. Now we choose A_l^k to be the Kronecker delta δ_l^k. Then applying (2.5), we find

$$\overline{\delta}_j^i = \frac{\Delta_{(i)}\overline{x}^{i\sigma_0\sigma_1\cdots\sigma_{k-1}}}{\Delta_k x^k}\frac{\Delta_l x^l}{\Delta_{(j)}\overline{x}^j}\delta_k^l$$

$$= \frac{\Delta_{(i)}\overline{x}^{i\sigma_0\sigma_1\cdots\sigma_{l-1}}}{\Delta_l x^l}\frac{\Delta_l x^l}{\Delta_{(j)}\overline{x}^j}$$

$$= \frac{\Delta_{(i)}\overline{x}^i}{\Delta_{(j)}\overline{x}^j}$$

$$= \delta_j^i.$$

Therefore the Kronecker delta is a mixed tensor of the second order whose components in any other system again form the Kronecker delta. This justifies the placing of one index as a subscript and the other as a superscript. Indeed, if we select the N^2 quantities

$$\delta_{ij} = \delta_j^i$$

as the components of a covariant tensor of the second order in a coordinate system x^i, then the components in the new system $\overline{x}^i$ are given by

$$\overline{\delta}_{ij} = \frac{\Delta_k x^k}{\Delta_{(i)}\overline{x}^i} \frac{\Delta_l x^l}{\Delta_{(j)}\overline{x}^j} \delta_{kl}$$

$$= \frac{\Delta_l x^l}{\Delta_{(i)}\overline{x}^i} \frac{\Delta_l x^l}{\Delta_{(j)}\overline{x}^j}.$$

Therefore the transformed components $\overline{\delta}_{ij}$ do not form the Kronecker delta. Let us now consider the N^2 quantities

$$\delta^{ij} = \delta_j^i.$$

Then using (1.6), we obtain

$$\overline{\delta}^{ij} = \frac{\Delta_{(i)}\overline{x}^{i\sigma_0\sigma_1\cdots\sigma_{k-1}}}{\Delta_k x^k} \frac{\Delta_{(j)}\overline{x}^j}{\Delta_l x^l} \delta^{kl}$$

$$= \frac{\Delta_{(i)}\overline{x}^{i\sigma_0\sigma_1\cdots\sigma_{l-1}}}{\Delta_l x^l} \frac{\Delta_{(j)}\overline{x}^j}{\Delta_l x^l}.$$

Again, the transformed components $\overline{\delta}^{ij}$ do not form the Kronecker delta.

Example 2.11. Now suppose that B^i and C^j are contravariant vectors and A_{ij} is a covariant tensor of the second order. We will show that

$$A_{ij}B^i C^j$$

is an invariant. Indeed, we have

$$A_{ij}B^iC^j = \frac{\Delta_{(l)}\bar{x}^{l\sigma_0\sigma_1\cdots\sigma_{i-1}}}{\Delta_i x^i}\frac{\Delta_{(m)}\bar{x}^{m\sigma_0\sigma_1\cdots\sigma_{j-1}}}{\Delta_j x^j}\frac{\Delta_i x^i}{\Delta_{(p)}\bar{x}^p}\bar{B}^p\frac{\Delta_j x^j}{\Delta_{(q)}\bar{x}^q}\bar{C}^q$$

$$= \left(\frac{\Delta_{(l)}\bar{x}^{l\sigma_0\sigma_1\cdots\sigma_{i-1}}}{\Delta_i x^i}\frac{\Delta_i x^i}{\Delta_{(p)}\bar{x}^p}\right)\left(\frac{\Delta_{(m)}\bar{x}^{m\sigma_0\sigma_1\cdots\sigma_{j-1}}}{\Delta_j x^j}\frac{\Delta_j x^j}{\Delta_{(q)}\bar{x}^q}\right)\bar{A}_{lm}\bar{B}^p\bar{C}^q$$

$$= \left(\frac{\Delta_{(l)}\bar{x}^l}{\Delta_{(p)}\bar{x}^p}\right)\left(\frac{\Delta_{(m)}\bar{x}^m}{\Delta_{(q)}\bar{x}^q}\right)\bar{A}_{lm}\bar{B}^p\bar{C}^q$$

$$= \delta_p^l\delta_q^m\bar{A}_{lm}\bar{B}^p\bar{C}^q$$

$$= \bar{A}_{pq}\bar{B}^p\bar{C}^q.$$

Example 2.12. Suppose the relation

$$a_j^i v^j = 0$$

holds for any contravariant vector v^j. We will show that

$$a_j^i = 0. \tag{2.6}$$

Since v^j are arbitrary, taking v^j as

$$(1, 0, \ldots, 0),$$

we find

$$a_1^i = 0,$$

taking v^j as

$$(0, 1, 0, \ldots, 0),$$

we find

$$a_2^i = 0,$$

and so on. Thus we get (2.6).

Example 2.13. Let λ_j be a given constant covariant vector. Suppose the relation

$$a_{ij}u^iu^j = 0 \tag{2.7}$$

holds for all vectors u^i such that

$$u^i\lambda_i = 0. \tag{2.8}$$

We will show that

$$a_{ij} + a_{ji} = \lambda_i v_j + \lambda_j v_i,$$

where v_j is some covariant vector.

Indeed, from relation (2.8) we get

$$u^i \lambda_i v_j u^j = 0$$

for some covariant vector v_j. Now applying (2.7), we arrive at

$$a_{ij} u^i u^j = u^i \lambda_i v_j u^j,$$

whereupon

$$(a_{ij} - \lambda_i v_j) u^i u^j = 0.$$

Since u^i is arbitrary, we find

$$a_{ij} = \lambda i_i v_j.$$

Interchanging the dummy suffices i and j, we get

$$a_{ji} = \lambda_j v_i.$$

Hence

$$a_{ij} + a_{ji} = \lambda_i v_j + \lambda_j v_i.$$

Example 2.14. If the equality

$$a_j^i u_i = v u_j \tag{2.9}$$

holds for any covariant vector u_i such that

$$u_i v^i = 0, \tag{2.10}$$

where v^i is a given contravariant vector, we will show that

$$a_j^i = v \delta_j^i + \lambda_j v^i,$$

where λ_i is some covariant vector.

Indeed, since (2.10) holds, we find

$$a_j^i u_i = v u_j + \lambda_j u_i v^i$$
$$= v \delta_j^i u_i + \lambda_j u_i v^i$$
$$= u_i(v \delta_j^i + \lambda_j u^i),$$

where λ_j is some covariant vector. Hence

$$(a_j^i - v \delta_j^i - \lambda_j v^i)u_i = 0.$$

Since u_i is arbitrary, we find

$$a_j^i - v \delta_j^i - \lambda_j v^i = 0.$$

Example 2.15. We will show that the Kronecker delta δ_j^i is an invariant. Indeed,

$$\overline{\delta}_j^i = \frac{\Delta_{(i)} \overline{x}^{i \sigma_0 \sigma_1 \cdots \sigma_{k-1}}}{\Delta_k x^k} \frac{\Delta_l x^l}{\Delta_{(j)} \overline{x}^j} \delta_l^k$$

$$= \frac{\Delta_{(i)} \overline{x}^{i \sigma_0 \sigma_1 \cdots \sigma_{k-1}}}{\Delta_k x^k} \left(\frac{\Delta_l x^l}{\Delta_{(j)} \overline{x}^j} \delta_l^k \right)$$

$$= \frac{\Delta_{(i)} \overline{x}^{i \sigma_0 \sigma_1 \cdots \sigma_{k-1}}}{\Delta_k x^k} \frac{\Delta_k x^k}{\Delta_{(j)} \overline{x}^j}$$

$$= \frac{\Delta_{(i)} \overline{x}^i}{\Delta_{(j)} \overline{x}^j}$$

$$= \delta_j^i.$$

2.2 Higher-order tensors

Definition 2.4. A set of N^{s+p} functions $A_{q_1 q_2 \cdots q_p}^{t_1 t_2 \cdots t_s}$ of the N coordinates x^i are said to be the components of a mixed tensor of the $(s + p)$th order, contravariant of the sth order and covariant of the pth order, if they transform according to the equation

$$\overline{A}_{r_1 r_2 \cdots r_p}^{u_1 u_2 \cdots u_s} = \frac{\Delta_{(u_1)} \overline{x}^{u_1 \sigma_0 \sigma_1 \cdots \sigma_{t_1-1}}}{\Delta_{t_1} x^{t_1}} \frac{\Delta_{(u_2)} \overline{x}^{u_2 \sigma_0 \sigma_1 \cdots \sigma_{t_2-1}}}{\Delta_{t_2} x^{t_2}} \cdots \frac{\Delta_{(u_s)} \overline{x}^{u_s \sigma_0 \sigma_1 \cdots \sigma_{t_s-1}}}{\Delta_{t_s} x^{t_s}}$$

$$\times \frac{\Delta_{q_1} x^{q_1}}{\Delta_{(r_1)} \overline{x}^{r_1}} \frac{\Delta_{q_2} x^{q_2}}{\Delta_{(r_2)} \overline{x}^{r_2}} \cdots \frac{\Delta_{q_p} x^{q_p}}{\Delta_{(r_p)} \overline{x}^{r_p}} A_{q_1 q_2 \cdots q_p}^{t_1 t_2 \cdots t_s} \quad \text{or}$$

$$A_{r_1 r_2 \cdots r_p}^{u_1 u_2 \cdots u_s} = \frac{\Delta_{u_1} x^{u_1}}{\Delta_{(t_1)} \overline{x}^{t_1}} \frac{\Delta_{u_2} x^{u_2}}{\Delta_{(t_2)} \overline{x}^{t_2}} \cdots \frac{\Delta_{u_s} x^{u_s}}{\Delta_{(t_s)} \overline{x}^{t_s}}$$

$$\times \frac{\Delta_{(q_1)} \overline{x}^{q_1 \sigma_0 \sigma_1 \cdots \sigma_{r_1-1}}}{\Delta_{r_1} x^{r_1}} \frac{\Delta_{(q_2)} \overline{x}^{q_2 \sigma_0 \sigma_1 \cdots \sigma_{r_2-1}}}{\Delta_{r_2} x^{r_2}} \cdots \frac{\Delta_{(q_p)} \overline{x}^{q_p \sigma_0 \sigma_1 \cdots \sigma_{r_p-1}}}{\Delta_{r_p} x^{r_p}} \overline{A}_{q_1 q_2 \cdots q_p}^{t_1 t_2 \cdots t_s}.$$

$$(2.11)$$

This formula is merely a combination of (1.6) to contravariant indices and (1.7) with respect covariant indices. The order of indices in a tensor is important. The tensor A^{ij} is not necessarily the same as the tensor A^{ji}.

Example 2.16. Suppose that

$$A_k^{ij}\lambda_i v_j v^k \tag{2.12}$$

is a scalar invariant and λ_i, v_j, and v^k are vectors. We will show that A_k^{ij} is a mixed tensor.
  Indeed, because (2.12) is an invariant, we have

$$\overline{A}_k^{ij}\overline{\lambda}_i\overline{v}_j\overline{v}^k = A_k^{ij}\lambda_i v_j v^k$$
$$= A_p^{\alpha\beta}\lambda_\alpha v_\beta v^p.$$

Now applying the transformation laws for covariant and contravariant vectors, we obtain

$$A_p^{\alpha\beta}\lambda_\alpha v_\beta v^p = \overline{A}_k^{ij}\frac{\Delta_\alpha x^\alpha}{\Delta_{(i)}\overline{x}_i}\lambda_\alpha\frac{\Delta_\beta x^\beta}{\Delta_{(j)}\overline{x}^j}v_\beta\frac{\Delta_{(k)}\overline{x}^{k\sigma_0\sigma_1\cdots\sigma_{p-1}}}{\Delta_p x^p}v^p$$
$$= \frac{\Delta_\alpha x^\alpha}{\Delta_{(i)}\overline{x}_i}\frac{\Delta_\beta x^\beta}{\Delta_{(j)}\overline{x}^j}\frac{\Delta_{(k)}\overline{x}^{k\sigma_0\sigma_1\cdots\sigma_{p-1}}}{\Delta_p x^p}\overline{A}_k^{ij}\lambda_\alpha v_\beta v^p.$$

Hence,

$$\left(A_p^{\alpha\beta} - \frac{\Delta_\alpha x^\alpha}{\Delta_{(i)}\overline{x}_i}\frac{\Delta_\beta x^\beta}{\Delta_{(j)}\overline{x}^j}\frac{\Delta_{(k)}\overline{x}^{k\sigma_0\sigma_1\cdots\sigma_{p-1}}}{\Delta_p x^p}\overline{A}_k^{ij}\right)\lambda_\alpha v_\beta v^p = 0.$$

Because λ_i, v_j and v^k are arbitrary, we arrive at

$$A_p^{\alpha\beta} - \frac{\Delta_\alpha x^\alpha}{\Delta_{(i)}\overline{x}_i}\frac{\Delta_\beta x^\beta}{\Delta_{(j)}\overline{x}^j}\frac{\Delta_{(k)}\overline{x}^{k\sigma_0\sigma_1\cdots\sigma_{p-1}}}{\Delta_p x^p}\overline{A}_k^{ij} = 0,$$

i. e., we get the transformation law for a mixed tensor. Thus A_k^{ij} is a mixed tensor.

Definition 2.5. If however two contravariant indices or two covariant indices can be interchanged without altering the tensor, then it is said to be symmetric with respect to these indices.

  Now we will list some of the properties of the symmetric tensors.
1. We will prove that if a tensor is symmetric with respect to two indices in any coordinate system, it remains symmetric with respect to these two indices in any other coordinate system. There is no loss of generality in proving this for the contravariant tensor of the second order

$$A^{ij} = A^{ji}.$$

Applying the transformation law for second-order contravariant tensors, we find

$$
\begin{aligned}
\overline{A}^{ij} &= \frac{\Delta_{(i)}\overline{x}^{\,i\sigma_0\sigma_1\cdots\sigma_{k-1}}}{\Delta_k x^k}\;\frac{\Delta_{(j)}\overline{x}^{\,j\sigma_0\sigma_1\cdots\sigma_{l-1}}}{\Delta_l x^l}A^{kl} \\[2mm]
&= \frac{\Delta_{(i)}\overline{x}^{\,i\sigma_0\sigma_1\cdots\sigma_{k-1}}}{\Delta_k x^k}\;\frac{\Delta_{(j)}\overline{x}^{\,j\sigma_0\sigma_1\cdots\sigma_{l-1}}}{\Delta_l x^l}A^{lk} \\[2mm]
&= \frac{\Delta_{(j)}\overline{x}^{\,j\sigma_0\sigma_1\cdots\sigma_{l-1}}}{\Delta_l x^l}\;\frac{\Delta_{(i)}\overline{x}^{\,i\sigma_0\sigma_1\cdots\sigma_{k-1}}}{\Delta_k x^k}A^{lk} \\[2mm]
&= \overline{A}^{ji},
\end{aligned}
$$

as required. We cannot usually define symmetry with respect to two indices, of which one denotes contravariance and the other denotes covariance, because this symmetry may not be preserved after a coordinate transformation. The Kronecker delta, however, is a mixed tensor, which possesses symmetry with respect to its two indices.

2. A symmetric tensor of the second order has at most $\frac{1}{2}N(N+1)$ different components. To see this, let A_{ij} be a symmetric covariant tensor of the second order. Then it has N^2 components

$$
\begin{matrix}
A_{11} & A_{12} & \cdots & A_{1N} \\
A_{21} & A_{22} & \cdots & A_{2N} \\
\vdots & \vdots & \vdots & \vdots \\
A_{N1} & A_{N2} & \cdots & A_{NN}
\end{matrix}.
$$

These components are of two types:

a. those in which the indices i and j are the same, i. e., the components along the diagonal. The maximum number of distinct components of this type is N.

b. those in which the indices i and j are different, i. e., the components along the nondiagonal. Hence the maximum number of such components is

$$N^2 - N = N(N-1).$$

Due to symmetry of A_{ij}, the components above and below the diagonal, the maximum number of distinct components of this type is

$$\frac{N(N-1)}{2}.$$

Therefore the maximum number of independent components is given by

$$N + \frac{N(N-1)}{2} = \frac{N(N+1)}{2}.$$

Example 2.17. Assume that

$$\phi = a_{jk}A^j A^k. \tag{2.13}$$

We will show that ϕ can be represented in the form

$$\phi = b_{jk}A^j A^k,$$

where b_{jk} is symmetric.

Indeed, in relation (2.13), interchanging the indices j and k, we find

$$\phi = a_{kj}A^k A^j$$
$$= a_{kj}A^j A^k.$$

Hence

$$2\phi = a_{jk}A^j A^k + a_{kj}A^j A^k$$
$$= (a_{jk} + a_{kj})A^j A^k,$$

whereupon

$$\phi = \frac{a_{jk} + a_{kj}}{2}A^j A^k.$$

Set

$$b_{jk} = \frac{a_{jk} + a_{kj}}{2}.$$

Note that b_{jk} is symmetric.

Example 2.18. Let the tensors a_{ij} and b_{ij} be symmetric, and let u^i and v^j be the components of contravariant vectors satisfying the equations

$$(a_{ij} - kb_{ij})u^i = 0,$$
$$(a_{ij} - k'b_{ij})v^j = 0, \quad i,j \in \{1,\ldots,N\}, \quad k \neq k'. \tag{2.14}$$

We will show that

$$b_{ij}u^i v^j = 0$$

or

$$a_{ij}u^i v^j = 0.$$

Indeed, since a_{ij} and b_{ij} are symmetric, we have

$$a_{ij} = a_{ji},$$
$$b_{ij} = b_{ji}.$$

Multiplying the first equation of (2.14) by v^j and its second equation by u^i, we find

$$(a_{ij} - kb_{ij})u^i v^j = 0,$$
$$(a_{ij} - k'b_{ij})v^j u^i = 0.$$

(2.15)

Subtracting both equations, we find

$$\begin{aligned}
0 &= (a_{ij} - kb_{ij})u^i v^j - (a_{ij} - k'b_{ij})v^j u^i \\
&= a_{ij}u^i v^j - kb_{ij}u^i v^j - a_{ij}v^j u^i + k'v^j u^i \\
&= a_{ij}u^i v^j - kb_{ij}u^i v^j - a_{ij}u^i v^j - kb_{ij}u^i v^j + k'b_{ij}u^i v^j \\
&= (k' - k)b_{ij}u^i v^j.
\end{aligned}$$

Since $k \neq k'$, we find

$$0 = b_{ij}u^i v^j.$$

Hence by the first equation of (2.15) we get

$$\begin{aligned}
0 &= a_{ij}u^i v^j - kb_{ij}u^i v^j \\
&= a_{ij}u^i v^j.
\end{aligned}$$

Example 2.19. Let a_{ij} be a symmetric tensor and let b_i be a vector such that

$$a_{ij}b_k + a_{jk}b_i + a_{ki}b_j = 0. \tag{2.16}$$

We will show that

$$a_{ij} = 0 \quad \text{or} \quad n_k = 0.$$

Indeed, by equation (2.16) we find

$$\overline{a}_{ij}\overline{b}_k + \overline{a}_{jk}\overline{b}_i + \overline{a}_{ki}\overline{b}_j = 0.$$

Now applying the transformation laws for a covariant tensor of the second order and a covariant vector, we find

$$0 = \frac{\Delta_p x^p}{\Delta_{(i)} \bar{x}^i} \frac{\Delta_q x^q}{\Delta_{(j)} \bar{x}^j} a_{pq} \frac{\Delta_r x^r}{\Delta_{(k)} \bar{x}^k} b_r + \frac{\Delta_p x^p}{\Delta_{(j)} \bar{x}^j} \frac{\Delta_q x^q}{\Delta_{(k)} \bar{x}^k} a_{pq} \frac{\Delta_r x^r}{\Delta_{(i)} \bar{x}^i} b_r$$

$$+ \frac{\Delta_p x^p}{\Delta_{(k)} \bar{x}^k} \frac{\Delta_q x^q}{\Delta_{(i)} \bar{x}^i} a_{pq} \frac{\Delta_r x^r}{\Delta_{(j)} \bar{x}^j} b_r$$

$$= \frac{\Delta_p x^p}{\Delta_{(i)} \bar{x}^i} \frac{\Delta_q x^q}{\Delta_{(j)} \bar{x}^j} \frac{\Delta_r x^r}{\Delta_{(k)} \bar{x}^k} a_{pq} b_r + \frac{\Delta_p x^p}{\Delta_{(j)} \bar{x}^j} \frac{\Delta_q x^q}{\Delta_{(k)} \bar{x}^k} \frac{\Delta_r x^r}{\Delta_{(i)} \bar{x}^i} a_{pq} b_r$$

$$+ \frac{\Delta_p x^p}{\Delta_{(k)} \bar{x}^k} \frac{\Delta_q x^q}{\Delta_{(i)} \bar{x}^i} \frac{\Delta_r x^r}{\Delta_{(j)} \bar{x}^j} a_{pq} b_r$$

$$= \left(\frac{\Delta_p x^p}{\Delta_{(i)} \bar{x}^i} \frac{\Delta_q x^q}{\Delta_{(j)} \bar{x}^j} \frac{\Delta_r x^r}{\Delta_{(k)} \bar{x}^k} + \frac{\Delta_p x^p}{\Delta_{(j)} \bar{x}^j} \frac{\Delta_q x^q}{\Delta_{(k)} \bar{x}^k} \frac{\Delta_r x^r}{\Delta_{(i)} \bar{x}^i} + \frac{\Delta_p x^p}{\Delta_{(k)} \bar{x}^k} \frac{\Delta_q x^q}{\Delta_{(i)} \bar{x}^i} \frac{\Delta_r x^r}{\Delta_{(j)} \bar{x}^j} \right) a_{pq} b_r.$$

Therefore

$$a_{pq} b_r = 0,$$

whereupon

$$a_{pq} = 0 \quad \text{or} \quad b_r = 0.$$

Definition 2.6. A tensor each component of which alters in sign but not in magnitude is said to be skew-symmetric with respect to these indices.

Now we will list some of the properties of the skew-symmetric tensors.

1. We will show that if a tensor is skew-symmetric with respect to two indices in any coordinate system, then it remains skew-symmetric with respect to these indices in any other coordinate system. Without loss of generality, consider the contravariant tensor of the second order

$$A^{ij} = -A^{ji}.$$

We have

$$\bar{A}^{ij} = \frac{\Delta_{(i)} \bar{x}^{i \sigma_0 \sigma_1 \cdots \sigma_{k-1}}}{\Delta_k x^k} \frac{\Delta_{(j)} \bar{x}^{j \sigma_0 \sigma_1 \cdots \sigma_{l-1}}}{\Delta_l x^l} A^{kl}$$

$$= -\frac{\Delta_{(i)} \bar{x}^{i \sigma_0 \sigma_1 \cdots \sigma_{k-1}}}{\Delta_k x^k} \frac{\Delta_{(j)} \bar{x}^{j \sigma_0 \sigma_1 \cdots \sigma_{l-1}}}{\Delta_l x^l} A^{lk}$$

$$= -\frac{\Delta_{(j)} \bar{x}^{j \sigma_0 \sigma_1 \cdots \sigma_{l-1}}}{\Delta_l x^l} \frac{\Delta_{(i)} \bar{x}^{i \sigma_0 \sigma_1 \cdots \sigma_{k-1}}}{\Delta_k x^k} A^{lk}$$

$$= -\bar{A}^{ji}.$$

Skew-symmetry, like symmetry cannot be defined with respect to two indices, of which one denotes contravariance and the other denotes covariance.

2. A skew-symmetry tensor A^{ij} of the second order has at most $\frac{1}{2}N(N-1)$ different arithmetical components.

   To see this, let A_{ij} be a skew-symmetric covariant tensor of the second order. Then it has N^2 components in V_N. These components are

$$\begin{matrix} 0 & A_{12} & \cdots & A_{1N} \\ A_{21} & 0 & \cdots & A_{2N} \\ \vdots & \vdots & \vdots & \vdots \\ A_{N1} & A_{N2} & \cdots & A_{NN} \end{matrix}.$$

They are of two types:

a. those in which two indices i and j are the same. In this case, we have

$$A_{ii} = -A_{ii}, \quad i \in \{1, 2, \ldots, N\},$$

   whereupon

$$A_{ii} = 0, \quad i \in \{1, 2, \ldots, N\}.$$

b. those in which the indices i and j are different, i. e., the components along the nondiagonal. Hence the maximum number of components of this type is

$$N^2 - N.$$

   Due to antisymmetry of A_{ij}, the components above and below the diagonal, the maximum number of distinct components of this type is

$$\frac{N(N-1)}{2}.$$

3. Note that if all components of a tensor in one coordinate system are zero at a point, then they are zero at this point in any coordinate system. If the components are identically zero in one coordinate system, then they are also identically zero in any coordinate system.

Definition 2.7. When a tensor is defined in all points of a curve or throughout the space V_N itself, we say that it constitutes a tensor field.

Example 2.20. Suppose that A_{ij} is a skew-symmetric tensor. We will prove that

$$(\delta_j^i \delta_l^k + \delta_l^i \delta_j^k)A_{ik} = 0.$$

Because A_{ij} is skew-symmetric, we have

$$A_{ij} = -A_{ji}.$$

Hence we obtain

$$(\delta_j^i \delta_l^k + \delta_l^i \delta_j^k)A_{ik} = \delta_j^i \delta_l^k A_{ik} + \delta_l^i \delta_j^k A_{ik}$$
$$= \delta_j^i A_{il} + \delta_l^i A_{ij}$$
$$= A_{jl} + A_{lj}$$
$$= A_{jl} - A_{jl}$$
$$= 0,$$

as required.

Example 2.21. Let a_{ijk} be symmetric with respect to the first two indices from the left and skew-symmetric with respect to the second and third indices from the left. Then

$$a_{ijk} = a_{jik}$$
$$= -a_{jki}$$
$$= -a_{kji}$$
$$= a_{kij}$$
$$= a_{ikj}$$
$$= -a_{ijk}.$$

Hence

$$2a_{ijk} = 0,$$

or

$$a_{ijk} = 0.$$

2.3 Algebraic operations with tensors

It is clear that we cannot expect to give any tensorial meaning to the equations

$$A^{ij} + B^i,$$

because it cannot satisfy the transformation law (2.11). It does, however, follow from this equation that any linear combination of tensors of the same type whose coefficients are invariants is a tensor of the same type. For instance, consider two mixed tensors A_{jk}^i and B_{jk}^i. We can form the tensor

$$\lambda A_k^{ij} + \nu B_k^{ij}, \tag{2.17}$$

which will satisfy the transformation law (2.11), provided that λ and ν are invariants.

Definition 2.8. In particular,

$$A^{ij}_k + B^{ij}_k$$

and

$$A^{ij}_k - B^{ij}_k$$

are called the sum and difference, respectively, of the tensors A^{ij}_k and B^{ij}_k.

Definition 2.9. Two tensors of the same type are said to be equal in the same coordinate system if they have the same covariant order and contravariant order and every component of one is equal to the corresponding component of the other.

Theorem 2.4. *Let λ and v be invariants, and let A^{ij}_k and B^{ij}_k be tensors. Then (2.17) is a tensor.*

Proof. Since λ and v are invariants, we have

$$\bar{\lambda} = \lambda,$$

$$\bar{v} = v.$$

Applying the transformation law, we get

$$\overline{\lambda A}^{ij}_k + \overline{v B}^{ij}_k = \lambda \frac{\Delta_{(i)} \bar{x}^{i\sigma_0 \sigma_1 \cdots \sigma_{p-1}}}{\Delta_p x^p} \frac{\Delta_{(j)} \bar{x}^{j\sigma_0 \sigma_1 \cdots \sigma_{q-1}}}{\Delta_q x^q} \frac{\Delta_r x^r}{\Delta_{(k)} \bar{x}^k} A^{pq}_r$$

$$+ v \frac{\Delta_{(i)} \bar{x}^{i\sigma_0 \sigma_1 \cdots \sigma_{p-1}}}{\Delta_p x^p} \frac{\Delta_{(j)} \bar{x}^{j\sigma_0 \sigma_1 \cdots \sigma_{q-1}}}{\Delta_q x^q} \frac{\Delta_r x^r}{\Delta_{(k)} \bar{x}^k} B^{pq}_r$$

$$= \frac{\Delta_{(i)} \bar{x}^{i\sigma_0 \sigma_1 \cdots \sigma_{p-1}}}{\Delta_p x^p} \frac{\Delta_{(j)} \bar{x}^{j\sigma_0 \sigma_1 \cdots \sigma_{q-1}}}{\Delta_q x^q} \frac{\Delta_r x^r}{\Delta_{(k)} \bar{x}^k} \lambda A^{pq}_r$$

$$+ \frac{\Delta_{(i)} \bar{x}^{i\sigma_0 \sigma_1 \cdots \sigma_{p-1}}}{\Delta_p x^p} \frac{\Delta_{(j)} \bar{x}^{j\sigma_0 \sigma_1 \cdots \sigma_{q-1}}}{\Delta_q x^q} \frac{\Delta_r x^r}{\Delta_{(k)} v \bar{x}^k} v B^{pq}_r$$

$$= \frac{\Delta_{(i)} \bar{x}^{i\sigma_0 \sigma_1 \cdots \sigma_{p-1}}}{\Delta_p x^p} \frac{\Delta_{(j)} \bar{x}^{j\sigma_0 \sigma_1 \cdots \sigma_{q-1}}}{\Delta_q x^q} \frac{\Delta_r x^r}{\Delta_{(k)} \bar{x}^k} (\lambda A^{pq}_r + v B^{pq}_r).$$

This shows that (2.17) satisfies the tensor law of transformation. This completes the proof. $\qquad\square$

Example 2.22. As another example, we can write

$$A_{ij} = \frac{1}{2}(A_{ij} + A_{ji}) + \frac{1}{2}(A_{ij} - A_{ji}).$$

Now

$$A_{ij} + A_{ji} = A_{ji} + A_{ij},$$

and

$$A_{ij} - A_{ji} = -(A_{ji} - A_{ij}),$$

i. e., $A_{ij} + A_{ji}$ is symmetric, and $A_{ij} - A_{ji}$ is skew-symmetric. Thus any covariant tensor of the second order is the sum of a symmetric and a skew-symmetric tensor. This is also true for a contravariant tensor of the second order.

Example 2.23. Let a_{ij} be the components of a covariant tensor of the second order such that

$$ba_{ij} + ca_{ij} = 0,$$

where b and c are nonzero constants. Then

$$ba_{ij} = -ca_{ij}.$$

Multiplying by b, we find

$$\begin{aligned}
b^2 a_{ij} &= -bc a_{ij} \\
&= -c(b a_{ij}) \\
&= -c(-c a_{ij}) \\
&= c^2 a_{ij}.
\end{aligned}$$

Hence

$$(b^2 - c^2)a_{ij} = 0.$$

Then

$$b^2 = c^2, \quad \text{and} \quad a_{ij} \neq 0,$$

or

$$b^2 \neq c^2, \quad \text{and} \quad a_{ij} \quad \text{is skew-symmetric.}$$

Let us take two tensors: one of contravariant order s and covariant order p, the other of contravariant order t and covariant order q. It then follows from (2.11) that the product of the components form a mixed tensor of contravariant order $s + t$ and covariant order $p + q$.

Definition 2.10. This tensor is called the outer product of the two tensors.

Example 2.24. For instance,

$$A^{ijl}_{kmnt} = B^{ij}_k C^l_{mnt}$$

is the outer product of the two tensors B^{ij}_k and C^l_{mnt} and is a tensor of the type indicated by its indices. Applying the transformation laws, we find

$$\overline{A}^{ijl}_{kmnt} = \overline{B}^{ij}_k \overline{C}^l_{mnt}$$

$$= \frac{\Delta_{(i)}\overline{x}^{i\sigma_0\sigma_1\cdots\sigma_{a-1}}}{\Delta_a x^a}\frac{\Delta_{(j)}\overline{x}^{j\sigma_0\sigma_1\cdots\sigma_{b-1}}}{\Delta_b x^b}\frac{\Delta_c x^c}{\Delta_{(k)}\overline{x}^k}B^{ab}_c\frac{\Delta_{(e)}\overline{x}^{e\sigma_0\sigma_1\cdots\sigma_{d-1}}}{\Delta_d x^d}\frac{\Delta_l x^l}{\Delta_{(m)}\overline{x}^m}\frac{\Delta_f x^f}{\Delta_{(n)}\overline{x}^n}\frac{\Delta_g x^g}{\Delta_{(t)}\overline{x}^t}C^d_{efg}$$

$$= \frac{\Delta_{(i)}\overline{x}^{i\sigma_0\sigma_1\cdots\sigma_{a-1}}}{\Delta_a x^a}\frac{\Delta_{(j)}\overline{x}^{j\sigma_0\sigma_1\cdots\sigma_{b-1}}}{\Delta_b x^b}\frac{\Delta_c x^c}{\Delta_{(k)}\overline{x}^k}\frac{\Delta_{(e)}\overline{x}^{e\sigma_0\sigma_1\cdots\sigma_{d-1}}}{\Delta_d x^d}\frac{\Delta_l x^l}{\Delta_{(m)}\overline{x}^m}\frac{\Delta_f x^f}{\Delta_{(n)}\overline{x}^n}\frac{\Delta_g x^g}{\Delta_{(t)}\overline{x}^t}B^{ab}_c C^d_{lfg}$$

$$= \frac{\Delta_{(i)}\overline{x}^{i\sigma_0\sigma_1\cdots\sigma_{a-1}}}{\Delta_a x^a}\frac{\Delta_{(j)}\overline{x}^{j\sigma_0\sigma_1\cdots\sigma_{b-1}}}{\Delta_b x^b}\frac{\Delta_{(e)}\overline{x}^{e\sigma_0\sigma_1\cdots\sigma_{d-1}}}{\Delta_d x^d}\frac{\Delta_c x^c}{\Delta_{(k)}\overline{x}^k}\frac{\Delta_l x^l}{\Delta_{(m)}\overline{x}^m}\frac{\Delta_f x^f}{\Delta_{(n)}\overline{x}^n}\frac{\Delta_g x^g}{\Delta_{(t)}\overline{x}^t}A^d_{efg}.$$

Remark 2.1. In taking the outer product of any number of tensors, care should be taken to use distinct indices. For example, it should be wrong to write the outer product of A^i_j, B_k, and C^{lm}_{kp} as

$$A^i_j B_k C^{lm}_{kp}.$$

Remark 2.2. The operation of outer product relates tensors of any two types at the same point.

Remark 2.3. The outer product of two tensors is a tensor whose order is the sum of the orders of the two tensors.

Remark 2.4. The division, in the usual sense, of one tensor by another is not defined.

2.4 Contractions of tensors

Now consider any mixed tensor A^{ij}_{lmn} and form the sum A^{ij}_{lmj}. By (2.11) we get

$$A^{ij}_{lmj} = \frac{\Delta_i x^i}{\Delta_{(p)}\overline{x}^p}\frac{\Delta_j x^j}{\Delta_{(q)}\overline{x}^q}\frac{\Delta_{(r)}\overline{x}^{r\sigma_0\sigma_1\cdots\sigma_{l-1}}}{\Delta_l x^l}\frac{\Delta_{(s)}\overline{x}^{s\sigma_0\sigma_1\cdots\sigma_{m-1}}}{\Delta_m x^m}\frac{\Delta_{(t)}\overline{x}^{t\sigma_0\sigma_1\cdots\sigma_{j-1}}}{\Delta_j x^j}\overline{A}^{pq}{}_{rst}$$

$$= \frac{\Delta_i x^i}{\Delta_{(p)}\overline{x}^p}\frac{\Delta_{(r)}\overline{x}^{r\sigma_0\sigma_1\cdots\sigma_{l-1}}}{\Delta_l x^l}\frac{\Delta_{(s)}\overline{x}^{s\sigma_0\sigma_1\cdots\sigma_{m-1}}}{\Delta_m x^m}\left(\frac{\Delta_{(t)}\overline{x}^{t\sigma_0\sigma_1\cdots\sigma_{j-1}}}{\Delta_j x^j}\frac{\Delta_j x^j}{\Delta_{(q)}\overline{x}^q}\right)\overline{A}^{pq}{}_{rst}$$

$$= \frac{\Delta_i x^i}{\Delta_{(p)} \bar{x}^p} \frac{\Delta_{(r)} \bar{x}^{r \sigma_0 \sigma_1 \cdots \sigma_{l-1}}}{\Delta_l x^l} \frac{\Delta_{(s)} \bar{x}^{s \sigma_0 \sigma_1 \cdots \sigma_{m-1}}}{\Delta_m x^m} \delta_q^t \bar{A}_{rst}^{pq}$$

$$= \frac{\Delta_i x^i}{\Delta_{(p)} \bar{x}^p} \frac{\Delta_{(r)} \bar{x}^{r \sigma_0 \sigma_1 \cdots \sigma_{l-1}}}{\Delta_l x^l} \frac{\Delta_{(s)} \bar{x}^{s \sigma_0 \sigma_1 \cdots \sigma_{m-1}}}{\Delta_m x^m} \bar{A}_{rst}^{pt}.$$

Thus we observe that A_{lmj}^{ij} is a mixed tensor, contravariant of the first order and covariant of the second order.

Definition 2.11. This process is called a contraction.

The contraction of tensors enables us to obtain a tensor of order $r - 2$ from a mixed tensor of order r. In the above computation, we could contract a stage further and arrive at the covariant vector A_{lij}^{ij}, i. e., we have

$$A_{lij}^{ij} = \frac{\Delta_i x^i}{\Delta_{(p)} \bar{x}^p} \frac{\Delta_j x^j}{\Delta_{(q)} \bar{x}^q} \frac{\Delta_{(r)} \bar{x}^{r \sigma_0 \sigma_1 \cdots \sigma_{l-1}}}{\Delta_l x^l} \frac{\Delta_{(s)} \bar{x}^{s \sigma_0 \sigma_1 \cdots \sigma_{i-1}}}{\Delta_i x^i} \frac{\Delta_{(t)} \bar{x}^{t \sigma_0 \sigma_1 \cdots \sigma_{j-1}}}{\Delta_j x^j} \bar{A}_{rst}^{pq}$$

$$= \frac{\Delta_{(r)} \bar{x}^{r \sigma_0 \sigma_1 \cdots \sigma_{l-1}}}{\Delta_l x^l} \left(\frac{\Delta_{(s)} \bar{x}^{s \sigma_0 \sigma_1 \cdots \sigma_{i-1}}}{\Delta_i x^i} \right) \left(\frac{\Delta_{(t)} \bar{x}^{t \sigma_0 \sigma_1 \cdots \sigma_{j-1}}}{\Delta_j x^j} \frac{\Delta_j x^j}{\Delta_{(q)} \bar{x}^q} \right) \bar{A}_{rst}^{pq}$$

$$= \frac{\Delta_{(r)} \bar{x}^{r \sigma_0 \sigma_1 \cdots \sigma_{l-1}}}{\Delta_l x^l} \delta_p^s \delta_q^t \bar{A}_{rst}^{pq}$$

$$= \frac{\Delta_{(r)} \bar{x}^{r \sigma_0 \sigma_1 \cdots \sigma_{l-1}}}{\Delta_l x^l} \bar{A}_{rst}^{st}.$$

Therefore we can form the following different tensors:

$$A_{lmj}^{ij}, \quad A_{ljm}^{ij}, \quad A_{jmn}^{ij}, \quad A_{lmi}^{ij}, \quad A_{lin}^{ij}, \quad A_{imn}^{ij},$$

$$A_{lij}^{ij}, \quad A_{imj}^{ij}, \quad A_{ijn}^{ij}, \quad A_{jin}^{ij}.$$

If the tensor A_{lmn}^{ij} possesses any symmetric properties, then there will be fewer tensors formed from it by contraction.

Consider A_j^i. We have

$$A_i^i = \frac{\Delta_i x^i}{\Delta_{(s)} \bar{x}^s} \frac{\Delta_{(p)} \bar{x}^{p \sigma_0 \sigma_1 \cdots \sigma_{i-1}}}{\Delta_i x^i} \bar{A}_p^s$$

$$= \delta_s^p \bar{A}_p^s$$

$$= \bar{A}_s^s.$$

This justifies us in calling an invariant a tensor of zeroth order.

We can also combine multiplication and contraction to produce new tensors. For instance, consider tensors A_k^{ij} and B_{mnt}^l. We have

$$A_r^{pq} B_{bcd}^r = \frac{\Delta_p x^p}{\Delta_{(i)}\overline{x}^i} \frac{\Delta_q x^q}{\Delta_{(j)}\overline{x}^j} \frac{\Delta_{(k)}\overline{x}^{k\sigma_0\sigma_1\cdots\sigma_{r-1}}}{\Delta_r x^r} \overline{A}_k^{ij}$$

$$\times \frac{\Delta_r x^r}{\Delta_{(l)}\overline{x}^l} \frac{\Delta_{(m)}\overline{x}^{m\sigma_0\sigma_1\cdots\sigma_{b-1}}}{\Delta_b x^b} \frac{\Delta_{(n)}\overline{x}^{n\sigma_0\sigma_1\cdots\sigma_{l-1}}}{\Delta_l x^l} \frac{\Delta_{(s)}\overline{x}^{s\sigma_0\sigma_1\cdots\sigma_{d-1}}}{\Delta_d x^d} \overline{B}_{mns}^l$$

$$= \frac{\Delta_p x^p}{\Delta_{(i)}\overline{x}^i} \frac{\Delta_q x^q}{\Delta_{(j)}\overline{x}^j} \frac{\Delta_{(m)}\overline{x}^{m\sigma_0\sigma_1\cdots\sigma_{b-1}}}{\Delta_b x^b} \frac{\Delta_{(n)}\overline{x}^{n\sigma_0\sigma_1\cdots\sigma_{l-1}}}{\Delta_l x^l} \frac{\Delta_{(s)}\overline{x}^{s\sigma_0\sigma_1\cdots\sigma_{d-1}}}{\Delta_d x^d}$$

$$\times \left(\frac{\Delta_{(k)}\overline{x}^{k\sigma_0\sigma_1\cdots\sigma_{r-1}}}{\Delta_r x^r} \frac{\Delta_r x^r}{\Delta_{(l)}\overline{x}^l} \right) \overline{A}_k^{ij}\overline{B}_{mns}^l$$

$$= \frac{\Delta_p x^p}{\Delta_{(i)}\overline{x}^i} \frac{\Delta_q x^q}{\Delta_{(j)}\overline{x}^j} \frac{\Delta_{(m)}\overline{x}^{m\sigma_0\sigma_1\cdots\sigma_{b-1}}}{\Delta_b x^b} \frac{\Delta_{(n)}\overline{x}^{n\sigma_0\sigma_1\cdots\sigma_{l-1}}}{\Delta_l x^l} \frac{\Delta_{(s)}\overline{x}^{s\sigma_0\sigma_1\cdots\sigma_{d-1}}}{\Delta_d x^d} \delta_l^k \overline{A}_k^{ij}\overline{B}_{mns}^l$$

$$= \frac{\Delta_p x^p}{\Delta_{(i)}\overline{x}^i} \frac{\Delta_q x^q}{\Delta_{(j)}\overline{x}^j} \frac{\Delta_{(m)}\overline{x}^{m\sigma_0\sigma_1\cdots\sigma_{b-1}}}{\Delta_b x^b} \frac{\Delta_{(n)}\overline{x}^{n\sigma_0\sigma_1\cdots\sigma_{k-1}}}{\Delta_k x^k} \frac{\Delta_{(s)}\overline{x}^{s\sigma_0\sigma_1\cdots\sigma_{d-1}}}{\Delta_d x^d} \overline{A}_k^{ij}\overline{B}_{mns}^k,$$

which is a mixed tensor, contravariant of second order and covariant of third order.

Definition 2.12. This process is called inner multiplication of two tensors, and the resulting tensor is called the inner product of the two tensors.

Note that we contract two indices of the same type as the resulting sum is not necessarily a tensor. Also, it should be clear that with our index notation, the summation convention applies to two indices one of which is superscript and the other one is a subscript.

2.5 Quotient law

Sometimes, it is necessary to ascertain whether a set of functions forms the components of a tensor. The direct method requires us to find out if they satisfy a tensor transformation equation of type (2.11). In practice, this is troublesome, and a simpler test is provided by the quotient law.

Theorem 2.5 (The quotient law). *N^p functions of x^i form the components of a tensor of order p, whose contravariant and covariant character can readily be determined, provided that an inner product of these functions with an arbitrary tensor is itself a tensor.*

Proof. It suffices to provide the proof in the following particular case. The set of N^3 functions A^{ijk} forms the components of a tensor of the type indicated by its indices if

$$A^{ijk} B_{ij}^p = C^{pk},$$

provided that B_{ij}^p is an arbitrary tensor and C^{pk} is a tensor. The transformed quantities, referred to a system of coordinates $\overline{x}^i$, satisfy the equations

$$\overline{A}^{ijk}\overline{B}^{p}_{ij} = \overline{C}^{pk}.$$

Applying (2.11), we find

$$A^{ijk}B^{p}_{ij} = A^{ijk}\frac{\Delta_p x^p}{\Delta_{(a)}\overline{x}^a}\frac{\Delta_{(b)}\overline{x}^{b\sigma_0\sigma_1\cdots\sigma_{i-1}}}{\Delta_i x^i}\frac{\Delta_{(c)}\overline{x}^{c\sigma_0\sigma_1\cdots\sigma_{j-1}}}{\Delta_j x^j}\overline{B}^{a}_{bc},$$

$$C^{pk} = \frac{\Delta_p x^p}{\Delta_{(d)}\overline{x}^d}\frac{\Delta_k x^k}{\Delta_{(e)}\overline{x}^e}\overline{C}^{de}.$$

Therefore

$$A^{ijk}\frac{\Delta_p x^p}{\Delta_{(a)}\overline{x}^a}\frac{\Delta_{(b)}\overline{x}^{b\sigma_0\sigma_1\cdots\sigma_{i-1}}}{\Delta_i x^i}\frac{\Delta_{(c)}\overline{x}^{c\sigma_0\sigma_1\cdots\sigma_{j-1}}}{\Delta_j x^j}\overline{B}^{a}_{bc} = \frac{\Delta_p x^p}{\Delta_{(d)}\overline{x}^d}\frac{\Delta_k x^k}{\Delta_{(e)}\overline{x}^e}\overline{C}^{de},$$

whereupon

$$A^{ijk}\left(\frac{\Delta_{(d)}\overline{x}^{d\sigma_0\sigma_1\cdots\sigma_{p-1}}}{\Delta_p x^p}\frac{\Delta_p x^p}{\Delta_{(a)}\overline{x}^a}\right)\frac{\Delta_{(b)}\overline{x}^{b\sigma_0\sigma_1\cdots\sigma_{i-1}}}{\Delta_i x^i}\frac{\Delta_{(c)}\overline{x}^{c\sigma_0\sigma_1\cdots\sigma_{j-1}}}{\Delta_j x^j}\frac{\Delta_{(e)}\overline{x}^{b\sigma_0\sigma_1\cdots\sigma_{k-1}}}{\Delta_k x^k}\overline{B}^{a}_{bc} = \overline{C}^{de},$$

or

$$A^{ijk}\delta^{d}_{a}\frac{\Delta_{(b)}\overline{x}^{b\sigma_0\sigma_1\cdots\sigma_{i-1}}}{\Delta_i x^i}\frac{\Delta_{(c)}\overline{x}^{c\sigma_0\sigma_1\cdots\sigma_{j-1}}}{\Delta_j x^j}\frac{\Delta_{(e)}\overline{x}^{b\sigma_0\sigma_1\cdots\sigma_{k-1}}}{\Delta_k x^k}\overline{B}^{a}_{bc} = \overline{C}^{de},$$

or

$$A^{ijk}\frac{\Delta_{(b)}\overline{x}^{b\sigma_0\sigma_1\cdots\sigma_{i-1}}}{\Delta_i x^i}\frac{\Delta_{(c)}\overline{x}^{c\sigma_0\sigma_1\cdots\sigma_{j-1}}}{\Delta_j x^j}\frac{\Delta_{(e)}\overline{x}^{b\sigma_0\sigma_1\cdots\sigma_{k-1}}}{\Delta_k x^k}\overline{B}^{d}_{bc} = \overline{C}^{de},$$

or

$$A^{ijk}\frac{\Delta_{(b)}\overline{x}^{b\sigma_0\sigma_1\cdots\sigma_{i-1}}}{\Delta_i x^i}\frac{\Delta_{(c)}\overline{x}^{c\sigma_0\sigma_1\cdots\sigma_{j-1}}}{\Delta_j x^j}\frac{\Delta_{(e)}\overline{x}^{b\sigma_0\sigma_1\cdots\sigma_{k-1}}}{\Delta_k x^k}\overline{B}^{d}_{bc} = \overline{A}^{bce}\overline{B}^{d}_{bc}.$$

Hence

$$\left(A^{ijk}\frac{\Delta_{(b)}\overline{x}^{b\sigma_0\sigma_1\cdots\sigma_{i-1}}}{\Delta_i x^i}\frac{\Delta_{(c)}\overline{x}^{c\sigma_0\sigma_1\cdots\sigma_{j-1}}}{\Delta_j x^j}\frac{\Delta_{(e)}\overline{x}^{b\sigma_0\sigma_1\cdots\sigma_{k-1}}}{\Delta_k x^k} - \overline{A}^{bce}\right)\overline{B}^{d}_{bc} = 0.$$

Because B^{p}_{ij} is an arbitrary tensor, we get

$$A^{ijk}\frac{\Delta_{(b)}\overline{x}^{b\sigma_0\sigma_1\cdots\sigma_{i-1}}}{\Delta_i x^i}\frac{\Delta_{(c)}\overline{x}^{c\sigma_0\sigma_1\cdots\sigma_{j-1}}}{\Delta_j x^j}\frac{\Delta_{(e)}\overline{x}^{b\sigma_0\sigma_1\cdots\sigma_{k-1}}}{\Delta_k x^k} - \overline{A}^{bce} = 0.$$

Thus A^{mnr} is a tensor of the third order and contravariant in all indices. This completes the proof. $\qquad\square$

Remark 2.5. In the above proof, it is important that the tensor B^l_{mn} will be arbitrary and must not possess any symmetric or skew-symmetric properties. Now suppose that B^l_{mn} is symmetrical in m and n. As in the proof of Theorem 2.5, we have

$$\overline{A}^{ijk}\frac{\Delta_m x^m\ \Delta_n x^n}{\Delta_{(i)}\overline{x}^i\ \Delta_{(j)}\overline{x}^j} = \frac{\Delta_{(k)}\overline{x}^{k\sigma_0\sigma_1\cdots\sigma_{r-1}}}{\Delta_r x^r}A^{mnr}$$

and

$$\overline{A}^{jik}\frac{\Delta_m x^m\ \Delta_n x^n}{\Delta_{(i)}\overline{x}^i\ \Delta_{(j)}\overline{x}^j} = \frac{\Delta_{(k)}\overline{x}^{k\sigma_0\sigma_1\cdots\sigma_{r-1}}}{\Delta_r x^r}A^{nmr}.$$

Then

$$(\overline{A}^{ijk}+\overline{A}^{jik})\frac{\Delta_m x^m\ \Delta_n x^n}{\Delta_{(i)}\overline{x}^i\ \Delta_{(j)}\overline{x}^j} = \frac{\Delta_{(k)}\overline{x}^{k\sigma_0\sigma_1\cdots\sigma_{r-1}}}{\Delta_r x^r}(A^{mnr}+A^{nmr})$$

and

$$\overline{A}^{ijk}+\overline{A}^{jik} = \frac{\Delta_{(i)}\overline{x}^{i\sigma_0\sigma_1\cdots\sigma_{m-1}}}{\Delta_m x^m}\ \frac{\Delta_{(j)}\overline{x}^{j\sigma_0\sigma_1\cdots\sigma_{n-1}}}{\Delta_n x^n}\ \frac{\Delta_{(k)}\overline{x}^{k\sigma_0\sigma_1\cdots\sigma_{r-1}}}{\Delta_r x^r}(A^{mnr}+A^{nmr}).$$

Therefore $A^{mnr}+A^{nmr}$ is a contravariant tensor of the third order.

Example 2.25. Let A^i and B^i be arbitrary contravariant vectors and suppose

$$C_{ij}A^i B^j \tag{2.18}$$

is an invariant. We will show that C_{ij} is a tensor of the second order.

Since (2.18) is an invariant, we get

$$\overline{C}_{ij}\overline{A}^i\overline{B}^j = C_{ij}A^i B^j.$$

Hence

$$\overline{C}_{ij}\frac{\Delta_{(i)}\overline{x}^{i\sigma_0\sigma_1\cdots\sigma_{m-1}}}{\Delta_m x^m}\ \frac{\Delta_{(j)}\overline{x}^{j\sigma_0\sigma_1\cdots\sigma_{n-1}}}{\Delta_n x^n}A^m B^n = C_{ij}A^i B^j,$$

and

$$\overline{C}_{ij}A^s B^t = \frac{\Delta_s x^s\ \Delta_t x^t}{\Delta_{(i)}\overline{x}^i\ \Delta_{(j)}\overline{x}^j}C_{st}A^s B^t.$$

Because A^s and B^t are arbitrary contravariant vectors, we find

$$\overline{C}_{ij} = \frac{\varDelta_s x^s}{\varDelta_{(i)} \overline{x}^i} \frac{\varDelta_t x^t}{\varDelta_{(j)} \overline{x}^j} C_{st}.$$

Therefore C_{st} is a covariant tensor of the second order.

Example 2.26. Let A^i be an arbitrary contravariant vector and suppose

$$C_{ij} A^i A^j \tag{2.19}$$

is an invariant. We will prove that

$$C_{ij} + C_{ji} \tag{2.20}$$

is a covariant tensor of the second order. Because (2.19) is an invariant, we get

$$\overline{C}_{ij} \overline{A}^i \overline{A}^j = C_{ij} A^i A^j.$$

Then

$$\overline{C}_{ij} \frac{\varDelta_{(i)} \overline{x}^{i\sigma_0 \sigma_1 \cdots \sigma_{t-1}}}{\varDelta_t x^t} \frac{\varDelta_{(j)} \overline{x}^{j\sigma_0 \sigma_1 \cdots \sigma_{s-1}}}{\varDelta_s x^s} A^t A^s = C_{ij} A^i A^j,$$

and

$$\overline{C}_{ji} \frac{\varDelta_{(i)} \overline{x}^{i\sigma_0 \sigma_1 \cdots \sigma_{t-1}}}{\varDelta_t x^t} \frac{\varDelta_{(j)} \overline{x}^{j\sigma_0 \sigma_1 \cdots \sigma_{s-1}}}{\varDelta_s x^s} A^t A^s = C_{ji} A^i A^j,$$

whereupon

$$(\overline{C}_{ij} + \overline{C}_{ji}) \frac{\varDelta_{(i)} \overline{x}^{i\sigma_0 \sigma_1 \cdots \sigma_{t-1}}}{\varDelta_t x^t} \frac{\varDelta_{(j)} \overline{x}^{j\sigma_0 \sigma_1 \cdots \sigma_{s-1}}}{\varDelta_s x^s} A^t A^s = (C_{ij} + C_{ji}) A^i A^j,$$

and

$$(\overline{C}_{ij} + \overline{C}_{ji}) A^m A^n = (C_{mn} + C_{nm}) \frac{\varDelta_m x^m}{\varDelta_{(i)} \overline{x}^i} \frac{\varDelta_n x^n}{\varDelta_{(j)} \overline{x}^j} A^m A^n.$$

Because A^m is an arbitrary contravariant vector, we find

$$(\overline{C}_{ij} + \overline{C}_{ji}) = (C_{mn} + C_{nm}) \frac{\varDelta_m x^m}{\varDelta_{(i)} \overline{x}^i} \frac{\varDelta_n x^n}{\varDelta_{(j)} \overline{x}^j}.$$

Therefore (2.20) is a covariant tensor of the second order.

Exercise 2.4. Let a quantity $A(p, q, r)$ be such that in the coordinate system x^i,

$$A(p, q, r)B_r^{qs} = C_p^s,$$

where B_r^{qs} and C_p^s are arbitrary tensors. Prove that $A(p, q, r)$ is a tensor.

2.6 Conjugate symmetric tensors of the second order

Consider a symmetric covariant tensor of the second order A_{ij} with determinant

$$\det(A_{ij}) \neq 0.$$

Let B^{ij} denote the expression formed by dividing the cofactor of A_{ij} in the determinant $\det(A_{ij})$ divided by $\det(A_{ij})$. We will show that B^{ij} is a contravariant tensor of the second order. By the theory of determinants we obtain

$$A_{ij}B^{ik} = \delta_j^k.$$

We cannot establish the tensor character of B^{ik} by applying the quotient law directly to this equation because A_{ij} is not arbitrary. Now choose an arbitrary contravariant vector C^i. Set

$$D_i = A_{ij}C^j.$$

We have

$$D_i = A_{ij}C^j$$

$$= \frac{\Delta_{(a)}\bar{x}^{a\sigma_0\sigma_1\cdots\sigma_{i-1}}}{\Delta_i x^i} \frac{\Delta_{(b)}\bar{x}^{b\sigma_0\sigma_1\cdots\sigma_{j-1}}}{\Delta_j x^j}\bar{A}_{ab}\frac{\Delta_j x^j}{\Delta_{(c)}\bar{x}^c}\bar{C}^c$$

$$= \frac{\Delta_{(a)}\bar{x}^{a\sigma_0\sigma_1\cdots\sigma_{i-1}}}{\Delta_i x^i}\left(\frac{\Delta_{(b)}\bar{x}^{b\sigma_0\sigma_1\cdots\sigma_{j-1}}}{\Delta_j x^j}\frac{\Delta_j x^j}{\Delta_{(c)}\bar{x}^c}\right)\bar{A}_{ab}\bar{C}^c$$

$$= \frac{\Delta_{(a)}\bar{x}^{a\sigma_0\sigma_1\cdots\sigma_{i-1}}}{\Delta_i x^i}\delta_c^b\bar{A}_{ab}\bar{C}^c$$

$$= \frac{\Delta_{(a)}\bar{x}^{a\sigma_0\sigma_1\cdots\sigma_{i-1}}}{\Delta_i x^i}\bar{A}_{ab}\bar{C}^b$$

$$= \frac{\Delta_{(a)}\bar{x}^{a\sigma_0\sigma_1\cdots\sigma_{i-1}}}{\Delta_i x^i}\bar{D}_a,$$

i. e., D_i is an arbitrary covariant vector. Hence

$$D_i B^{ik} = A_{ij} C^j B^{ik}$$
$$= C^j A_{ij} B^{ik}$$
$$= C^j \delta_j^k$$
$$= C^k.$$

Now we apply the quotient law to the equation

$$D_i B^{ik} = C^k,$$

and we find

$$\overline{D}_i \overline{B}^{ik} = \overline{C}^k,$$

or

$$\frac{\Delta_m x^m}{\Delta_{(i)} \overline{x}^i} D_m \overline{B}^{ik} = \frac{\Delta_{(k)} \overline{x}^{k\sigma_0 \sigma_1 \cdots \sigma_{n-1}}}{\Delta_n x^n} C_n.$$

Inner multiplication by

$$\frac{\Delta_s \overline{x}^{s\sigma_0 \sigma_1 \cdots \sigma_{m-1}}}{\Delta_m x^m}$$

gives

$$D_m \overline{B}^i k = \frac{\Delta_{(s)} \overline{x}^{s\sigma_0 \sigma_1 \cdots \sigma_{m-1}}}{\Delta_m x^m} \frac{\Delta_{(k)} \overline{x}^{k\sigma_0 \sigma_1 \cdots \sigma_{n-1}}}{\Delta_n x^n} C_n$$
$$= \frac{\Delta_{(s)} \overline{x}^{s\sigma_0 \sigma_1 \cdots \sigma_{m-1}}}{\Delta_m x^m} \frac{\Delta_{(k)} \overline{x}^{k\sigma_0 \sigma_1 \cdots \sigma_{n-1}}}{\Delta_n x^n} D_m B^{mn},$$

whereupon

$$\overline{B}^{ik} = \frac{\Delta_{(s)} \overline{x}^{s\sigma_0 \sigma_1 \cdots \sigma_{m-1}}}{\Delta_m x^m} \frac{\Delta_{(k)} \overline{x}^{k\sigma_0 \sigma_1 \cdots \sigma_{n-1}}}{\Delta_n x^n} B^{mn}.$$

Therefore B^{ij} is a tensor of the second order. It is symmetric because A_{ij} is symmetric.

Let now E_{ij} denote the cofactor of V^{ij} in the determinant $\det(B^{ij})$ divided by $\det(B^{ij})$. Form the theory of determinants we have

$$\det(A_{ij}) \det(B_{ij}) = 1.$$

Therefore

$$\det(B_{ij}) \neq 0.$$

Next,

$$E_{ij}B^{ik} = \delta_j^k.$$

Inner multiplication by A_{kl} yields

$$E_{ij}A_{kl}B^{ik} = A_{kl}\delta_j^k,$$

or

$$E_{ij}A_{kl}B^{ki} = A_{kl}\delta_j^k,$$

or

$$E_{ij}\delta_l^i = A_{kl}\delta_j^k,$$

or

$$\begin{aligned} E_{ij} &= A_{ij} \\ &= A_{ji}. \end{aligned}$$

Thus this process only leads back to the original covariant tensor of the second order.

Example 2.27. We say that A_{ij} and B^{ij} are conjugate tensors if

$$A_{ij}B^{ik} = \delta_j^k.$$

It is important to note that a tensor of the second order has a conjugate tensor only if its determinant is different than zero.

Example 2.28. Suppose that A_{ij} is a covariant tensor of the second order and that

$$A_{ij} = 0 \quad \text{if } i \neq j$$

and

$$\det(A_{ij}) \neq 0.$$

Let B^{ij} be its conjugate tensor. Then

$$A_{ij}B^{ik} = \delta_j^k$$

and

$$A_{ii}B^{ii} = 1,$$

whereupon

$$B^{ii} = \frac{1}{A_{ii}}.$$

For $i \neq j$, any component of any cofactor of A_{ij} in the $\det(A_{ij})$ contains at least one A_{lk}, $l \neq k$. Then

$$B_{ij} = 0, \quad i \neq j.$$

2.7 Advanced practical problems

Problem 2.1. Let the components of a contravariant tensor A^{ij} in the x^i coordinate system be as follows:

$$A^{11} = 2, \quad A^{12} = -1, \quad A^{21} = 2, \quad A^{22} = 1.$$

Find its components in the $\bar{x}^i$ coordinate system, where

$$\bar{x}^1 = (x^2)^2,$$
$$\bar{x}^2 = x^1 + x^2, \quad x^1 > 0, \ x^2 > 0.$$

Problem 2.2. Let the components of a contravariant tensor A_{ij} of the second order in the x^i coordinate system be as follows:

$$A_{11} = 1, \quad A_{12} = -2, \quad A_{21} = 1, \quad A_{22} = 2.$$

Find its components in the $\bar{x}^i$ coordinate system, where

$$\bar{x}^1 = x^1,$$
$$\bar{x}^2 = (x^2)^2, \quad x^1 > 0, \ x^2 > 0.$$

Problem 2.3. Let the components of a mixed tensor A_j^i of the second order in the x^i coordinate system be as follows:

$$A_1^1 = -1, \quad A_2^1 = 1, \quad A_1^2 = 2, \quad A_2^2 = -2.$$

Find its components in the $\bar{x}^i$ coordinate system, where

$$\bar{x}^1 = x^1,$$
$$\bar{x}^2 = (x^2)^2, \quad x^2 > 0.$$

Problem 2.4. If the relation

$$a_{ij}v^iv^j = b_{ij}v^iv^j$$

holds for any arbitrary values of v^i, then prove that

$$a_{ij} + a_{ji} = b_{ij} + b_{ji}.$$

Hint 2.6. Set

$$c_{ij} = a_{ij} - b_{ij}.$$

Then the given relation takes the form

$$c_{ij}v^iv^j = 0.$$

Prove that

$$c_{ij} + c_{ji} = 0.$$

Problem 2.5. If the equality

$$a^i_j v_i = \beta v_j$$

holds for any covariant vector v_i, where β is a scalar, then prove that

$$a^i_j = \beta \delta^i_j.$$

Solution. Set

$$b^i_j = a^i_j - \beta \delta^i_j.$$

The given relation can be written in the form

$$a^i_j v_i = \beta \delta^i_j v_i.$$

Hence

$$(a^i_j - \beta \delta^i_j)v_i = 0,$$

or

$$b^i_j v_i = 0.$$

Because v_i is arbitrary, we find

$$\beta_j^i = 0.$$

Problem 2.6. If the equality

$$a_j^i v^j = \beta v^i$$

holds for any contravariant vector v^i, where β is a scalar, then prove that

$$a_j^i = \beta \delta_j^i.$$

Problem 2.7. Let A_{jk}^i be a mixed tensor. Prove that A_{ji}^i is a covariant vector.

Problem 2.8. If the relation

$$a_{hijk}\lambda^h v^i \lambda^j v^k = 0,$$

where λ^i and v^j are components of two arbitrary contravariant vectors, then prove that

$$a_{hijk} + a_{hkji} + a_{jihk} + a_{jkhi} = 0.$$

Problem 2.9. If the relation

$$a_{ijk}\lambda^i \lambda^j \lambda^k = 0$$

holds for any arbitrary contravariant vector λ^i, then prove that

$$a_{ijk} + a_{jki} + a_{kij} + a_{jik} + a_{kji} + a_{ikj} = 0.$$

Problem 2.10. If

$$a^{ij} u_i u_j$$

is an invariant for an arbitrary covariant vector u_i, then prove that

$$a^{ij} + a^{ji}$$

is a contravariant tensor of the second order.

Problem 2.11. If a_{ij} are components of a covariant tensor of the second order and λ^i and v^j are the components of two contravariant vectors, then prove that

$$a_{ij}\lambda^i \lambda^j$$

is an invariant.

Problem 2.12. If

$$a_{ij} u^i u^j$$

is an invariant, where u^i is an arbitrary contravariant vector, a_{ij} is a symmetric tensor, and

$$u^i = \lambda^i + v^i,$$

then prove that

$$a_{ij} \lambda^i v^j$$

is an invariant.

Problem 2.13. Prove that any contraction of a mixed tensor A^i_j is an invariant.

Problem 2.14. Prove that any contraction of a tensor A^i_{jk} is a covariant vector.

Problem 2.15. If A^{ij}_{kl} are tensor components, then prove that A^{ij}_{ij} is an invariant.

Problem 2.16. Prove that the outer product of a contravariant vector and a covariant vector is a mixed tensor.

Problem 2.17. Let A_{ij} be a covariant tensor, and let B^i be a contravariant vector. Prove that $A_{ij} B^i$ is a covariant vector.

Problem 2.18. If a_{ij} is a symmetric tensor and b_i is a vector such that

$$a_{ij} b_k + a_{jk} b_i + a_{ki} b_j = 0,$$

then prove that

$$a_{ij} = 0 \quad \text{or} \quad b_k = 0.$$

Problem 2.19. If a_{ij} is a component of a covariant symmetric tensor and b_i is a nonzero covariant vector such that

$$a_{ij} b_k + a_{ik} b_j + a_{ki} b_j = 0,$$

then prove that

$$a_{ij} = 0.$$

Problem 2.20. Prove:

1.

$$e_{ij} = \frac{j - i}{|j - i|}.$$

2.

$$e_{ijk} = \frac{(j - i)(k - i)(k - j)}{|j - i||k - i||k - j|}.$$

3.

$$e_{i_1 i_2 \cdots e_n} = \frac{(i_2 - i_1)(i_3 - i_1) \cdots (i_n - i_1)(i_3 - i_2) \cdots (i_n - i_2) \cdots (i_n - i_{n-1})}{|i_2 - i_1||i_3 - i_1| \cdots |i_n - i_1||i_3 - i_2| \cdots |i_n - i_2| \cdots |i_n - i_{n-1}|}.$$

3 The Riemann metric

In this chapter, an N-dimensional Riemannian space has been chosen for the development of tensor calculus on time scales. Metric and associated tensors are defined, and some of their properties are explored. Affine and curvilinear coordinates are introduced.

3.1 The metric tensor

Consider a space of N dimensions and a displacement vector $\Delta_i x^i$, $i \in \{1, 2, \ldots, N\}$, determined by a pair of neighboring points x^i and $x^i + \Delta_i x^i$. The distance Δs between the two adjacent points whose coordinates in any system are x^i and $x^i + \Delta_i x^i$, $i \in \{1, 2, \ldots, N\}$, is given by the quadratic formulas

$$\Delta s^2 = g_{ij} \Delta_i x^i \Delta_j x^j, \quad i, j \in \{1, \ldots, N\}, \tag{3.1}$$

where the coefficients g_{ij} are arbitrary functions of the coordinates x^i such that

$$g = |g_{ij}| \neq 0.$$

Definition 3.1. The quadratic dynamic form

$$g_{ij} \Delta_i x^i \Delta_j x^j, \tag{3.2}$$

which expresses the distance between two neighboring points, is called a metric, or a Riemannian metric, or a line element. The coefficients g_{ij} in equation (3.1) are called the metric tensor or fundamental tensor of the Riemannian metric.

The quadratic form (3.1) is positive definite if it is positive for all the values of the dynamic differentials $\Delta_i x^i$, not all equal to zero.

Since the distance between two pints is independent of the coordinate system, the line element is an invariant.

Definition 3.2. The element Δs is called the element of arc in V_N.

Definition 3.3. If Δs is real, then it is called an elementary distance.

Every three-dimensional Euclidean space E^3 referred to an orthogonal Cartesian system can be written as

$$\Delta s^2 = \delta_{ij} \Delta_i x^i \Delta_j x^j, \quad i, j \in \{1, 2, \ldots, N\}, \quad \delta_{ij} = g_{ij},$$

where δ_i^j is the Kronecker delta. Sometimes,

https://doi.org/10.1515/9783112228494-003

$$\Delta s^2 = g_{ij}\Delta_i x^i \Delta_j x^j,$$

where the numerical factor e, called the indicator, is equal to $+1$ or -1, so that Δs^2 is always nonnegative.

Definition 3.4. An N-dimensional space characterized by this metric is called the Riemannian space of N dimensions and is denoted by V_N. Geometry based on this metric is called Riemannian geometry of N dimensions.

Now we establish the nature of g_{ij}.

Theorem 3.1. *In a Riemannian space the fundamental tensor is a covariant symmetric tensor.*

Proof. Let us consider a covariant transformation from x^i to $\overline{x}^i$, $i \in \{1, 2, \ldots, N\}$, given by

$$\overline{x}^i = \overline{x}^i(x^1, x^2, \ldots, x^N),$$

so that the metric (3.1) transforms to

$$\Delta s^2 = \overline{g}_{ij}\Delta_{(i)}\overline{x}^i \Delta_{(j)}\overline{x}^j. \tag{3.3}$$

Now

$$\Delta_{(i)}\overline{x}^i = \frac{\Delta_{(i)}\overline{x}^{i\sigma_0\sigma_1\cdots\sigma_{p-1}}}{\Delta_p x^p}\Delta_p x^p.$$

This is the law of transformation of a covariant vector. Thus $\Delta_i x^i$ is a covariant vector. Hence, using (3.3), we get

$$g_{ij}\Delta_i x^i \Delta_j x^j = \overline{g}_{ij}\frac{\Delta_{(i)}\overline{x}^{i\sigma_0\sigma_1\cdots\sigma_{p-1}}}{\Delta_p x^p}\frac{\Delta_{(j)}\overline{x}^{j\sigma_0\sigma_1\cdots\sigma_{q-1}}}{\Delta_q x^q}\Delta_p x^p \Delta_q x^q,$$

or

$$g_{pq}\Delta_p x^p \Delta_q x^q = \overline{g}_{ij}\frac{\Delta_{(i)}\overline{x}^{i\sigma_0\sigma_1\cdots\sigma_{p-1}}}{\Delta_p x^p}\frac{\Delta_{(j)}\overline{x}^{j\sigma_0\sigma_1\cdots\sigma_{q-1}}}{\Delta_q x^q}\Delta_p x^p \Delta_q x^q$$

Therefore

$$\left(g_{pq} - \overline{g}_{ij}\frac{\Delta_{(i)}\overline{x}^{i\sigma_0\sigma_1\cdots\sigma_{p-1}}}{\Delta_p x^p}\frac{\Delta_{(j)}\overline{x}^{j\sigma_0\sigma_1\cdots\sigma_{q-1}}}{\Delta_q x^q}\right)\Delta_p x^p \Delta_q x^q = 0.$$

Since $\Delta_p x^p$ and $\Delta_q x^q$ are arbitrary, we find

$$g_{pq} = \overline{g}_{ij} \frac{\Delta_{(i)}\overline{x}^{i\sigma_0\sigma_1\cdots\sigma_{p-1}} \Delta_{(j)}\overline{x}^{j\sigma_0\sigma_1\cdots\sigma_{q-1}}}{\Delta_p x^p \; \Delta_q x^q}.$$

This is the second-order covariant tensor law of transformation. Therefore g_{ij} is a covariant tensor of the second order.

Finally, we will show that g_{ij} is symmetric. Now g_{ij} can be written as

$$g_{ij} = \frac{1}{2}(g_{ij} + g_{ji}) + \frac{1}{2}(g_{ij} - g_{ji})$$
$$= A_{ij} + B_{ij},$$

where

$$A_{ij} = \frac{1}{2}(g_{ij} + g_{ji}),$$
$$B_{ij} = \frac{1}{2}(g_{ij} - g_{ji}).$$

Note that A_{ij} is a symmetric covariant tensor of the second order and B_{ij} is a skew-symmetric covariant tensor of the second order. Therefore

$$g_{ij}\Delta_i x^i \Delta_j x^j = (A_{ij} + B_{ij})\Delta_i x^i \Delta_j x^j$$
$$= A_{ij}\Delta_i x^i \Delta_j x^j + B_{ij}\Delta_i x^i \Delta_j x^j,$$

or

$$(g_{ij} - A_{ij})\Delta_i x^i \Delta_j x^j = B_{ij}\Delta_i x^i \Delta_j x^j. \tag{3.4}$$

Interchanging the dummy indices in

$$B_{ij}\Delta_i x^i \Delta_j x^j$$

and using that B_{ij} is skew-symmetric, we get

$$B_{ij}\Delta_i x^i \Delta_j x^j = B_{ji}\Delta_j x^j \Delta_i x^i$$
$$= -B_{ij}\Delta_j x^j \Delta_i x^i$$
$$= -B_{ij}\Delta_i x^i \Delta_j x^j,$$

whereupon

$$2B_{ij}\Delta_i x^i \Delta_j x^j = 0,$$

or

$$B_{ij}\Delta_i x^i \Delta_j x^j = 0.$$

Because $\Delta_i x^i$ and $\Delta_j x^j$ are arbitrary, we conclude that

$$B_{ij} = 0.$$

Therefore

$$g_{ij} = A_{ij},$$

and g_{ij} is symmetric. This completes the proof. $\qquad\square$

Definition 3.5. In an N-dimensional space, a coordinate system in terms of which

$$g_{ij} = 0, \quad i \neq j,$$

is called an orthogonal coordinate system.

Definition 3.6. In an N dimensional space, a coordinate system in terms of which

$$g_{ii} = 1, \quad g_{ij} = 0, \quad i \neq j,$$

is called a Cartesian coordinate system.

Theorem 3.2. *The line element*

$$g_{ij}\Delta_i x^i \Delta_j x^j$$

is an invariant.

Proof. Let us consider a coordinate transformation from x^i to $\bar{x}^i$ given by

$$\bar{x}^i = \bar{x}^i(x^1, x^2, \ldots, x^N).$$

Since g_{ij} is a covariant tensor of the second order, we have

$$\bar{g}_{pq} = \frac{\Delta_i x^i}{\Delta_{(p)}\bar{x}^p} \frac{\Delta_j x^j}{\Delta_{(q)}\bar{x}^q} g_{ij},$$

or

$$\bar{g}_{pq} - \frac{\Delta_i x^i}{\Delta_{(p)}\bar{x}^p} \frac{\Delta_j x^j}{\Delta_{(q)}\bar{x}^q} g_{ij} = 0,$$

and

$$\left(\bar{g}_{pq} - \frac{\Delta_i x^i}{\Delta_{(p)}\bar{x}^p} \frac{\Delta_j x^j}{\Delta_{(q)}\bar{x}^q} g_{ij} \right) \Delta_{(p)}\bar{x}^p \Delta_{(q)}\bar{x}^q = 0,$$

or

$$\overline{g}_{pq}\Delta_{(p)}\overline{x}^p\Delta_{(q)}\overline{x}^q = \frac{\Delta_i x^i}{\Delta_{(p)}\overline{x}^p}\frac{\Delta_j x^j}{\Delta_{(q)}\overline{x}^q}g_{ij}\Delta_{(p)}\overline{x}^p\Delta_{(q)}\overline{x}^q,$$

or

$$\overline{g}_{pq}\Delta_{(p)}\overline{x}^p\Delta_{(q)}\overline{x}^q = \frac{\Delta_i x^i}{\Delta_{(p)}\overline{x}^p}\Delta_{(p)}\overline{x}^p\frac{\Delta_j x^j}{\Delta_{(q)}\overline{x}^q}g_{ij}\Delta_{(q)}\overline{x}^q$$

$$= g_{ij}\Delta_i x^i\Delta_j x^j.$$

This completes the proof. $\qquad\square$

Example 3.1. We will show that the volume element ΔV is an invariant, where

$$V = \int_R \sqrt{g}\,\Delta_1 x^1\Delta_2 x^2\cdots\Delta_N x^n,$$

and R is a finite region of V_N bounded by a closed V_{N-1}.

Since g_{ij} is a covariant tensor of the second order, we have

$$\overline{g}_{ij} = \frac{\Delta_p x^p}{\Delta_{(i)}\overline{x}^i}\frac{\Delta_q x^q}{\Delta_{(j)}\overline{x}^j}g_{pq},$$

whereupon

$$|\overline{g}_{ij}| = \left|\frac{\Delta_p x^p}{\Delta_{(i)}\overline{x}^i}\right|\left|\frac{\Delta_q x^q}{\Delta_{(j)}\overline{x}^j}\right||g_{pq}|,$$

or

$$\overline{g} = J^2 g,$$

where

$$J = \left|\frac{\Delta_p x^p}{\Delta_{(i)}\overline{x}^i}\right|.$$

Hence

$$J = \sqrt{\frac{\overline{g}}{g}}.$$

Now using that

$$\Delta_1 x^1\Delta_2 x^2\cdots\Delta_N x^N = \left|\frac{\Delta_p x^p}{\Delta_{(i)}\overline{x}^i}\right|\Delta_{(1)}\overline{x}^1\Delta_{(2)}\overline{x}^2\cdots\Delta_{(N)}\overline{x}^N,$$

we find

$$\frac{\Delta_1 x^1 \Delta_2 x^2 \cdots \Delta_N x^N}{\Delta_{(1)} \bar{x}^1 \Delta_{(2)} \bar{x}^2 \cdots \Delta_{(N)} \bar{x}^N} = \left| \frac{\Delta_p x^p}{\Delta_{(i)} \bar{x}^i} \right|$$
$$= J$$
$$= \sqrt{\frac{\bar{g}}{g}},$$

or

$$\sqrt{g}\,\Delta_1 x^1 \Delta_2 x^2 \cdots \Delta_N x^N = \sqrt{\bar{g}}\,\Delta_{(1)} \bar{x}^1 \Delta_{(2)} \bar{x}^2 \cdots \Delta_{(N)} \bar{x}^N,$$

or

$$\Delta V = \Delta \bar{V}.$$

Definition 3.7. The covariant tensor ε of the second order is defined by

$$\varepsilon_{11} = 0,$$
$$\varepsilon_{12} = \sqrt{g},$$
$$\varepsilon_{21} = -\sqrt{g},$$
$$\varepsilon_{22} = 0.$$

This tensor is skew-symmetric. Applying the tensor law of transformation, we find

$$\bar{\varepsilon}_{ij} = \frac{\Delta_p x^p}{\Delta_{(i)} \bar{x}^i} \frac{\Delta_q x^q}{\Delta_{(j)} \bar{x}^j} \varepsilon_{pq},$$

or

$$\bar{\varepsilon}_{ij} = \frac{\Delta_1 x^1}{\Delta_{(i)} \bar{x}^i} \frac{\Delta_1 x^1}{\Delta_{(j)} \bar{x}^j} \varepsilon_{11} + \frac{\Delta_1 x^1}{\Delta_{(i)} \bar{x}^i} \frac{\Delta_2 x^2}{\Delta_{(j)} \bar{x}^j} \varepsilon_{12}$$

$$+ \frac{\Delta_2 x^2}{\Delta_{(i)} \bar{x}^i} \frac{\Delta_1 x^1}{\Delta_{(j)} \bar{x}^j} \varepsilon_{21} + \frac{\Delta_2 x^2}{\Delta_{(i)} \bar{x}^i} \frac{\Delta_2 x^2}{\Delta_{(j)} \bar{x}^j} \varepsilon_{22}$$

$$= \sqrt{g}\,\frac{\Delta_1 x^1}{\Delta_{(i)} \bar{x}^i} \frac{\Delta_2 x^2}{\Delta_{(j)} \bar{x}^j} - \sqrt{g}\,\frac{\Delta_2 x^2}{\Delta_{(i)} \bar{x}^i} \frac{\Delta_1 x^1}{\Delta_{(j)} \bar{x}^j}$$

$$= \sqrt{g} \det \begin{pmatrix} \dfrac{\Delta_1 x^1}{\Delta_{(i)} \bar{x}^i} & \dfrac{\Delta_1 x^1}{\Delta_{(j)} \bar{x}^j} \\ \dfrac{\Delta_2 x^2}{\Delta_{(i)} \bar{x}^i} & \dfrac{\Delta_2 x^2}{\Delta_{(j)} \bar{x}^j} \end{pmatrix}$$

$$= \sqrt{g}\,\frac{\Delta(x^1, x^2)}{\Delta(\bar{x}^i, \bar{x}^j)}.$$

Therefore

$$\bar{\varepsilon}_{11} = 0,$$

$$\bar{\varepsilon}_{12} = \sqrt{g}\frac{\Delta(x^1, x^2)}{\Delta(\bar{x}^i, \bar{x}^j)},$$

$$\bar{\varepsilon}_{21} = -\sqrt{g}\frac{\Delta(x^1, x^2)}{\Delta(\bar{x}^i, \bar{x}^j)},$$

$$\bar{\varepsilon}_{22} = 0.$$

Definition 3.8. Let g^{ij} be the components of the conjugate tensor of g_{ij}, i. e.,

$$g^{ij} = \frac{\text{cofactor of } g_{ij} \text{ in } |g|}{|g|},$$

where $|g| \neq 0$. The tensor g^{ij} is called the fundamental contravariant tensor of V_N, g_{ij} is called the first fundamental tensor, and g^{ij} is called the second fundamental tensor.

Theorem 3.3. *The second fundamental tensor g^{ij} has the following properties:*
1. $g_{ij}g^{kj} = \delta_i^k$.
2. $g^{ij}g_{ij} = N$.
3. g^{ij} *is a symmetric contravariant tensor of the second order.*

Proof. 1. Let the cofactor of g_{ij} in g be denoted by $\xi(i,j)$. By the properties of determinants we get

$$g_{ij}\xi(i,j) = g,$$

whereupon

$$g_{ij}\frac{\xi(i,j)}{g} = 1,$$

or

$$g_{ij}g^{ij} = 1.$$

Now

$$g_{ij}\xi(k,j) = 0, \quad i \neq k.$$

Hence

$$g_{ij}\frac{\xi(k,j)}{g} = 0, \quad i \neq k,$$

or

$$g_{ij}g^{kj} = 0, \quad k \neq i.$$

Applying the definition of the Kronecker delta, we arrive at

$$g_{ij}g^{kj} = \delta_i^k.$$

2. Using the preceding property, we obtain

$$\begin{aligned}
g_{ij}g^{ij} &= \delta_i^i \\
&= \delta_1^1 + \delta_2^2 + \cdots + \delta_N^N \\
&= N.
\end{aligned}$$

3. By the first property of g^{ij} we find

$$\begin{aligned}
|g_{ij}||g^{kj}| &= |\delta_i^k| \\
&= 1.
\end{aligned}$$

Thus

$$|g^{kj}| \neq 0.$$

Now define the tensor

$$r_{ij} = \frac{\text{cofactor of } g^{ij} \text{ in } |g^{ij}|}{|g^{ij}|},$$

i. e., r_{ij} is the conjugate tensor of g^{ij}. Then

$$r_{ij}g^{ik} = \delta_j^k.$$

Multiplying both sides by g_{lk}, we find

$$\begin{aligned}
r_{ij}g^{ik}g_{lk} &= \delta_j^k g_{lk} \\
&= g_{lj},
\end{aligned}$$

whereupon

$$r_{ij}\delta_l^i = g_{lj},$$

or

$$r_{ij} = g_{ij}.$$

Since r_{ij} is the conjugate tensor of g^{ij}, we conclude that g_{ij} is the conjugate tensor of g^{ij}. Therefore g^{ij} is a symmetric contravariant tensor of the second order. This completes the proof. $\qquad\square$

Example 3.2. Let E^3 be covered by orthogonal Cartesian coordinates y^i and consider the transformation

$$
\begin{aligned}
y^1 &= x^1 \sin_1(x^2, x_0^2) \cos_1(x^3, x_0^3), \\
y^2 &= x^1 \sin_1(x^2, x_0^2) \sin_1(x^3, x_0^3), \\
y^3 &= x^1 \cos_1(x^2, x_0^2),
\end{aligned}
$$

where $(x_0^1, x_0^2, x_0^3) \in \Lambda^3$.

The metric in the Euclidean space E^3, referred to Cartesian coordinates, is given by

$$
\tilde{d}s^2 = \left(\varDelta_{(1)} y^1\right)^2 + \left(\varDelta_{(2)} y^2\right)^2 + \left(\varDelta_{(3)} y^3\right)^2.
$$

We have

$$
\begin{aligned}
g_{ii} &= 1, \\
g_{ij} &= 0, \quad i, j \in \{1, 2, 3\}, \ i \neq j.
\end{aligned}
$$

We will find $\overline{g}_{ij}$. First, we will compute the partial derivatives

$$
\frac{\varDelta_{(j)} y^j}{\varDelta_l x^l}, \quad j, l \in \{1, 2, 3\}.
$$

We have

$$
\frac{\varDelta_{(1)} y^1}{\varDelta_1 x^1}(x^1, x^2, x^3) = \sin_1(x^2, x_0^2) \cos_1(x^3, x_0^3),
$$

$$
\frac{\varDelta_{(1)} y^1}{\varDelta_2 x^2}(x^1, x^2, x^3) = x^1 \cos_1(x^2, x_0^2) \cos_1(x^3, x_0^3),
$$

$$
\frac{\varDelta_{(1)} y^1}{\varDelta_3 x^3}(x^1, x^2, x^3) = -x^1 \sin_1(x^2, x_0^2) \sin_1(x^3, x_0^3),
$$

$$
\frac{\varDelta_{(2)} y^2}{\varDelta_1 x_1}(x^1, x^2, x^3) = \sin_1(x^2, x_0^2) \sin_1(x^3, x_0^3),
$$

$$
\frac{\varDelta_{(2)} y^2}{\varDelta_2 x^2}(x^1, x^2, x^3) = x^1 \cos_1(x^2, x_0^2) \sin_1(x^3, x_0^3),
$$

$$
\frac{\varDelta_{(2)} y^2}{\varDelta_3 x^3}(x^1, x^2, x^3) = x^1 \sin_1(x^2, x_0^2) \cos_1(x^3, x_0^3),
$$

$$
\frac{\varDelta_{(3)} y^3}{\varDelta_1 x^1}(x^1, x^2, x^3) = \cos_1(x^2, x_0^2),
$$

$$\frac{\Delta_{(3)}y^3}{\Delta_2 x^2}(x^1, x^2, x^3) = -x^1 \sin_1(x^2, x_0^2),$$

$$\frac{\Delta_{(3)}y^3}{\Delta_3 x^3}(x^1, x^2, x^3) = 0.$$

Then

$$\overline{g}_{11} = \left(\frac{\Delta_{(1)}y^1}{\Delta_1 x^1}(x^1, x^2, x^3)\right)^2 + \left(\frac{\Delta_{(2)}y^2}{\Delta_1 x^1}(x^1, x^2, x^3)\right)^2 + \left(\frac{\Delta_{(3)}y^3}{\Delta_1 x^1}(x^1, x^2, x^3)\right)^2$$

$$= (\sin_1(x^2, x_0^2)\cos_1(x^3, x_0^3))^2 + (\sin_1(x^2, x_0^2)\sin_1(x^3, x_0^3))^2 + (\cos_1(x^2, x_0^2))^2$$

$$= (\sin_1(x^2, x_0^2))^2(\cos_1(x^3, x_0^3))^2 + (\sin_1(x^2, x_0^2))^2(\sin_1(x^3, x_0^3))^2 + (\cos_1(x^2, x_0^2))^2$$

$$= (\sin_1(x^2, x_0^2))^2((\cos_1(x^3, x_0^3))^2 + (\sin_1(x^3, x_0^3))^2) + (\cos_1(x^2, x_0^2))^2$$

$$= (\sin_1(x^2, x_0^2))^2 e_{\mu_3}(x^3, x_0^3) + (\cos_1(x^2, x_0^2)),$$

$$\overline{g}_{22} = \left(\frac{\partial y^1}{\Delta_2 x^2}(x^1, x^2, x^3)\right)^2 + \left(\frac{\partial y^2}{\Delta_2 x^2}(x^1, x^2, x^3)\right)^2 + \left(\frac{\partial y^3}{\Delta_2 x^2}(x^1, x^2, x^3)\right)^2$$

$$= (x^1 \cos_1(x^2, x_0^2)\cos_1(x^3, x_0^3))^2$$
$$\quad + (x^1 \cos_1(x^2, x_0^2)\sin_1(x^3, x_0^3))^2 + (\cos_1(x^2, x_0^2))^2$$

$$= (x^1)^2(\cos_1(x^2, x_0^2))^2(\cos_1(x^3, x_0^3))^2$$
$$\quad + (x^1)^2(\cos_1(x^2, x_0^2))^2(\sin_1(x^3, x_0^3))^2 + (\cos_1(x^2, x_0^2))^2$$

$$= (x^1)^2(\cos_1(x^2, x_0^2))^2((\cos_1(x^3, x_0^3))^2 + (\sin_1(x^3, x_0^3))^2)$$
$$\quad + (\cos_1(x^2, x_0^2))^2$$

$$= (x^1)^2(\cos_1(x^2, x_0^2))^2 e_{\mu_3}(x^3, x_0^3) + (\cos_1(x^2, x_0^2))^2$$

$$= ((x^1)^2 e_{\mu_3}(x^3, x_0^3) + 1)(\cos_1(x^2, x_0^2))^2,$$

$$\overline{g}_{33} = \left(\frac{\partial y^1}{\Delta_3 x^3}(x^1, x^2, x^3)\right)^2 + \left(\frac{\partial y^2}{\Delta_3 x^3}(x^1, x^2, x^3)\right)^2 + \left(\frac{\partial y^3}{\Delta_3 x^3}(x^1, x^2, x^3)\right)^2$$

$$= (-x^1 \sin_1(x^2, x_0^2)\sin_1(x^3, x_0^3))^2$$
$$\quad + (x^1 \sin_1(x^2, x_0^2)\cos_1(x^3, x_0^3))^2$$

$$= (x^1)^2(\sin_1(x^2, x_0^2))^2(\sin_1(x^3, x_0^3))^2$$
$$\quad + (x^1)^2(\sin_1(x^2, x_0^2))^2(\cos_1(x^3, x_0^3))^2$$

$$= (x^1)^2(\sin_1(x^2, x_0^2))^2((\sin_1(x^3, x_0^3))^2 + (\cos_1(x^3, x_0^3))^2)$$

$$= (x^1)^2(\sin_1(x^2, x_0^2))^2 e_{\mu_3}(x^3, x_0^3),$$

and

$$\overline{g}_{jl} = 0, \quad j, l \in \{1, 2, 3\}, \ j \neq l.$$

Therefore

$$\Delta s^2 = ((\sin_1(x^2, x_0^2))^2 e_{\mu_3}(x^3, x_0^3) + (\cos_1(x^2, x_0^2))^2)^2 (\Delta_1 x^1)^2$$
$$+ ((x^1)^2 e_{\mu_3}(x^3, x_0^3) + 1)(\cos_1(x^2, x_0^2))^2 (\Delta_2 x^2)^2$$
$$+ (x^1)^2 (\sin_1(x^2, x_0^2))^2 e_{\mu_3}(x^3, x_0^3)(\Delta_3 x^3)^2.$$

Example 3.3. Let y^i be orthogonal Cartesian coordinates. Consider the transformation

$$y^1 = \sin_1(x^1, x_0^1) \cos_1(x^2, x_0^2),$$
$$y^2 = \sin_1(x^1, x_0^1) \sin_1(x^2, x_0^2),$$
$$y^3 = \cos_1(x^1, x_0^1),$$

where $(x_0^1, x_0^2) \in \Lambda^2$. We will find $\overline{g}_{ij}$. We have

$$\frac{\Delta_{(1)} y^1}{\Delta_1 x^1} = \cos_1(x^1, x_0^1) \cos_1(x^2, x_0^2),$$

$$\frac{\Delta_{(1)} y^1}{\Delta_2 x^2} = -\sin_1(x^1, x_0^1) \sin_1(x^2, x_0^2),$$

$$\frac{\Delta_{(2)} y^2}{\Delta_1 x^1} = \cos_1(x^1, x_0^1) \sin_1(x^2, x_0^2),$$

$$\frac{\Delta_{(2)} y^2}{\Delta_2 x^2} = \sin_1(x^1, x_0^1) \cos_1(x^2, x_0^2),$$

$$\frac{\Delta_{(3)} y^3}{\Delta_1 x^1} = -\sin_1(x^1, x_0^1),$$

$$\frac{\Delta_{(3)} y^3}{\Delta_2 x^2} = 0.$$

The components of the system $\overline{g}_{mn}$ are obtained as follows:

$$\overline{g}_{11} = \left(\frac{\Delta_{(1)} y^1}{\Delta_1 x^1}\right)^2 + \left(\frac{\Delta_{(2)} y^2}{\Delta_2 x^2}\right)^2 + \left(\frac{\Delta_{(3)} y^3}{\Delta_1 x^1}\right)^2$$
$$= (\cos_1(x^1, x_0^1))^2 (\cos_1(x^2, x_0^2))^2 + (\cos_1(x^1, x_0^1))^2 (\sin_1(x^2, x_0^2))^2$$
$$+ (\sin_1(x^1, x_0^1))^2$$
$$= (\cos_1(x^1, x_0^1))^2 ((\sin_1(x^2, x_0^2))^2 + (\cos_1(x^2, x_0^2))^2)$$
$$+ (\sin_1(x^1, x_0^1))^2$$
$$= e_{\mu_2}(x^2, x_0^2)(\cos_1(x^1, x_0^1))^2 + (\sin_1(x^1, x_0^1))^2,$$

$$\overline{g}_{12} = \overline{g}_{21}$$

$$= \frac{\Delta_{(1)}y^1}{\Delta_1 x^1}\frac{\Delta_{(1)}y^1}{\Delta_2 x^2} + \frac{\Delta_{(2)}y^2}{\Delta_1 x^1}\frac{\Delta_{(2)}y^2}{\Delta_2 x^2} + \frac{\Delta_{(3)}y^3}{\Delta_1 x^1}\frac{\Delta_{(3)}y^3}{\Delta_2 x^2}$$

$$= -\cos_1(x^1, x_0^1)\cos_1(x^2, x_0^2)\sin_1(x^2, x_0^2)\sin_1(x^1, x_0^1)$$
$$\quad + \cos_1(x^1, x_0^1)\cos_1(x^2, x_0^2)\sin_1(x^2, x_0^2)\sin_1(x^1, x_0^1)$$

$$= 0,$$

$$\overline{g}_{22} = \left(\frac{\Delta_{(1)}y^1}{\Delta_2 x^2}\right)^2 + \left(\frac{\Delta_{(2)}y^2}{\Delta_2 x^2}\right)^2 + \left(\frac{\Delta_{(3)}y^3}{\Delta_2 x^2}\right)^2$$

$$= (\sin_1(x^1, x_0^1))^2(\sin_1(x^2, x_0^2))^2 + (\sin_1(x^1, x_0^1))^2(\cos_1(x^2, x_0^2))^2$$
$$= (\sin_1(x^1, x_0^1))^2((\sin_1(x^2, x_0^2))^2 + (\cos_1(x^2, x_0^2))^2)$$
$$= e_{\mu_2}(x^2, x_0^2)(\sin_1(x^1, x_0^1))^2.$$

Thus we obtain the metric of the subspace R_2 in the form

$$\Delta s^2 = e_{\mu_2}(x^2, x_0^2)(\cos_1(x^1, x_0^1))^2 + (\sin_1(x^1, x_0^1))^2(\Delta_1 x^1)^2$$
$$\quad + e_{\mu_2}(x^2, x_0^2)(\sin_1(x^1, x_0^1))^2(\Delta_2 x^2)^2.$$

Example 3.4. Since

$$g_{mn}g^{mn} = N,$$

differentiating with respect to x^s, we obtain

$$0 = \frac{\partial}{\Delta_s x^s}(g_{mn}g^{mn})$$

$$= g_{mn}^{\sigma_s}\frac{\partial g^{mn}}{\Delta_s x^s} + g^{mn}\frac{\partial g_{mn}}{\Delta_s x^s}$$

$$= g_{mn}\frac{\partial g^{mn}}{\Delta_s x^s} + g^{mn\sigma_s}\frac{\partial g_{mn}}{\Delta_s x^s}.$$

Example 3.5. We will prove that

$$(g_{hj}g_{ik} - g_{hk}g_{ij})g^{hj} = (N-1)g_{ik},$$

where g_{ij} and g^{ij} have their usual meaning. We have

$$(g_{hj}g_{ik} - g_{hk}g_{ij})g^{hj} = g_{hj}g_{ik}g^{hj} - g_{hk}g_{ij}g^{hj}$$
$$= g_{hj}g^{hj}g_{ik} - g_{hk}g^{hj}g_{ij}$$
$$= Ng_{ik} - \delta_k^j g_{ij}$$
$$= Ng_{ik} - g_{ik}$$
$$= (N-1)g_{ik}.$$

Example 3.6. For V_2 in which

$$g_{11} = E, \quad g_{12} = F, \quad g_{22} = G,$$

we will find g and g^{ij}, $i, j \in \{1, 2\}$. Since g_{ij} is symmetric, we have

$$g_{21} = F$$

and then

$$g = \det \begin{pmatrix} g_{11} & g_{12} \\ g_{21} & g_{22} \end{pmatrix}$$

$$= \det \begin{pmatrix} E & F \\ F & G \end{pmatrix}$$

$$= EG - F^2.$$

Hence

$$g^{11} = \frac{\text{cofactor of } g_{11} \text{ in } g}{g}$$

$$= \frac{g_{22}}{g}$$

$$= \frac{G}{EG - F^2},$$

$$g^{12} = g^{21}$$

$$= \frac{\text{cofactor of } g_{12} \text{ in } g}{g}$$

$$= \frac{g_{21}}{g}$$

$$= -\frac{F}{EG - F^2},$$

and

$$g^{22} = \frac{\text{cofactor of } g_{22} \text{ in } g}{g}$$

$$= \frac{g_{11}}{g}$$

$$= \frac{G}{EG - F^2}.$$

Consider a continuous curve in a Riemannian space V_N. Then the coordinates of any current point on it are expressed as functions of $t \in \mathbb{T}$. Let s denote the arc length of the curve measured from a fixed point P_0 on the curve. Then the length Δs of the arc between the points whose coordinates are x^i and $x^i + \Delta_i x^i$ is given by the quadratic formula

$$\Delta s^2 = g_{ij}\Delta_i x^i \Delta_j x^j, \quad i,j \in \{1, 2, \ldots, N\}.$$

Let L denote the arc length of the curve between the points P_1 and P_2 on the curve that corresponds to two values $a \in \mathbb{T}$ and $b \in \mathbb{T}$ of the parameter t. Then

$$L = \int_{P_1}^{P_2} \Delta s$$

$$= \int_a^b \sqrt{g_{ij}\frac{\Delta_i x^i}{\Delta t}\frac{\Delta_j x^j}{\Delta t}}\,\Delta t.$$

If

$$g_{ij}\frac{\Delta_i x^i}{\Delta t}\frac{\Delta_j x^j}{\Delta t} = 0 \tag{3.5}$$

along the curve, then the points P_1 and P_2 are of zero distance despite the fact that they do not coincide.

Definition 3.9. Such a curve is called minimal curve or null curve.

If Δs^2 is positive definite, then null curves do not exist. A curve is null if it or its subarcs have zero length. Here a subarc is understood to be nontrivial, i. e., it consists of more than one point and corresponds to an interval $[c, d]$, $c, d \in \mathbb{T}$. A curve is null at a point if for some value of the parameter t, the tangent vector is a null vector, i. e., (3.5) holds.

Definition 3.10. The set of t values at which the curve is null is said to be the null set of the curve.

Let now

$$x^i = x^i(\tilde{t}), \quad \tilde{a} \le \tilde{t} \le \tilde{b}, \ \tilde{a}, \tilde{t}, \tilde{b} \in \tilde{T},$$

be a different parameterization, where

$$\tilde{t} = \phi(t),$$
$$\phi^{\Delta}(t) > 0,$$
$$\tilde{\sigma}(\tilde{t}) = \phi(\sigma(t)), \quad t \in \mathbb{T}, \ \tilde{t} \in \tilde{T},$$

and

$$\tilde{a} = \phi(a),$$
$$\tilde{b} = \phi(b).$$

Then

$$
\begin{aligned}
L &= \int_a^b \sqrt{g_{ij}\frac{\varDelta_i x^i}{\varDelta t}\frac{\varDelta_j x^j}{\varDelta t}}\,\varDelta t \\[2mm]
&= \int_a^b \sqrt{g_{ij}\frac{\varDelta_i x^i}{\widetilde{\varDelta t}}\frac{\widetilde{\varDelta t}}{\varDelta t}\frac{\varDelta_j x^j}{\widetilde{\varDelta t}}\frac{\widetilde{\varDelta t}}{\varDelta t}}\,\varDelta t \\[2mm]
&= \int_a^b \sqrt{g_{ij}\frac{\varDelta_i x^i}{\varDelta t}\frac{\varDelta_j x^j}{\varDelta t}}\,\phi^{\varDelta}(t)\varDelta t \\[2mm]
&= \int_{\tilde{a}}^{\tilde{b}} \sqrt{g_{ij}\frac{\varDelta_i x^i}{\widetilde{\varDelta t}}\frac{\varDelta_j x^j}{\widetilde{\varDelta t}}}\,\widetilde{\varDelta t}.
\end{aligned}
$$

Example 3.7. We will find the length of arc for $\mathbb{T} = \mathbb{Z}, 1 \le ty \le 3$, for the curve

$$x^1 = 1,$$
$$x^2 = t, \quad t \in \mathbb{T},$$

if the metric is

$$\varDelta s^2 = \frac{1}{(x^2)^2}\left((\varDelta_1 x^1)^2 + (\varDelta_2 x^2)^2\right).$$

Here

$$\sigma(t) = t + 1,$$
$$\mu(t) = 1, \quad t \in \mathbb{T},$$

and

$$
\begin{aligned}
g_{11} &= \frac{1}{(x^2)^2}, \\[1mm]
g_{12} &= 0, \\[1mm]
g_{21} &= 0, \\[1mm]
g_{22} &= \frac{1}{(x^2)^2}.
\end{aligned}
$$

Then

$$\Delta s^2 = g_{11}(\Delta_1 x^1)^2 + g_{22}(\Delta_2 x^2)^2$$
$$= \frac{1}{t^2}(\Delta t)^2.$$

Hence

$$L = \int\limits_1^3 \frac{1}{t}\Delta t$$
$$= \mu(t)\frac{1}{t}\Big|_{t=1} + \mu(t)\frac{1}{t}\Big|_{t=2}$$
$$= \mu(1) + \mu(2)\frac{1}{2}$$
$$= 1 + \frac{1}{2}$$
$$= \frac{3}{2}.$$

Example 3.8. Under the metric

$$\Delta s^2 = (\Delta_1 x^1)^2 + (\Delta_2 x^2)^2 + (\Delta_3 x^3)^2 - (\Delta_4 x^4)^2,$$

consider the curve given by

$$x^1 = 3\cos_1(t, 1),$$
$$x^2 = 3\sin_1(t, 1),$$
$$x^3 = 4t,$$
$$x^4 = 5t, \quad t \in \mathbb{T},$$

where $\mathbb{T} = 2^{\mathbb{N}_0}$. We will find the length of arc for $t \in [1, 4]$. We have $N = 4$ and

$$g_{11} = 1,$$
$$g_{12} = 0,$$
$$g_{13} = 0,$$
$$g_{14} = 0,$$
$$g_{21} = 0,$$
$$g_{22} = 1,$$
$$g_{23} = 0,$$
$$g_{24} = 0,$$
$$g_{31} = 0,$$
$$g_{32} = 0,$$
$$g_{33} - 1,$$

$$g_{34} = 0,$$
$$g_{41} = 0,$$
$$g_{42} = 0,$$
$$g_{43} = 0,$$
$$g_{44} = -1$$

and

$$\sigma(t) = 2t,$$
$$\mu(t) = t, \quad t \in \mathbb{T}.$$

Then

$$\frac{\Delta_1 x^1}{\Delta t} = -3 \sin_1(t, 1),$$

$$\frac{\Delta_2 x^2}{\Delta t} = 3 \cos_1(t, 1),$$

$$\frac{\Delta_3 x^3}{\Delta t} = 4,$$

$$\frac{\Delta_4 x^4}{\Delta t} = 5, \quad t \in \mathbb{T},$$

and

$$g_{11}\left(\frac{\Delta_1 x^1}{\Delta t}\right)^2 + g_{22}\left(\frac{\Delta_2 x^2}{\Delta t}\right)^2 + g_{33}\left(\frac{\Delta_3 x^3}{\Delta t}\right)^2 + g_{44}\left(\frac{\Delta_4 x^4}{\Delta t}\right)^2$$
$$= \left(-3 \sin_1(t, 1)\right)^2 + \left(3 \cos_1(t, 1)\right)^2 + 4^2 - 5^2$$
$$= 9\left(\sin_1(t, 1)\right)^2 + 9\left(\cos_1(t, 1)\right)^2 + 16 - 25$$
$$= 9\left(\left(\sin_1(t, 1)\right)^2 + \left(\cos_1(t, 1)\right)^2\right) - 9$$
$$= 9 e_\mu(t, 1) - 9, \quad t \in \mathbb{T}.$$

Therefore

$$L = \int_1^4 \sqrt{9 e_\mu(t, 1) - 9} \, \Delta t$$
$$= 3\mu(t) \sqrt{e_\mu(t, 1) - 1}\Big|_{t=1} + 3\mu(t) \sqrt{e_\mu(t, 1) - 1}\Big|_{t=2}$$
$$= 6\sqrt{1 - 1} + 12\sqrt{e_\mu(4, 1) - 1}$$
$$= 12\sqrt{e_\mu(4, 1) - 1}.$$

Exercise 3.1. Under the metric

$$\Delta s^2 = 2(\Delta_1 x^1)^2 + x^2(\Delta_2 x^2)^2 - (\Delta_3 x^3)^2,$$

consider the curve

$$x^1 = t,$$
$$x^2 = t^2,$$
$$x^3 = t, \quad t \in \mathbb{T},$$

where $\mathbb{T} = 2^{\mathbb{N}_0}$. Find the length of arc for $t \in [1, 8]$.

Answer 3.4.

$$30\sqrt{3} + 8\sqrt{2307}.$$

3.2 Associated tensors

One of the fundamental concepts of tensor calculus resides in the "raising" and "lowering" the indices of a tensor.

In E^N the covariance and contravariance of a tensor was a quality that was impossible to change. With the introduction of a metric in E^N, this barrier falls in V_N.

Let A^i and B_i be a contravariant and a covariant vector, respectively, in the x^i coordinate system. Define two vectors

$$A_i = g_{ij} A^j \quad \text{and} \quad B^i = g^{ij} B_j.$$

Then the associate to a contravariant vector A^j is formed by lowering its index by the fundamental metric tensor g_{ij}, and the associate to a covariant vector B_j is formed by raising its index by the conjugate metric tensor. Now we see that

$$\begin{aligned}
g^{ij} A_i &= g^{ij} g_{ik} A^k \\
&= \delta^j_k A^k \\
&= A^j.
\end{aligned} \tag{3.6}$$

Thus the procedure of raising and lowering indices is clearly reversible. By equations (3.6) it follows that the associate to A_i is A^i. Therefore if A^i is the associate to A_i, then A_i is the associate to A^i. In other words, A_i and A^i are mutually associated, and so they are associate vectors.

Next, consider tensors of order greater than one. Any index of such a tensor can be lowered or raised by the fundamental tensors as in the case of vectors. Consider a tensor A_{ij} and form the following inner products:

$$A^i_{\bullet j} = g^{ik} A_{kj},$$
$$A^j_{i\bullet} = g^{jk} A_{ik},$$

and

$$A^{ij} = g^{ik} g^{jl} A_{kl}.$$

Definition 3.11. The tensors

$$A^i_{\bullet j}, \quad A^j_{i\bullet}, \quad \text{and} \quad A^{ij} \tag{3.7}$$

are called associates to the tensor A_{ij}.

Note that tensors (3.7) may be formed from each other by lowering and raising indices. For instance,

$$A^i_{\bullet j} = g^{ik} A_{kj}$$
$$= g^{ik} g_{mj} A^m_{k\bullet}.$$

Similarly, consider a tensor A_{ijk} and form the following inner products:

$$g^{mi} A_{ijk} = A^m_{\bullet jk},$$
$$g^{mj} A_{ijk} = A^m_{i\bullet k},$$

and

$$g^{mk} A_{ijk} = A^m_{ij\bullet}.$$

All these tensors are associated with the tensor A_{ijk}. Operating on these tensors with g^{ij}, we can form other associated tensors.

Definition 3.12. Let $A^{i_1 i_2 \cdots i_p}_{j_1 j_2 \cdots j_q}$ be a given tensor of order $p + q$. A tensor obtained by the process of inner multiplication of $A^{i_1 i_2 \cdots i_p}_{j_1 j_2 \cdots j_q}$ with either of the fundamental metric tensor g_{ij} or its conjugate g^{ij} is called an associated tensor of the given tensor.

For a particular case, from the fundamental tensor g_{ij} we get

$$g^{ik} g_{kj} = g^i_{\bullet j}$$
$$= \delta^i_j$$

and

$$g^{ik} g_{jk} = g^i_{j\bullet}$$
$$= \delta^i_j.$$

Thus

$$g^i_{\cdot j} = g^i_{j\cdot},$$

and it is not necessary to distinguish them, so we may write δ^i_j.

Example 3.9. Let A_i and B_j be two covariant vectors. We will prove that

$$g^{ij}(A_iB_j - A_jB_i) = 0.$$

Indeed, we have

$$\begin{aligned}
g^{ij}(A_iB_j - A_jB_i) &= g^{ij}A_iB_j - g^{ij}A_jB_i \\
&= A^jB_j - A^iB_i \\
&= A^iB_i - A^iB_i \\
&= 0.
\end{aligned}$$

Example 3.10. Let

$$g_{pq}A^q = B_p.$$

Then using the definition of an associated vector, we find

$$A_p = g_{pq}A^q.$$

Hence

$$A_p = B_p,$$

and then

$$\begin{aligned}
g^{pq}B_q &= g^{pq}A_q \\
&= A^p.
\end{aligned}$$

Example 3.11. We will prove that

$$A_{ij}B^{ij} = A^{ij}B_{ij}.$$

Indeed, by the definition for an associated tensor we have

$$A_{ij} = g_{ik}g_{jl}A^{kl}$$

and

$$B^{ij} = g^{im} g^{jn} B_{mn}.$$

Then

$$
\begin{aligned}
A_{ij} B^{ij} &= g_{ik} g_{jl} A^{kl} g^{im} g^{jn} B_{mn} \\
&= g_{ik} g_{jl} g^{im} g^{jn} A^{kl} B_{mn} \\
&= g_{ik} g^{im} g_{jl} g^{jn} A^{kl} B_{mn} \\
&= \delta_k^m \delta_l^n A^{kl} B_{mn} \\
&= A^{kl} B_{kl} \\
&= A^{ij} B_{ij}.
\end{aligned}
$$

Example 3.12. We have

$$B^{jkl} = g^{jm} g^{kn} g^{lp} B_{mnp}$$

and

$$B_{mnp} = g_{jm} g_{kn} g_{lp} B^{jkl}.$$

Example 3.13. We have

$$B^{\bullet k \bullet}_{j \bullet l} = g_{jq} g_{lr} B^{qkr}$$

and

$$B^{qkr} = g^{jq} g^{lr} B^{\bullet k \bullet}_{j \bullet l}.$$

Example 3.14. We have

$$B^{p \bullet rs \bullet}_{\bullet q \bullet\bullet l} = g^{pj} g^{rk} g_{lt} B^{\bullet\bullet\bullet st}_{jqk \bullet\bullet}$$

and

$$B^{\bullet\bullet\bullet st}_{jqk \bullet\bullet} = g_{pj} g_{rk} g^{lt} B^{p \bullet rs \bullet}_{\bullet q \bullet\bullet l}.$$

Example 3.15. Let

$$u^i = g^{ij} u_j$$

and

$$v^i = g^{ij} v_j.$$

Then

$$g_{ik}u^i = g_{ik}g^{ij}u_j$$
$$= \delta_k^j u_j$$
$$= u_k,$$
$$u^i v_i = (g^{ij}u_j)v_i$$
$$= (g^{ij}v_i)u_j$$
$$= v^j u_j,$$

and

$$u^i g_{ij}u^j = (g^{ik}u_k)g_{ij}u^j$$
$$= u_k g^{ik}g_{ij}u^j$$
$$= u_k \delta_j^k u^j$$
$$= u_k u^k$$
$$= u_k g^{kj}u_j$$
$$= u_i g^{ij}u_j.$$

Exercise 3.2. Prove that:

1.

$$A^{pq} = g^{pj}A_j^{\cdot q}.$$

2.

$$A_{\cdot q}^{p\cdot r} = g^{pj}g^{rl}A_{jql}.$$

Definition 3.13. The magnitude or length A of a contravariant vector A^i is defined as

$$A = \sqrt{g_{ij}A^i A^j}.$$

The magnitude or length B of a covariant vector B_i is defined as

$$B = \sqrt{g^{ij}B_i B_j}.$$

We have

$$A^2 = g_{ij}A^i A^j$$
$$= A_j A^j$$
$$= A_j g^{ij}A_i$$
$$= g^{ij}A_i A_j$$

and

$$B^2 = g^{ij} B_i B_j$$
$$= B^j B_j$$
$$= B^j g_{ij} B^i$$
$$= g_{ij} B^i B^j.$$

Example 3.16. Consider $\frac{\Delta_i x^i}{\Delta s}$. We have

$$\Delta s^2 = g_{ij} \Delta_i x^i \Delta_j x^j,$$

whereupon

$$1 = g_{ij} \frac{\Delta_i x^i}{\Delta s} \frac{\Delta_j x^j}{\Delta s}.$$

Thus $\frac{\Delta_i x^i}{\Delta s}$ is a unit contravariant vector.

Example 3.17. We will show that in V_4 with the line element

$$\Delta s^2 = -(\Delta_1 x^1)^2 - (\Delta_2 x^2)^2 - (\Delta_3 x^3)^2 + c^2(\Delta_4 x^4)^2,$$

the vector $(-1, 0, 0, \frac{1}{c})$ is a null vector.

Here

$$g_{11} = -1,$$
$$g_{22} = -1,$$
$$g_{33} = -1,$$
$$g_{44} = c^2,$$
$$g_{ij} = 0, \quad i, j \in \{1, \ldots, 4\},\ i \neq j,$$

and

$$A^1 = -1,$$
$$A^2 = 0,$$
$$A^3 = 0,$$
$$A^4 = \frac{1}{c}.$$

Then

$$g_{11}(A^1)^2 + g - 22(A^2)^2 + g_{33}(A^3)^2 + g_{44}(A^4)^2$$
$$= -1(-1)^2 - 1(0)^2 - 1(0)^2 + c^2\left(\frac{1}{c}\right)^2$$
$$= -1 + 1$$
$$= 0.$$

Example 3.18. We will show that in V_4 with the line element

$$\Delta s^2 = -(\Delta_1 x^1)^2 - (\Delta_2 x^2)^2 - (\Delta_3 x^3)^2 + c^2(\Delta_4 x^4)^2,$$

the vector $(1, 0, 0, \frac{\sqrt{2}}{c})$ is a null vector.

Here

$$g_{11} = -1,$$
$$g_{22} = -1,$$
$$g_{33} = -1,$$
$$g_{44} = c^2,$$
$$g_{ij} = 0, \quad i, j \in \{1, \ldots, 4\}, \ i \neq j,$$

and

$$A^1 = 1,$$
$$A^2 = 0,$$
$$A^3 = 0,$$
$$A^4 = \frac{\sqrt{2}}{c}.$$

Then

$$g_{11}(A^1)^2 + g - 22(A^2)^2 + g_{33}(A^3)^2 + g_{44}(A^4)^2$$
$$= -1(1)^2 - 1(0)^2 - 1(0)^2 + c^2\left(\frac{\sqrt{2}}{c}\right)^2$$
$$= -1 + 2$$
$$= 1.$$

Example 3.19. We will show that the length of a vector is invariant. Indeed, using the transformation formula for a contravariant vector and the transformation formula for a covariant tensor of the second order, we get

$$A^2 = g_{ij}A^i A^j$$

$$= \bar{g}_{pq}\frac{\Delta_{(p)}\bar{x}^p}{\Delta_i x^i}\frac{\Delta_{(q)}\bar{x}^q}{\Delta_j x^j}\bar{A}^s\frac{\Delta_i x^i}{\Delta_{(s)}\bar{x}^s}\bar{A}^m\frac{\Delta_j x^j}{\Delta_{(m)}\bar{x}^m}$$

$$= \bar{g}_{pq}\delta^p_s \delta^q_m \bar{A}^s \bar{A}^m$$

$$= \bar{g}_{sm}\bar{A}^s \bar{A}^m$$

$$= (\bar{A})^2.$$

Example 3.20. Let

$$A^j = \frac{1}{\sqrt{g_{pq}B^p B^q}}B^j,$$

where B^j is a contravariant vector, and g_{ij} is the fundamental tensor. Then

$$A^2 = g_{ij}A^i A^j$$

$$= g_{ij}\frac{1}{\sqrt{g_{pq}B^p B^q}}B^i \frac{1}{\sqrt{g_{pq}B^p B^q}}B^j$$

$$= g_{ij}B^i B^j \frac{1}{g_{pq}B^p B^q}$$

$$= \frac{B^2}{B^2}$$

$$= 1.$$

Exercise 3.3. Prove that in V_4 with the line element

$$\Delta s^2 = -(\Delta_1 x^1)^2 - (\Delta_2 x^2)^2 - (\Delta_3 x^3)^2 + c^2(\Delta_4 x^4)^2,$$

the vector $(1, 1, 0, \frac{\sqrt{3}}{c})$ is an unit vector.

Definition 3.14. Let A^i and B^i be two nonnull contravariant vectors. Then the angle θ between them is defined by

$$\cos\theta = \frac{g_{ij}A^i B^j}{\sqrt{g_{ij}A^i A^j}\sqrt{g_{ij}B^i B^j}}.$$

The angle θ between two nonnull covariant vectors A_i and B_i is defined by

$$\cos\theta = \frac{g^{ij}A_i B_j}{\sqrt{g^{ij}A_i A_j}\sqrt{g^{ij}B_i B_j}}.$$

If A and B are two nonnull vectors with contravariant and covariant components A^i, A_i and B^i, B_i, respectively, then the angle θ between them is defined by

$$\cos \theta = \frac{A^i B_i}{\sqrt{A^i A_i}\,\sqrt{B^i B_i}}.$$

Example 3.21. Consider a space with a line element

$$\varDelta s^2 = -(\varDelta_1 x^1)^2 - (\varDelta_2 x^2)^2 - (\varDelta_3 x^3)^2 + c^2(\varDelta_4 x^4)^2,$$

where c is a nonzero constant, and two vectors

$$(1,0,0,0) \quad \text{and} \quad \left(\sqrt{2}, 0, 0, \frac{\sqrt{3}}{c} \right).$$

Here

$$g_{11} = -1,$$
$$g_{22} = -1,$$
$$g_{33} = -1,$$
$$g_{44} = c^2,$$
$$g_{mn} = 0, \quad m, n \in \{1,\ldots,4\},\ m \neq n,$$

and

$$A^1 = 1,$$
$$A^2 = 0,$$
$$A^3 = 0,$$
$$A^4 = 0$$

and

$$B^1 = \sqrt{2},$$
$$B^2 = 0,$$
$$B^3 = 0,$$
$$B^4 = \frac{\sqrt{3}}{c}.$$

Then

$$g_{11}A^1 B^1 + g_{22}A^2 B^2 + g - 33A^3 B^3 + g_{44}A^4 B^4$$

$$= (-1) \cdot 1 \cdot \sqrt{2} + (-1) \cdot 0 \cdot 0 + (-1) \cdot 0 \cdot 0 + c^2 \cdot 0 \cdot \frac{\sqrt{3}}{c}$$

$$= -\sqrt{2},$$

$$g_{11}(A^1)^2 + g_{22}(A^2)^2 + g_{33}(A^3)^2 + g_{44}(A^4)^2$$
$$= (-1) \cdot 1^2 + (-1) \cdot 0^2 + (-1) \cdot 0^2 + c^2 \cdot 0^2$$
$$= -1,$$

and

$$g_{11}(B^1)^2 + g_{22}(B^2)^2 + g_{33}(B^3)^2 + g_{44}(B^4)^2$$
$$= (-1) \cdot (\sqrt{2})^2 + (-1) \cdot 0^2 + (-1) \cdot 0^2 + c^2 \cdot \left(\frac{\sqrt{3}}{c}\right)^2$$
$$= -2 + 3$$
$$= 1.$$

Thus

$$\cos \theta = \frac{g_{ij}A^i B^j}{\sqrt{g_{ij}A^i A^j}\sqrt{g_{ij}B^i B^j}}$$
$$= \frac{-\sqrt{2}}{\sqrt{-1}\sqrt{1}}$$
$$= \frac{-\sqrt{2}}{i}$$
$$= i\sqrt{2}$$

and the angle between the two vectors is not real.

Definition 3.15. If $\theta = \frac{\pi}{2}$, then the corresponding vectors are said to be orthogonal.

Therefore two nonnull contravariant vectors A^i and B^i are orthogonal if

$$g_{ij}A^i B^j = 0.$$

Similarly, two nonnull covariant vectors A_i and B_i are orthogonal if

$$g^{ij}A_i B_j = 0,$$

and two nonnull vectors A and B with contravariant and covariant components A^i, A_i and B^i, B_i, respectively, are orthogonal if

$$A^i B_i = 0.$$

Example 3.22. Let u^i and v^i be orthogonal unit vectors. We will show that

$$(g_{hj}g_{ki} - g_{hk}g_{ji})u^h v^i u^j v^k = 1.$$

Since u^i and v^i are orthogonal unit vectors, we have

$$u^i u_i = 1,$$
$$v^i v_i = 1,$$

and

$$u^i v_i = 0.$$

Then

$$
\begin{aligned}
(g_{hj}g_{ki} - g_{hk}g_{ji})u^h v^i u^j v^k &= g_{hj}g_{ki}u^h v^i u^j v^k - g_{hk}g_{ji}u^h v^i u^j v^k \\
&= (g_{hj}u^h)u^j(g_{ki}v^i)v^k - (g_{hk}u^h)u^j(g_{ji}v^i)v^k \\
&= (u_j u^j)(v_k v^k) - (u_k v^k)(u^j v_j) \\
&= 1 \cdot 1 - 0 \cdot 0 \\
&= 1.
\end{aligned}
$$

Example 3.23. Let A^i and B^i be two nonnull vectors, and let

$$U^i = A^i + B^i,$$
$$V^i = A^i - B^i.$$

Let also,

$$g_{ij}U^i U^j = g_{ij}V^i V^j. \tag{3.8}$$

We will prove that A^i and B^i are orthogonal. Indeed, we have

$$
\begin{aligned}
g_{ij}U^i U^j &= g_{ij}(A^i + B^i)(A^j + B^j) \\
&= g_{ij}(A^i A^j + A^i B^j + B^i A^j + B^i B^j) \\
&= g_{ij}A^i A^j + g_{ij}A^i B^j + g_{ij}A^j B^i + g_{ij}B^i B^j \\
&= g_{ij}A^i A^j + 2g_{ij}A^i B^j + g_{ij}B^i B^j
\end{aligned}
$$

and

$$
\begin{aligned}
g_{ij}V^i V^j &= g_{ij}(A^i - B^i)(A^j - B^j) \\
&= g_{ij}(A^i A^j - B^i A^j - A^i B^j + B^i B^j) \\
&= g_{ij}A^i A^j - g_{ij}A^j B^i - g_{ij}A^i B^j + g_{ij}B^i B^j \\
&= g_{ij}A^i A^j - 2g_{ij}A^i B^j + g_{ij}B^i B^j.
\end{aligned}
$$

Now applying (3.8), we find

$$g_{ij}A^iA^j - 2g_{ij}A^iB^j + g_{ij}B^iB^j = g_{ij}A^iA^j - 2g_{ij}A^iB^j + g_{ij}B^iB^j,$$

whereupon

$$4g_{ij}A^iB^j = 0,$$

or

$$g_{ij}A^iB^j = 0.$$

Thus A^i and B^i are orthogonal.

Example 3.24. Let a_{ij} be a symmetric tensor, and let A^i and B^i be unit vectors that are orthogonal to C^j and satisfy the conditions

$$(a_{ij} - k_1 g_{ij})A^i + \lambda_1 g_{ij}C^i = 0 \tag{3.9}$$

and

$$(a_{ij} - k_2 g_{ij})B^i + \lambda_2 g_{ij}C^i = 0, \tag{3.10}$$

where $k_1 \neq k_2$. We will prove that A^i and B^i are orthogonal and

$$a_{ij}A^iB^j = 0. \tag{3.11}$$

Since A^i and B^i are unit vectors and they are orthogonal to C^i, we have

$$g_{ij}A^iA^j = 1,$$
$$g_{ij}B^iB^j = 1$$

and

$$g_{ij}A^iC^j = 0,$$
$$g_{ij}B^iC^j = 0.$$

Now multiplying (3.9) by B^i, we find

$$0 = (a_{ij} - k_1 g_{ij})A^iB^j + \lambda_1 g_{ij}C^iB^j$$
$$= a_{ij}A^iB^j - k_1 g_{ij}A^iB^j,$$

whereupon

$$a_{ij}A^iB^j = k_1 g_{ij}A^iB^j. \tag{3.12}$$

Multiplying (3.10) by A^j, we arrive at

$$\begin{aligned}
0 &= (a_{ij} - k_2 g_{ij})B^iA^j + \lambda_2 g_{ij}C^iA^j \\
&= a_{ij}B^iA^j - k_2 g_{ij}B^iA^j \\
&= a_{ij}A^iB^j - k_2 g_{ij}A^iB^j,
\end{aligned}$$

from which

$$a_{ij}A^iB^j = k_2 g_{ij}A^iB^j. \tag{3.13}$$

Now subtracting from equation (3.12) equation (3.13), we find

$$0 = (k_1 - k_2)g_{ij}A^iB^j,$$

or

$$g_{ij}A^iB^j = 0.$$

Therefore A^i and B^j are orthogonal. Hence, using (3.12) and (3.13), we obtain (3.11).

Exercise 3.4. Let A and B be orthogonal vectors of length l. Prove that

$$(g_{hj}g_{ik} - g_{hk}g_{ij})A^hA^jB^iB^k = -l^4.$$

Definition 3.16. A curve along which only one coordinate varies is called a coordinate curve. If only the particular coordinate x^i with particular integer i varies, then the curve is called an x^i curve.

Definition 3.17. The angle between two coordinate curves is defined as the angle between their tangents.

The x^j coordinate curve is defined as

$$x^i = c^i, \quad i \in \{1,\ldots,N\}, \ i \neq j,$$

where $c^i \in \mathbb{T}_i$. Then

$$\Delta_i x^i = 0, \quad i \neq j,$$

and

$$\Delta_j x^j \neq 0.$$

Hence

$$\Delta x^i = (0, 0, \ldots, 0, \Delta_j x^j, 0, \ldots, 0).$$

Let A^i and B^i be the tangent vectors to the x^p and x^q coordinate curves, respectively. Then

$$A^i = \Delta x^i$$
$$= (0, 0, \ldots, 0, A^p, 0, \ldots, 0),$$
$$B^i = \Delta x^i$$
$$= (0, 0, \ldots, 0, B^q, 0, \ldots, 0).$$

Let θ_{pq} be the angle between the two coordinate curves. Then

$$\cos \theta_{pq} = \frac{g_{ij} A^i B^j}{\sqrt{g_{ij} A^i A^j} \sqrt{g_{ij} B^i B^j}}$$
$$= \frac{g_{pq} A^p B^q}{\sqrt{g_{pp} (A^p)^2} \sqrt{g_{qq} (A^q)^2}}$$
$$= \frac{g_{pq} A^p B^q}{\sqrt{g_{pp}} \sqrt{g_{qq}} A^p B^q}$$
$$= \frac{g_{pq}}{\sqrt{g_{pp}} \sqrt{g_{qq}}}.$$

If the coordinate curves x^p and x^q are orthogonal, then

$$\theta_{pq} = \frac{\pi}{2}$$

and

$$\frac{g_{pq}}{\sqrt{g_{pp}} \sqrt{g_{qq}}} = 0,$$

i. e.,

$$g_{pq} = 0.$$

Example 3.25. Consider the surface

$$x^1 = a \sin_1(u^1, u^{10}) \cos_1(u^2, u^{20}),$$
$$x^2 = a \sin_1(u^1, u^{10}) \sin_1(u^2, u^{20}),$$
$$x^3 = a \cos_1(u^1, u^{10}),$$

where $u^{10} \in \mathbb{T}_{(1)}$ and $u^{20} \in \mathbb{T}_{(2)}$, x^i are orthogonal Cartesian coordinates, and a is a constant. We will prove that the coordinate curves are orthogonal. The metric in the Euclidean space E^3, referred to Cartesian coordinates, is given by

$$\Delta s^2 = \left(\Delta_1 x^1\right)^2 + \left(\Delta_2 x^2\right)^2 + \left(\Delta_3 x^3\right)^2.$$

We have

$$g_{11} = 1,$$
$$g_{22} = 1,$$
$$g_{33} = 1,$$
$$g_{ij} = 0, \quad i,j \in \{1,2,3\}, \ i \neq j,$$

and

$$\frac{\Delta_1}{\Delta_{(1)} u^1} = a \cos_1(u^1, u^{10}) \cos_1(u^2, u^{20}),$$

$$\frac{\Delta_1 x^1}{\Delta_{(2)} u^2} = -a \sin_1(u^1, u^{10}) \sin_1(u^2, u^{20}),$$

$$\frac{\Delta_1 x^1}{\Delta_{(3)} u^3} = 0,$$

$$\frac{\Delta_2 x^2}{\Delta_{(1)} u^1} = a \cos_1(u^1, u^{10}) \cos_1(u^2, u^{20}),$$

$$\frac{\Delta_2 x^2}{\Delta_{(2)} u^2} = a \sin_1(u^1, u^{10}) \cos_1(u^2, u^{20}),$$

$$\frac{\Delta_2 x^2}{\Delta_{(3)} u^3} = 0,$$

$$\frac{\Delta_3 x^3}{\Delta_{(1)} u^1} = -a \sin_1(u^1, u^{10}),$$

$$\frac{\Delta_3 x^3}{\Delta_{(2)} u^2} = 0,$$

$$\frac{\Delta_3 x^3}{\Delta_{(3)} u^3} = 0.$$

Hence

$$\overline{g}_{12} = \frac{\Delta_i x^i}{\Delta_{(1)} u^1} \frac{\Delta_j x^j}{\Delta_{(2)} u^2} g_{ij}$$

$$= \frac{\Delta_1 x^1}{\Delta_{(1)} u^1} \frac{\Delta_1 x^1}{\Delta_{(2)} u^2} g_{11} + + \frac{\Delta_2 x^2}{\Delta_{(1)} u^1} \frac{\Delta_2 x^2}{\Delta_{(2)} u^2} g_{22} + \frac{\Delta_3 x^3}{\Delta_{(1)} u^1} \frac{\Delta_3 x^3}{\Delta_{(2)} u^2} g_{33}$$

$$\begin{aligned}
&= (a\cos_1(u^1,u^{10})\cos_1(u^2,u^{20}))(-a\sin_1(u^1,u^{10})\sin_1(u^2,u^{20})) \\
&\quad + (a\cos_1(u^1,u^{10})\sin_1(u^2,u^{20}))(a\sin_1(u^1,u^{10})\cos_1(u^2,u^{20})) \\
&= 0,
\end{aligned}$$

$$\overline{g}_{13} = \frac{\Delta_i x^i}{\Delta_{(1)} u^1}\frac{\Delta_j x^j}{\Delta_{(3)} u^3}g_{ij}$$
$$= 0,$$

$$g_{23} = \frac{\Delta_i x^i}{\Delta_{(2)} u^2}\frac{\Delta_j x^j}{\Delta_{(3)} u^3}g_{ij}$$
$$= 0.$$

Let $t^1, t^2, \ldots, t^M$ be M parameters, $M < N$. The N equations

$$x^i = x^i(t^1, t^2, \ldots, t^M), \quad i \in \{1, 2, \ldots, N\},$$

define the M-dimensional subspace V_M of V_N. If we eliminate the M parameters t^1, $t^2, \ldots, t^M$ from these N equations, we obtain $N - M$ equations in x^j, which represent an M-dimensional curve in V_N. In particular, if $M = N - 1$, then we obtain only one equation in x^j, which represents an $(N - 1)$-dimensional curve in V_N.

Definition 3.18. This particular curve is called a hypersurface in V_N.

Thus a family of hypersurfaces of V_N is determined by

$$\phi(x^j) = \phi(x^1, x^2, \ldots, x^N) = \text{constant},$$

where $\phi(x^j)$ is a scalar function of coordinates x^j. Moreover, a hypersurface is obtained if x^j are functions of $N - 1$ independent variables.

Definition 3.19. A parametric hypersurface is a hypersurface on which one particular coordinate x^j, say, is a constant, while the others vary. Let us call it the x^j hypersurface with

$$x^j = c = \text{constant}.$$

Definition 3.20. The angle between two hypersurfaces is defined as the angle between their normals at their points of intersections.

Let

$$\phi(x^i) = \phi(x^1, x^2, \ldots, x^N) = \text{constant} \tag{3.14}$$

and

$$\psi(x^i) = \psi(x^1, x^2, \ldots, x^N) = \text{constant} \tag{3.15}$$

represent two families of hypersurfaces. Now differentiating (3.14) and (3.15), we obtain

$$\frac{\partial \phi}{\Delta_j x^j} \Delta_j x^j = \phi_j \Delta_j x^j$$
$$= 0$$

(3.16)

and

$$\frac{\partial \psi}{\Delta_j x^j} \Delta_j x^j = \psi_j \Delta_j x^j$$
$$= 0,$$

(3.17)

where

$$\phi_j = \frac{\partial \phi}{\Delta_j x^j},$$
$$\psi_j = \frac{\partial \psi}{\Delta_j x^j}.$$

These partial derivatives are components of covariant vectors. Relations (3.16) and (3.17) show that ϕ_j and ψ_j are orthogonal to all displacements $\Delta_j x^j$ at a point P on the surface. Hence $\Delta_j x^j$ is in the tangential of the hypersurfaces (3.14) and (3.15). Thus the gradient vectors ϕ_j and ψ_j at any point of the hypersurfaces are normals to the hypersurfaces at the point. Let ω be the angle between hypersurfaces (3.14) and (3.15). Then we have

$$\cos \omega = \frac{g^{ij} \phi_i \psi_j}{\sqrt{g^{ij} \phi_i \phi_j} \sqrt{g^{ij} \psi_i \psi_j}}.$$

In particular, let hypersurfaces (3.14) and (3.15) be taken as coordinate hypersurfaces, say x^p and x^q, respectively. Then

$$\phi = x^p = \text{constant}$$

and

$$\psi = x^q = \text{constant}.$$

Then

$$\phi_p = 1,$$
$$\psi_q = 1,$$

and

$$\phi_i = 0, \quad i \neq p,$$
$$\psi_i = 0, \quad i \neq q.$$

In this case, we have

$$\cos \omega_{pq} = \frac{g^{pq}}{\sqrt{g^{pp}}\sqrt{g^{qq}}}.$$

Thus, if

$$g^{ij} = 0, \quad i \neq j,$$

then the parametric hypersurfaces are orthogonal to each other.

Definition 3.21. If in V_N, there are N families of hypersurfaces such that at any point each hypersurface is orthogonal to the $N-1$ hypersurfaces of the other families passing through that point, then they are said to form an Nply orthogonal system of hypersurfaces.

Theorem 3.5. *The necessary and sufficient condition for the existence of an Nply orthogonal system of coordinate hypersurfaces is that the fundamental form is of the form*

$$\Delta s^2 = g_{11}\left(\Delta_1 x^1\right)^2 + g_{22}\left(\Delta_2 x^2\right)^2 + \cdots + g_{nn}\left(\Delta_n x^n\right)^2. \tag{3.18}$$

Proof. 1.　Suppose that coordinate surfaces form an Nply orthogonal system of hypersurfaces. Then

$$g^{ij} = 0, \quad i,j \in \{1, 2, \ldots, N\}, \ i \neq j.$$

If g denotes $|g_{ij}|$ and l denotes $|g^{ij}|$, then

$$l = \frac{1}{g}.$$

Hence $l \neq 0$. Because

$$l = g^{11} g^{22} \cdots g^{NN},$$

we conclude that

$$g^{jj} \neq 0, \quad j \in \{1, 2, \ldots, N\}.$$

Now using that

$$g_{ij} = \frac{\text{cofactor of } g^{ij} \text{ in } l}{l},$$

or

$$lg_{ij} = \text{cofactor of } g^{ij} \quad \text{in } l,$$

or

$$\frac{g_{ij}}{g} = \text{cofactor of } g^{ij} \quad \text{in } l,$$

we conclude that

$$g_{ij} = 0, \quad i \neq j.$$

Thus the fundamental form is given by (3.18).

2. Let the fundamental form be given by (3.18). Then

$$g_{ij} = 0, \quad i \neq j.$$

Hence

$$g^{ij} = \frac{\text{cofactor of } g_{ij} \text{ in } g}{g}$$
$$= 0, \quad i \neq j.$$

Therefore the coordinate hypersurfaces form an Nply orthogonal system of coordinate hypersurfaces. This completes the proof. $\qquad\square$

Now suppose that some V_N admits an Nply orthogonal system of hypersurfaces. The fundamental form in this case is

$$\Delta s^2 = g_{jj}\left(\Delta_j x^j\right)^2.$$

Let an Nply orthogonal system of hypersurfaces be given by the hypersurfaces

$$\phi_j = c_j, \quad j \in \{1, 2, \ldots, N\},$$

where c_j are constants. Now the condition that the hypersurfaces is

$$g_{ij} \frac{\partial \phi_p}{\Delta_i x^i} \frac{\partial \phi_q}{\Delta_j x^j} = 0.$$

These equations admit $\frac{N(N-1)}{2}$ simultaneous dynamic equations to find N unknowns. If the equations admit N solutions, then we have

$$\frac{N(N-1)}{2} = N.$$

These equations are consistent if

$$\frac{N(N-1)}{2} \leq N \quad \text{or} \quad N \leq 3.$$

Therefore an arbitrary V_N does not admit an Nply orthogonal system of hypersurfaces.

Definition 3.22. A congruence of curves in a V_N is a family of curves, one of which passes through each point of V_N.

Definition 3.23. An orthogonal ennuple in a Riemannian space V_N consists of N mutually orthogonal congruences of curves.

Consider N tangents λ_h^i, $h \in \{1, 2, \ldots, N\}$, to congruences λ_h, $h \in \{1, 2, \ldots, N\}$, of an orthogonal ennuple in V_N. The quantities λ_h^i and λ_{hi} denote the contravariant and covariant components of λ_h, respectively.

Suppose that any two congruences of the orthogonal ennuple are λ_h and λ_k, so that

$$\lambda_h \lambda_k = \delta_k^h,$$
$$g_{ij} \lambda_h^i \lambda_k^j = \delta_k^h.$$

Then

$$g_{ij} \lambda_h^i \lambda_h^j = \delta_h^h$$
$$= 1,$$
$$g_{ij} \lambda_k^i \lambda_k^j = 1,$$

and

$$g_{ij} \lambda_h^i \lambda_k^j = 0, \quad k \neq j.$$

Define

$$\lambda_h^i = \frac{\text{cofactor of } \lambda_{hi} \text{ in } |\lambda_{hi}|}{|\lambda_{hi}|}.$$

From the determinant property we have

$$\sum_{h=1}^{N} \lambda_h^i \lambda_{hj} = \delta_j^i. \tag{3.19}$$

Multiplying this equation by g^{jk}, we get

$$\sum_{h=1}^{N} \lambda_h^i \lambda_{hj} g^{jk} = \delta_j^i g^{jk},$$

whereupon

$$g^{ik} = \sum_{h=1}^{N} \lambda_h^i \lambda_h^k,$$

or

$$g^{ij} = \sum_{h=1}^{N} \lambda_h^i \lambda_h^j. \tag{3.20}$$

Multiplying equation (3.19) by g_{ik}, we get

$$\sum_{h=1}^{N} \lambda_h^i \lambda_{hj} g_{ik} = \delta_j^i g_{ik},$$

or

$$\sum_{h=1}^{N} \lambda_{hk} \lambda_{hj} = g_{jk},$$

or

$$g_{ij} = \sum_{h=1}^{N} \lambda_{hi} \lambda_{hj}. \tag{3.21}$$

Equations (3.20) and (3.21) give the values of the fundamental tensors g^{ij} and g_{ij}, respectively.

Now consider the vector u given by

$$u^i = \sum_{h=1}^{N} C_h \lambda_h^i, \tag{3.22}$$

where C_h are constants that will be determined below. Multiplying equation (3.22) by λ_{ki}, we get

$$u^i \lambda_{ki} = \sum_{h=1}^{N} C_h \lambda_h^i \lambda_{ki}$$

$$= \sum_{h=1}^{N} C_h \delta_h^k$$

$$= C_k.$$

Therefore

$$u^i = \sum_{h=1}^{N} (u^j \lambda_{hj}) \lambda_h^i.$$

Let m denotes the magnitude of u. Then

$$m = u^i u_i$$

$$= \left(\sum_{h=1}^{N} C_h \lambda_h^i \right) \left(\sum_{k=1}^{N} C_k \lambda_{ki} \right)$$

$$= \sum_{h=1}^{N} \sum_{k=1}^{N} C_h C_k \lambda_h^i \lambda_{ki}$$

$$= \sum_{h=1}^{N} \sum_{k=1}^{N} C_h C_k \delta_k^h$$

$$= \sum_{h=1}^{N} C_h C_h$$

$$= \sum_{h=1}^{N} C_h^2.$$

From the last equation it follows that

$$u = 0 \quad \Longleftrightarrow \quad m = 0 \quad \Longleftrightarrow \quad C_h = 0.$$

3.3 Affine coordinates

Let us consider the transformation from a given coordinate system x^i to a rectangular system $\bar{x}^i$. The Jacobian matrix for such a transformation is

$$J = \left(\frac{\Delta_{(i)} \bar{x}^{i\sigma_0 \sigma_1 \cdots \sigma_{j-1}}}{\Delta_j x^j} \right).$$

Then the matrix $G = (g_{ij})$ of the Euclidean metric tensor in the x^i system is

$$G = J^T J.$$

Our task is to find $G = (g_{ij})$ for the three-dimensional affine coordinates x^i. Let

$$u = (\delta_1^i),$$
$$v = (\delta_2^j),$$
$$w = (\delta_3^j)$$

and

$$\alpha = \angle(u, v),$$
$$\beta = \angle(v, w),$$
$$\gamma = \angle(u, w).$$

Then

$$\cos \alpha = \frac{g_{ij}\delta_1^i\delta_2^j}{\sqrt{g_{ij}\delta_1^i\delta_1^j}\sqrt{g_{ij}\delta_2^i\delta_2^j}}$$
$$= \frac{g_{12}}{\sqrt{g_{11}}\sqrt{g_{22}}}$$
$$= g_{12},$$

as obviously

$$g_{11} = g_{22} = g_{33} = 1.$$

Likewise,

$$\cos \beta = g_{13},$$
$$\cos \gamma = g_{23}.$$

Then the complete symmetric matrix is

$$G = (g_{ij})$$
$$= \begin{pmatrix} 1 & \cos \alpha & \cos \beta \\ \cos \alpha & 1 & \cos \gamma \\ \cos \beta & \cos \gamma & 1 \end{pmatrix}.$$

The corresponding metric is given by

$$\Delta s^2 = \left(\Delta_1 x^1\right)^2 + \left(\Delta_2 x^2\right)^2 + \left(\Delta_3 x^3\right)^2 + 2\cos \alpha \Delta_1 x^1 \Delta_2 x^2$$
$$+ 2\cos \beta \Delta_1 x^1 \Delta_3 x^3 + 2\cos \gamma \Delta_2 x^2 \Delta_3 x^3.$$

3.4 Curvalinear coordinates

In this section, we study a type of coordinates in E^3, of which two coordinates will be specific. Consider a general functional transformation

$$T : x^j = x^j(y^1, y^2, y^3), \quad j \in \{1, 2, 3\}, \tag{3.23}$$

such that x^j is continuously differentiable function of (y^1, y^2, y^3) in some region R of E^3, $x^j \in \mathbb{T}_j, y^j \in \mathbb{T}_{(j)}$, and

$$x^j(\sigma_{(1)}(y^1), y^2, y^3) = x^j(y^1, \sigma_{(2)}(y^2), y^3)$$
$$= x^j(y^1, y^2, \sigma_{(3)}(y^3))$$
$$= \sigma_j(x^j(y^1, y^2, y^3)), \quad j \in \{1, 2, 3\}.$$

Since x^1, x^2, and x^3 have to be independent, the Jacobian

$$J = \frac{\Delta(x^1, x^2, x^3)}{\Delta(y^1, y^2, y^3)}$$

$$= \det \begin{pmatrix} \dfrac{\Delta_1 x^1}{\Delta_{(1)} y^1} & \dfrac{\Delta_1 x^1}{\Delta_{(2)} y^2} & \dfrac{\Delta_1 x^1}{\Delta_{(3)} y^3} \\[2mm] \dfrac{\Delta_2 x^2}{\Delta_{(1)} y^1} & \dfrac{\Delta_2 x^2}{\Delta_{(2)} y^2} & \dfrac{\Delta_2 x^2}{\Delta_{(3)} y^3} \\[2mm] \dfrac{\Delta_3 x^3}{\Delta_{(1)} y^1} & \dfrac{\Delta_3 x^3}{\Delta_{(2)} y^2} & \dfrac{\Delta_3 x^3}{\Delta_{(3)} y^3} \end{pmatrix}$$

$$\neq 0.$$

Under these conditions, y^1, y^2, and y^3 can be obtained as single-valued functions of x^1, x^2, and x^3 with continuous first-order partial derivatives. Thus the inverse transformation

$$T^{-1} : y^j = y^j(x^1, x^2, x^3), \quad j \in \{1, 2, 3\},$$

will be single valued, and the transformations T and T^{-1} establish one-to-one correspondence between the sets of (x^1, x^2, x^3) and (y^1, y^2, y^3).

Definition 3.24. The coordinates (y^1, y^2, y^3) are said to be a curvilinear coordinate system in R of E^3.

The curvilinear coordinates y^1, y^2, and y^3 in E^3 are related to the rectangular Cartesian coordinates x^1, x^2, and x^3 by formula (3.23).

Let us now fix one of y^1, y^2, or y^3, say

$$y^1 = c^1 = \text{constant} \in \mathbb{T}_{(1)},$$

and let y^2 and y^3 be allowed to vary. Then a point $P(x^1, x^2, x^3)$ will satisfy the relations

$$x^j = x^j(c^1, y^2, y^3), \quad j \in \{1, 2, 3\},$$

which define a surface, and the point $P(x^1, x^2, x^3)$ will lie on the surface, which we will denote by

$$y^1 = c^1.$$

If the constant is now allowed to assume various values, then we get a one-parameter family of surfaces. Similarly,

$$x^j = x^j(y^1, c^2, y^3), \quad j \in \{1, 2, 3\},$$

and

$$x^j = x^j(y^1, y^2, c^3), \quad j \in \{1, 2, 3\},$$

define the families of surfaces

$$y^2 = c^2 \quad \text{and} \quad y^3 = c^3,$$

respectively.

Definition 3.25. Each of the surfaces

$$y^1 = c^1, \quad y^2 = c^2, \quad \text{and} \quad y^3 = c^3$$

is called a coordinate surface of the curvilinear coordinate system, and their intersections pair-by-pair are the coordinate lines.

There will be three families of such surfaces corresponding to different values of c^1, c^2, and c^3. Through a given point $P(x^1, x^2, x^3)$, there pass three coordinate surfaces corresponding to fixed values of c^1, c^2, and c^3. The condition that the Jacobian $J \neq 0$ in the region under consideration expresses the fact that the surfaces

$$y^1 = c^1, \quad y^2 = c^2, \quad \text{and} \quad y^3 = c^3$$

intersect at a unique point.

Let us fix two of the coordinates y^1, y^2, and y^3 in T, say

$$y^2 = c^2 \in \mathbb{T}_{(2)},$$
$$y^3 = c^3 \in \mathbb{T}_{(3)},$$

and allow y^1 to vary. Then $P(x^1, x^2, x^3)$ will satisfy the relations

$$x^j = x^j(y^1, c^2, c^3), \quad j \in \{1, 2, 3\}.$$

Definition 3.26. Since x^1, x^2, and x^3 are functions of one variable, it follows that $P(x^1, x^2, x^3)$ will be on a curve, called a coordinate curve. This coordinate curve

$$y^2 = c^2 \quad \text{and} \quad y^3 = c^3$$

is called the y^1 curve.

Thus the line of intersection

$$y^2 = c^2,$$
$$y^3 = c^3,$$

is the y^1 coordinate line, because along this line, the variable y^1 is the only changing one. Therefore the y^1 line lies on both surfaces

$$y^2 = c^2 \quad \text{and} \quad y^3 = c^3.$$

Similarly, we can define y^2 and y^3 curves. Note that through a given point $P(x^1, x^2, x^3)$, there pass three coordinate curves corresponding to fixed values c^1, c^2, and c^3.

Example 3.26. Consider the transformation

$$\begin{aligned}
x^1 &= y^3, \\
x^2 &= (y^1)^3, \\
x^3 &= (y^2)^2, \quad x^3 \geq 0.
\end{aligned}$$

We have

$$\frac{\Delta_1 x^1}{\Delta_{(1)} y^1} = 0,$$

$$\frac{\Delta_1 x^1}{\Delta_{(2)} y^2} = 0,$$

$$\frac{\Delta_1 x^1}{\Delta_{(3)} y^3} = 1,$$

$$\frac{\Delta_2 x^2}{\Delta_{(1)} y^1} = (\sigma_{(1)}(y^1))^2 + y^2 \sigma_{(1)}(y^1) + (y^1)^2,$$

$$\frac{\Delta_2 x^2}{\Delta_{(2)} y^2} = 0,$$

$$\frac{\Delta_2 x^2}{\Delta_{(3)} y^3} = 0,$$

$$\frac{\Delta_3 x^3}{\Delta_{(1)} y^1} = 0,$$

$$\frac{\Delta_3 x^3}{\Delta_{(2)} y^2} = \sigma_{(2)}(y^2) + y^2,$$

$$\frac{\Delta_3 x^3}{\Delta_{(3)} y^3} = 0.$$

Hence the Jacobian of the transformation is

$$J = \det \begin{pmatrix}
\dfrac{\Delta_1 x^1}{\Delta_{(1)} y^1} & \dfrac{\Delta_1 x^1}{\Delta_{(2)} y^2} & \dfrac{\Delta_1 x^1}{\Delta_{(3)} y^3} \\[2ex]
\dfrac{\Delta_2 x^2}{\Delta_{(1)} y^1} & \dfrac{\Delta_2 x^2}{\Delta_{(2)} y^2} & \dfrac{\Delta_2 x^2}{\Delta_{(3)} y^3} \\[2ex]
\dfrac{\Delta_3 x^3}{\Delta_{(1)} y^1} & \dfrac{\Delta_3 x^3}{\Delta_{(2)} y^2} & \dfrac{\Delta_3 x^3}{\Delta_{(3)} y^3}
\end{pmatrix}$$

$$= \det \begin{pmatrix} 0 & 0 & 1 \\ (\sigma_{(1)}(y^1))^2 + y^1\sigma_{(1)}(y^1) + (y^1)^2 & 0 & 0 \\ 0 & \sigma_{(2)}(y^2) + y^2 & 0 \end{pmatrix}$$

$$= \left((\sigma_{(1)}(y^1))^2 + y^1\sigma_{(1)}(y^1) + (y^1)^2\right)(\sigma_{(2)}(y^2) + y^2)$$

$$\neq 0.$$

Hence the inverse transformation exists and is given by

$$y^1 = \sqrt[3]{x^2},$$
$$y^2 = \sqrt{x^3},$$
$$y^3 = x^1, \quad y^2 \geq 0.$$

If we take

$$\mathbb{T}_1 = 2^{\mathbb{N}_0},$$
$$\mathbb{T}_2 = 3^{\mathbb{N}_0},$$
$$\mathbb{T}_3 = 4\mathbb{N}_0,$$

then

$$\mathbb{T}_{(1)} = \left(\sqrt[3]{3}\right)^{\mathbb{N}_0},$$
$$\mathbb{T}_{(2)} = 2\sqrt{\mathbb{N}_0},$$
$$\mathbb{T}_{(3)} = 2^{\mathbb{N}_0}.$$

We have

$$\sigma_{(1)}(y^1) = \sqrt[3]{3}y^1,$$
$$\sqrt[3]{\sigma_2(x^2)} = \sqrt[3]{3x^2}$$
$$= \sqrt[3]{3}\sqrt[3]{x^2}$$
$$= \sqrt[3]{3}y^1$$
$$= \sigma_{(1)}(y^1),$$
$$\sigma_{(2)}(y^2) = 2\sqrt{(y^2)^2 + 1},$$
$$\sqrt{\sigma_3(x^3)} = \sqrt{4x^3 + 4}$$
$$= 2\sqrt{x^3 + 1}$$
$$= 2\sqrt{(y^2)^2 + 1}$$
$$= \sigma_{(2)}(y^2),$$

$$\sigma_{(3)}(y^3) = 2y^3$$
$$= 2x^1$$
$$= \sigma_1(x^1).$$

Now we have to obtain the line element of E^3 in curvilinear coordinates. Let

$$P(y^1, y^2, y^3) \quad \text{and} \quad Q(y^1 + \Delta_{(1)}y^1, y^2 + \Delta_{(2)}y^2, y^3 + \Delta_{(3)}y^3)$$

be two neighboring points in the region R of E^3 in which a curvilinear coordinate system (3.23) is defined. The Euclidean distance between a pair of such points is determined by the quadratic form

$$\Delta s^2 = \left((y^1 + \Delta_{(1)}y^1) - y^1\right)^2 + \left((y^2 + \Delta_{(2)}y^2) - y^2\right)^2 + \left((y^3 + \Delta_{(3)}y^3) = -y^3\right)^2$$
$$= \left(\Delta_{(1)}y^1\right)^2 + \left(\Delta_{(2)}y^2\right)^2 + \left(\Delta_{(3)}y^3\right)^2$$
$$= \left(\Delta_{(j)}y^j\right)^2$$
$$= \frac{\Delta_{(j)}y^j}{\Delta_k x^k} \frac{\Delta_{(j)}y^j}{\Delta_l x^l} \Delta_k x^k \Delta_l x^l$$
$$= g_{kl} \Delta_k x^k \Delta_l x^l,$$

where

$$g_{kl} = \frac{\Delta_{(j)}y^j}{\Delta_k x^k} \frac{\Delta_{(j)}y^j}{\Delta_l x^l}.$$

This is the elementary arc length in the curvilinear coordinate system. Obviously, g_{ij} are symmetric.

Definition 3.27. We call g_{ij} the fundamental metric tensor.

Let

$$g = |g_{ij}|.$$

Since

$$g_{ij} \Delta_i x^i \Delta_j x^j$$

is positive definite, $g > 0$ in R. Then we can introduce the conjugate tensor g^{ij} as

$$g^{ij} = \frac{\text{cofactor of } g_{ij} \text{ in } g}{g}, \quad g \neq 0.$$

Hence g^{ij} is a second-order tensor conjugate to g_{ij}.

3.5 Advanced practical problems

Problem 3.1. Under the metric

$$\Delta s^2 = \left(\Delta_1 x^1\right)^2 + \left(x^2\right)^2\left(\Delta_2 x^2\right)^2 + \left(\Delta_3 x^3\right)^2,$$

consider the curve

$$x^1 = t^2,$$
$$x^2 = 1 - t,$$
$$x^3 = 2 - 3t, \quad t \in \mathbb{T},$$

where $\mathbb{T} = 2\mathbb{Z}$. Find the length of arc for $t \in [0, 4]$.

Answer 3.6.

$$12.$$

Problem 3.2. Let $\mathbb{T} = 2^{\mathbb{N}_0}$. Find the length of the curve

$$x^1 = 3 - t,$$
$$x^2 = 6t + 3,$$
$$x^3 = t^2, \quad t \in [1, 4],$$

for the metric

$$\begin{pmatrix} 12 & 4 & 0 \\ 4 & 1 & 1 \\ 0 & 1 & (x^1)^2 \end{pmatrix}.$$

Answer 3.7.

$$3\sqrt{5} + 36\sqrt{2}.$$

Problem 3.3. Prove that

$$A^{\bullet\bullet r}_{pq} = g_{pj}g_{qk}g^{rl}A^{jk}_{\bullet\bullet l}.$$

Problem 3.4. Prove that in V_4 with the line element

$$\Delta s^2 = -\left(\Delta_1 x^1\right)^2 - \left(\Delta_2 x^2\right)^2 - \left(\Delta_3 x^3\right)^2 + c^2\left(\Delta_4 x^4\right)^2,$$

the vector $\left(\sqrt{2}, 0, 0, \frac{\sqrt{3}}{c}\right)$ is an unit vector.

Problem 3.5. Let

$$A_i = \frac{1}{\sqrt{g^{pq}B_pB_q}}B_i,$$

where B_i is a covariant vector, and g^{ij} is the fundamental tensor. Prove that A_i is a unit vector.

Problem 3.6. Let A^p be a vector field. Prove that the corresponding unit vectors are

$$\frac{A^p}{\sqrt{A^pA_q}} \quad \text{and} \quad \frac{A^p}{\sqrt{g_{pq}A^pA^q}}.$$

Problem 3.7. If

$$A_i = \frac{1}{\sqrt{g^{pq}B_pB_q}},$$

where B_i is a covariant vector, then prove that A_i is a unit vector.

Problem 3.8. Prove that the magnitude of two associated vectors is invariant.

Problem 3.9. Prove that the angle between two contravariant vectors is real when the Riemannian metric is positive definite.

Problem 3.10. Let θ be the angle between two nonnull vectors A^j and B^j. Prove that:
1.

$$(\sin \theta)^2 = \frac{(g_{ij}g_{pq} - g_{ip}g_{jq})A^iB^pA^jB^q}{(g_{ij}A^iA^j)(g_{ij}B^iB^j)}.$$

2. if A^i and B^j are orthogonal unit vectors, then

$$(g_{ij}g_{pq} - g_{ip}g_{jq})A^iB^pA^jB^q = 1.$$

4 Covariant differentiation

In this chapter, the Christoffel symbols of the first and second kinds are defined. Some of their basic properties are established. Covariant derivatives are introduced. The divergence, Laplace operator, and curl are defined and explored. Intrinsic differentiation is studied.

4.1 Christoffel symbols

In this section, we consider two expressions for the components of g_{ij} due to Christoffel, which will be useful in the development of tensor calculus.

Definition 4.1. The Christoffel symbol of the first kind is defined as

$$\Gamma_{ij,k} = \frac{1}{2}\left(\frac{\partial g_{ik}}{\Delta_j x^j} + \frac{\partial g_{jk}}{\Delta_i x^i} - \frac{\partial g_{ij}}{\Delta_k x^k} \right), \quad i,j \in \{1,2,\ldots,N\}.$$

Definition 4.2. The Christoffel symbol of the second kind is defined as

$$\Gamma_{ij}^k = g^{mk}\Gamma_{ij,m}$$
$$= \frac{1}{2}g^{mk}\left(\frac{\partial g_{im}}{\Delta_j x^j} + \frac{\partial g_{jm}}{\Delta_i x^i} - \frac{\partial g_{ij}}{\Delta_m x^m} \right),$$

where g^{ij} is the conjugate tensor for the fundamental metric tensor g_{ij}.

Both these symbols in V_N are sets of functions of coordinates x^j. There are N distinct Christoffel symbols of each kind for each independent g_{ij}. Because g_{ij} is a symmetric tensor of the second kind and has $\frac{1}{2}N(N+1)$ independent components, the number of independent components of the Christoffel symbols is

$$N \cdot \frac{1}{2}N(N+1) = \frac{1}{2}N^2(N+1).$$

Example 4.1. We will find the Christoffel symbols of the second kind in V_2 with line element

$$\Delta s^2 = a^2(\Delta_1 x^1)^2 + a^2(\sin_1(x^1, x_0^1))^2(\Delta_2 x^2)^2,$$

where a is a constant, and $x_0^1 \in \mathbb{T}_1$ is fixed. In this case, we have

$$g_{11} = a^2,$$
$$g_{12} = 0,$$

https://doi.org/10.1515/9783112228494-004

$$g_{21} = 0,$$
$$g_{22} = a^2(\sin_1(x^1, x_0^1))^2.$$

Let

$$l(x^1) = \int_0^1 \sin_1(h\sigma_1(x^1) + (1-h)x^1, x_0^1)\,dh.$$

We have

$$\frac{\partial g_{11}}{\Delta_1 x^1} = 0,$$

$$\frac{\partial g_{11}}{\Delta_2 x^2} = 0,$$

and using the Pötzsche chain rule, we find

$$\frac{\partial g_{22}}{\Delta_1 x^1} = 2a^2\left(\int_0^1 \sin_1(h\sigma_1(x^1) + (1-h)x^1, x_0^1)\,dh\right)\cos_1(x^1, x_0^1)$$

$$= 2a^2 l(x^1)\cos_1(x^1, x_0^1),$$

$$\frac{\partial g_{22}}{\Delta_2 x^2} = 0.$$

Then

$$g = \det\begin{pmatrix} g_{11} & g_{12} \\ g_{21} & g_{22} \end{pmatrix}$$

$$= \det\begin{pmatrix} a^2 & 0 \\ 0 & a^2(\sin_1(x^1, x_0^1))^2 \end{pmatrix}$$

$$= a^4(\sin_1(x^1, x_0^1))^2.$$

The conjugate tensors g^{ij} of the tensors g_{ij} are as follows:

$$g^{11} = \frac{\text{cofactor of } g_{11} \text{ in } g}{g}$$

$$= \frac{a^2(\sin_1(x^1, x_0^1))^2}{a^4(\sin_1(x^1, x_0^1))^2}$$

$$= \frac{1}{a^2},$$

$$g^{12} = \frac{\text{cofactor of } g_{12} \text{ in } g}{g}$$

$$= 0,$$

$$g^{21} = \frac{\text{cofactor of } g_{21} \text{ in } g}{g}$$

$$= 0,$$

$$g^{22} = \frac{\text{cofactor of } g_{22} \text{ in } g}{g}$$

$$= \frac{a^2}{a^4(\sin_1(x^1, x_0^1))^2}$$

$$= \frac{1}{a^2(\sin_1(x^1, x_0^1))^2}.$$

Hence, for the Christoffel symbols of the first kind, we find

$$\Gamma_{11,1} = \frac{1}{2}\left(\frac{\partial g_{11}}{\Delta_1 x^1} + \frac{\partial g_{11}}{\Delta_1 x^1} - \frac{\partial g_{11}}{\Delta_1 x^1}\right)$$

$$= \frac{1}{2}\frac{\partial g_{11}}{\Delta_1 x^1}$$

$$= 0,$$

$$\Gamma_{11,2} = \frac{1}{2}\left(\frac{\partial g_{12}}{\Delta_1 x^1} + \frac{\partial g_{12}}{\Delta_1 x^1} - \frac{\partial g_{11}}{\Delta_2 x^2}\right)$$

$$= 0,$$

$$\Gamma_{12,1} = \frac{1}{2}\left(\frac{\partial g_{11}}{\Delta_2 x^2} + \frac{\partial g_{21}}{\Delta_1 x^1} - \frac{\partial g_{12}}{\Delta_1 x^1}\right)$$

$$= 0,$$

$$\Gamma_{12,2} = \frac{1}{2}\left(\frac{\partial g_{12}}{\Delta_2 x^2} + \frac{\partial g_{22}}{\Delta_1 x^1} - \frac{\partial g_{12}}{\Delta_2 x^2}\right)$$

$$= \frac{1}{2}(2a^2 l(x^1)\cos_1(x^1, x_0^1))$$

$$= a^2 l(x^1)\cos_1(x^1, x_0^1),$$

$$\Gamma_{21,1} = \frac{1}{2}\left(\frac{\partial g_{21}}{\Delta_1 x^1} + \frac{\partial g_{11}}{\Delta_2 x^2} - \frac{\partial g_{21}}{\Delta_1 x^1}\right)$$

$$= 0,$$

$$\Gamma_{21,2} = \frac{1}{2}\left(\frac{\partial g_{22}}{\Delta_1 x^1} + \frac{\partial g_{12}}{\Delta_2 x^2} - \frac{\partial g_{21}}{\Delta_2 x^2}\right)$$

$$= \frac{1}{2}(2a^2 l(x^1)\cos_1(x^1, x_0^1))$$

$$= a^2 l(x^1)\cos_1(x^1, x_0^1),$$

$$\Gamma_{22,1} = \frac{1}{2}\left(\frac{\partial g_{21}}{\Delta_2 x^2} + \frac{\partial g_{21}}{\Delta_2 x^2} - \frac{\partial g_{22}}{\Delta_1 x^1}\right)$$

$$= \frac{1}{2}(-2a^2 l(x^1)\cos_1(x^1, x_0^1))$$

$$= -a^2 l(x^1)\cos_1(x^1, x_0^1),$$

$$\Gamma_{22,2} = \frac{1}{2}\left(\frac{\partial g_{22}}{\Delta_2 x^2} + \frac{\partial g_{22}}{\Delta_2 x^2} - \frac{\partial g_{22}}{\Delta_2 x^2}\right)$$
$$= 0.$$

Therefore, for the Christoffel symbols of the second kind, we obtain

$$\begin{aligned}
\Gamma_{11}^1 &= g^{1m}\Gamma_{11,m} \\
&= g^{11}\Gamma_{11,1} + g^{12}\Gamma_{11,2} \\
&= \frac{1}{a^2}\cdot 0 + 0\cdot 0 \\
&= 0, \\
\Gamma_{12}^1 &= g^{1m}\Gamma_{12,m} \\
&= g^{11}\Gamma_{12,1} + g^{12}\Gamma_{12,2} \\
&= \frac{1}{a^2}\cdot 0 + 0\cdot 0 \\
&= 0, \\
\Gamma_{21}^1 &= g^{1m}\Gamma_{21,m} \\
&= g^{11}\Gamma_{21,1} + g^{12}\Gamma_{21,2} \\
&= \frac{1}{a^2}\cdot 0 + 0\cdot (a^2 l(x^1)\cos_1(x^1, x_0^1)) \\
&= 0, \\
\Gamma_{22}^1 &= g^{1m}\Gamma_{22,m} \\
&= g^{11}\Gamma_{22,1} + g^{12}\Gamma_{22,2} \\
&= \frac{1}{a^2}(-a^2 l(x^1)\cos_1(x^1, x_0^1)) + 0\cdot 0 \\
&= -l(x^1)\cos_1(x^1, x_0^1), \\
\Gamma_{11}^2 &= g^{2m}\Gamma_{11,m} \\
&= g^{21}\Gamma_{11,1} + g^{22}\Gamma_{11,2} \\
&= 0\cdot 0 + \frac{1}{a^2(\sin_1(x^1, x_0^1))^2}\cdot 0 \\
&= 0, \\
\Gamma_{12}^2 &= g^{2m}\Gamma_{12,m} \\
&= g^{21}\Gamma_{12,1} + g^{22}\Gamma_{12,2} \\
&= 0\cdot 0 + \frac{1}{a^2(\sin_1(x^1, x_0^1))^2}(a^2 l(x^1)\cos_1(x^1, x_0^1)) \\
&= \frac{l(x^1)\cos_1(x^1, x_0^1)}{(\sin_1(x^1, x_0^1))^2}, \\
\Gamma_{21}^2 &= g^{2m}\Gamma_{21,m} \\
&= g^{21}\Gamma_{21,1} + g^{22}\Gamma_{21,2}
\end{aligned}$$

$$= 0 \cdot 0 + \frac{1}{a^2(\sin_1(x^1, x_0^1))^2}\left(a^2 l(x^1)\cos_1(x^1, x_0^1)\right)$$

$$= \frac{l(x^1)\cos_1(x^1, x_0^1)}{(\sin_1(x^1, x_0^1))^2},$$

$$\Gamma_{22}^2 = g^{2m}\Gamma_{22,m}$$

$$= g^{21}\Gamma_{22,1} + g^{22}\Gamma_{22,2}$$

$$= 0 \cdot \left(-a^2 l(x^1)\cos_1(x^1, x_0^1)\right) + \frac{1}{a^2(\sin_1(x^1, x_0^1))^2}\cdot 0$$

$$= 0.$$

Example 4.2. We will find the Christoffel symbols of the second kind in V_2 with line element

$$\Delta s^2 = (\Delta_1 x^1)^2 + (f(x^1, x^2))^2(\Delta_2 x^2)^2,$$

where f is a continuously differentiable function of x^1 and x^2. In this case, we have

$$g_{11} = 1,$$
$$g_{12} = 0,$$
$$g_{21} = 0,$$
$$g_{22} = (f(x^1, x^2))^2.$$

Let

$$l_1(x^1, x^2) = \int_0^1 f(h\sigma_1(x^1) + (1-h)x^1, x^2)\,dh,$$

$$l_2(x^1, x^2) = \int_0^1 f(x^1, h\sigma_2(x^2) + (1-h)x^2)\,dh.$$

We have

$$\frac{\partial g_{11}}{\Delta_1 x^1} = 0,$$

$$\frac{\partial g_{11}}{\Delta_2 x^2} = 0,$$

and using the Pötzsche chain rule, we find

$$\frac{\partial g_{22}}{\Delta_1 x^1} = 2\left(\int_0^1 f(h\sigma_1(x^1) + (1-h)x^1, x^2)\,dh\right)f_{x_1}^{\Delta_1}(x^1, x^2)$$

$$= 2l_1(x^1, x^2)f_{x_1}^{\Delta_1}(x^1, x^2),$$

$$\frac{\partial g_{22}}{\Delta_2 x^2} = 2\left(\int_0^1 f(h\sigma_1(x^1) + (1-h)x^1, x^2)\,dh \right) f_{x_2}^{\Delta_2}(x^1, x^2)$$

$$= 2l_2(x^1, x^2) f_{x_2}^{\Delta_2}(x^1, x^2).$$

Then

$$g = \det \begin{pmatrix} g_{11} & g_{12} \\ g_{21} & g_{22} \end{pmatrix}$$

$$= \det \begin{pmatrix} 1 & 0 \\ 0 & (f(x^1, x^2))^2 \end{pmatrix}$$

$$= (f(x^1, x^2))^2.$$

The conjugate tensors g^{ij} of the tensors g_{ij} are as follows:

$$g^{11} = \frac{\text{cofactor of } g_{11} \text{ in } g}{g}$$

$$= \frac{(f(x^1, x^2))^2}{(f(x^1, x^2))^2}$$

$$= 1,$$

$$g^{12} = \frac{\text{cofactor of } g_{12} \text{ in } g}{g}$$

$$= 0,$$

$$g^{21} = \frac{\text{cofactor of } g_{21} \text{ in } g}{g}$$

$$= 0,$$

$$g^{22} = \frac{\text{cofactor of } g_{22} \text{ in } g}{g}$$

$$= \frac{1}{(f(x^1, x^2))^2}.$$

Hence, for the Christoffel symbols of the first kind, we find

$$\Gamma_{11,1} = \frac{1}{2}\left(\frac{\partial g_{11}}{\Delta_1 x^1} + \frac{\partial g_{11}}{\Delta_1 x^1} - \frac{\partial g_{11}}{\Delta_1 x^1} \right)$$

$$= \frac{1}{2}\frac{\partial g_{11}}{\Delta_1 x^1}$$

$$= 0,$$

$$\Gamma_{11,2} = \frac{1}{2}\left(\frac{\partial g_{12}}{\Delta_1 x^1} + \frac{\partial g_{12}}{\Delta_1 x^1} - \frac{\partial g_{11}}{\Delta_2 x^2} \right)$$

$$= 0,$$

$$\Gamma_{12,1} = \frac{1}{2}\left(\frac{\partial g_{11}}{\Delta_2 x^2} + \frac{\partial g_{21}}{\Delta_1 x^1} - \frac{\partial g_{12}}{\Delta_1 x^1}\right)$$
$$= 0,$$
$$\Gamma_{12,2} = \frac{1}{2}\left(\frac{\partial g_{12}}{\Delta_2 x^2} + \frac{\partial g_{22}}{\Delta_1 x^1} - \frac{\partial g_{12}}{\Delta_2 x^2}\right)$$
$$= \frac{1}{2}\left(2l_1(x^1, x^2)f_{x_1}^{\Delta_1}(x^1, x^2)\right)$$
$$= l_1(x^1, x^2)f_{x_1}^{\Delta_1}(x^1, x^2),$$
$$\Gamma_{21,1} = \frac{1}{2}\left(\frac{\partial g_{21}}{\Delta_1 x^1} + \frac{\partial g_{11}}{\Delta_2 x^2} - \frac{\partial g_{21}}{\Delta_1 x^1}\right)$$
$$= 0,$$
$$\Gamma_{21,2} = \frac{1}{2}\left(\frac{\partial g_{22}}{\Delta_1 x^1} + \frac{\partial g_{12}}{\Delta_2 x^2} - \frac{\partial g_{21}}{\Delta_2 x^2}\right)$$
$$= \frac{1}{2}\left(2l_1(x^1, x^2)f_{x_1}^{\Delta_1}(x^1, x^2)\right)$$
$$= l_1(x^1, x^2)f_{x_1}^{\Delta_1}(x^1, x^2),$$
$$\Gamma_{22,1} = \frac{1}{2}\left(\frac{\partial g_{21}}{\Delta_2 x^2} + \frac{\partial g_{21}}{\Delta_2 x^2} - \frac{\partial g_{22}}{\Delta_1 x^1}\right)$$
$$= \frac{1}{2}\left(-2l_1(x^1, x^2)f_{x_1}^{\Delta_1}(x^1, x^2)\right)$$
$$= -l_1(x^1, x^2)f_{x_1}^{\Delta_1}(x^1, x^2),$$
$$\Gamma_{22,2} = \frac{1}{2}\left(\frac{\partial g_{22}}{\Delta_2 x^2} + \frac{\partial g_{22}}{\Delta_2 x^2} - \frac{\partial g_{22}}{\Delta_2 x^2}\right)$$
$$= \frac{1}{2}\left(-2l_2(x^1, x^2)f_{x_2}^{\Delta_2}(x^1, x^2)\right)$$
$$= -l_2(x^1, x^2)f_{x_2}^{\Delta_2}(x^1, x^2).$$

Therefore, for the Christoffel symbols of the second kind, we obtain

$$\Gamma_{11}^1 = g^{1m}\Gamma_{11,m}$$
$$= g^{11}\Gamma_{11,1} + g^{12}\Gamma_{11,2}$$
$$= 1 \cdot 0 + 0 \cdot 0$$
$$= 0,$$
$$\Gamma_{12}^1 = g^{1m}\Gamma_{12,m}$$
$$= g^{11}\Gamma_{12,1} + g^{12}\Gamma_{12,2}$$
$$= 1 \cdot 0 + 0 \cdot 0$$
$$= 0,$$
$$\Gamma_{21}^1 = g^{1m}\Gamma_{21,m}$$
$$= g^{11}\Gamma_{21,1} + g^{12}\Gamma_{21,2}$$

$$= 1 \cdot 0 + 0 \cdot \left(l_1(x^1, x^2) f_{x_1}^{\Delta_1}(x^1, x^2) \right)$$
$$= 0,$$
$$\Gamma_{22}^1 = g^{1m}\Gamma_{22,m}$$
$$= g^{11}\Gamma_{22,1} + g^{12}\Gamma_{22,2}$$
$$= 1\left(-l_1(x^1, x^2) f_{x_1}^{\Delta_1}(x^1, x^2)\right) + 0 \cdot \left(l_2(x^1, x^2) f_{x_2}^{\Delta_2}(x^1, x^2) \right)$$
$$= -l_1(x^1, x^2) f_{x_1}^{\Delta_1}(x^1, x^2),$$
$$\Gamma_{11}^2 = g^{2m}\Gamma_{11,m}$$
$$= g^{21}\Gamma_{11,1} + g^{22}\Gamma_{11,2}$$
$$= 0 \cdot 0 + \frac{1}{(f(x^1, x^2))^2} \cdot 0$$
$$= 0,$$
$$\Gamma_{12}^2 = g^{2m}\Gamma_{12,m}$$
$$= g^{21}\Gamma_{12,1} + g^{22}\Gamma_{12,2}$$
$$= 0 \cdot 0 + \frac{1}{(f(x^1, x^2))^2}\left(l_1(x^1, x^2) f_{x_1}^{\Delta_1}(x^1, x^2) \right)$$
$$= \frac{l_1(x^1, x^2) f_{x_1}^{\Delta_1}(x^1, x^2)}{(f(x^1, x^2))^2},$$
$$\Gamma_{21}^2 = g^{2m}\Gamma_{21,m}$$
$$= g^{21}\Gamma_{21,1} + g^{22}\Gamma_{21,2}$$
$$= 0 \cdot 0 + \frac{1}{(f(x^1, x^2))^2}\left(l_1(x^1, x^2) f_{x_1}^{\Delta_1}(x^1, x^2) \right)$$
$$= \frac{l_1(x^1, x^2) f_{x_1}^{\Delta_1}(x^1, x^2)}{(f(x^1, x^2))^2},$$
$$\Gamma_{22}^2 = g^{2m}\Gamma_{22,m}$$
$$= g^{21}\Gamma_{22,1} + g^{22}\Gamma_{22,2}$$
$$= 0 \cdot \left(-l_1(x^1, x^2) f_{x_1}^{\Delta_1}(x^1, x^2)\right) + \frac{1}{(f(x^1, x^2))^2} \cdot \left(-l_2(x^1, x^2) f_{x_2}^{\Delta_2}(x^1, x^2)\right)$$
$$= -\frac{l_2(x^1, x^2) f_{x_2}^{\Delta_2}(x^1, x^2)}{(f(x^1, x^2))^2}.$$

Exercise 4.1. Compute the nonzero Christoffel symbols corresponding to the metric:

1.

$$\Delta s^2 = \left(\Delta_1 x^1\right)^2 + \left(\left(x^1\right)^2 - \left(x^2\right)^2\right)\left(\Delta_2 x^2\right)^2.$$

2.

$$\Delta s^2 = f(x^1, x^2)\left(\Delta_1 x^1\right)^2 + \left(\Delta_2 x^2\right)^2,$$

where f is continuously differentiable with respect to x^1 and x^2.

3.

$$\Delta s^2 = (\Delta_1 x^1)^2 + f(x^1)(\Delta_2 x^2)^2 + (\Delta_3 x^3)^2,$$

where f is continuously differentiable with respect to x^1.

4.

$$\Delta s^2 = (\Delta_1 x^1)^2 + f(x^1, x^2)(\Delta_2 x^2)^2 + (\Delta_3 x^3)^2,$$

where f is continuously differentiable with respect to x^1 and x^2.

5.

$$\Delta s^2 = (\Delta_1 x^1)^2 + f(x^1, x^2, x^3)(\Delta_2 x^2)^2 + (\Delta_3 x^3)^2,$$

where f is continuously differentiable with respect to x^1, x^2, and x^3.

Now we will deduce some important properties of the Christoffel symbols.

1. The Christoffel symbols $\Gamma_{ij,k}$ and Γ_{ij}^k are symmetric with respect to the indices i and j.

Proof. Since g_{ij} are symmetric, we have

$$\frac{\partial g_{ij}}{\Delta_k x^k} = \frac{\partial g_{ji}}{\Delta_k x^k},$$

and then

$$\begin{aligned}
\Gamma_{ji,k} &= \frac{1}{2}\left(\frac{\partial g_{jk}}{\Delta_i x^i} + \frac{\partial g_{ik}}{\Delta_j x^j} - \frac{\partial g_{ji}}{\Delta_k x^k}\right) \\
&= \frac{1}{2}\left(\frac{\partial g_{jk}}{\Delta_i x^i} + \frac{\partial g_{ik}}{\Delta_j x^j} - \frac{\partial g_{ij}}{\Delta_k x^k}\right) \\
&= \Gamma_{ij,k}.
\end{aligned}$$

Hence

$$\begin{aligned}
\Gamma_{ij}^k &= g^{km}\Gamma_{ij,m} \\
&= g^{km}\Gamma_{ji,m} \\
&= \Gamma_{ji}^k.
\end{aligned}$$

This completes the proof. □

2. The necessary and sufficient condition that all the Christoffel symbols vanish at a point is that g_{ij} are constants.

Proof. Let g_{ij} vanish at a point P. Then

$$\frac{\partial g_{ik}}{\Delta_j x^j} = 0,$$

$$\frac{\partial g_{jk}}{\Delta_i x^i} = 0,$$

and

$$\frac{\partial g_{ij}}{\Delta_k x^k} = 0.$$

Hence

$$\Gamma_{ij,k} = \frac{1}{2}\left(\frac{\partial g_{jk}}{\Delta_i x^i} + \frac{\partial g_{ik}}{\Delta_j x^j} - \frac{\partial g_{ij}}{\Delta_k x^k} \right)$$
$$= 0.$$

Now suppose that

$$\Gamma_{ij,k} = 0.$$

Then

$$\frac{1}{2}\left(\frac{\partial g_{jk}}{\Delta_i x^i} + \frac{\partial g_{ik}}{\Delta_j x^j} - \frac{\partial g_{ij}}{\Delta_k x^k} \right) = 0.$$

Because x^i are independent, we find

$$\frac{\partial g_{ik}}{\Delta_j x^j} = 0,$$

$$\frac{\partial g_{jk}}{\Delta_i x^i} = 0,$$

$$\frac{\partial g_{ij}}{\Delta_k x^k} = 0.$$

Thus g_{ij} are constants. This completes the proof. $\qquad\square$

3. We have

$$\Gamma_{ij,m} = g_{km}\Gamma_{ij}^{k}.$$

Proof. By the definition of Γ_{ij}^{k} we have

$$\Gamma^k_{ij} = g^{km}\Gamma_{ij,m}.$$

Multiplying the last equation by g_{kp}, we obtain

$$g_{kp}\Gamma^k_{ij} = g_{kp}g^{km}\Gamma_{ij,m}$$
$$= \delta^m_p \Gamma_{ij,m}$$
$$= \Gamma_{ij,p}.$$

This completes the proof. $\qquad\square$

4. We have

$$\Gamma_{ij,k} + \Gamma_{jk,i} = \frac{\partial g_{ik}}{\Delta_j x^j}.$$

Proof. By the definition of the Christoffel indices we find

$$\Gamma_{ij,k} = \frac{1}{2}\left(\frac{\partial g_{ik}}{\Delta_j x^j} + \frac{\partial g_{jk}}{\Delta_i x^i} - \frac{\partial g_{ij}}{\Delta_k x^k}\right),$$
$$\Gamma_{jk,i} = \frac{1}{2}\left(\frac{\partial g_{ji}}{\Delta_k x^k} + \frac{\partial g_{ki}}{\Delta_j x^j} - \frac{\partial g_{jk}}{\Delta_i x^i}\right)$$
$$= \frac{1}{2}\left(\frac{\partial g_{ij}}{\Delta_k x^k} + \frac{\partial g_{ik}}{\Delta_j x^j} - \frac{\partial g_{jk}}{\Delta_i x^i}\right).$$

Adding both equations, we find

$$\Gamma_{ij,k} + \Gamma_{jk,i} = \frac{\partial g_{ik}}{\Delta_j x^j}.$$

This completes the proof. $\qquad\square$

5. We have

$$\frac{\partial g^{ki}}{\Delta_j x^j} = -g^{pk\sigma_j}\Gamma^i_{jp} - g^{pk\sigma_j}g^{mi}\Gamma_{jm,p}$$
$$= -g^{pk}g^{mi\sigma_j}\Gamma_{jp,m} - g^{mi\sigma_j}\Gamma^k_{jm}.$$

Proof. Note that

$$\delta^i_p = g_{pm}g^{mi}.$$

Then

$$0 = \frac{\partial}{\Delta_j x^j}\left(\delta_p^i\right)$$

$$= \frac{\partial}{\Delta_j x^j}\left(g_{pm}g^{mi}\right)$$

$$= \frac{\partial g_{pm}}{\Delta_j x^j}g^{mi} + g_{pm}^{\sigma_j}\frac{\partial g^{mi}}{\Delta_j x^j},$$

whereupon

$$-\frac{\partial g_{pm}}{\Delta_j x^j}g^{mi} = g_{pm}^{\sigma_j}\frac{\partial g^{mi}}{\Delta_j x^j}.$$

Multiplying the last equation by $g^{pk\sigma_j}$, we find

$$-g^{pk\sigma_j}\frac{\partial g_{pm}}{\Delta_j x^j}g^{mi} = g^{pk\sigma_j}g_{pm}^{\sigma_j}\frac{\partial g^{mi}}{\Delta_j x^j}$$

$$= \delta_m^{k\sigma_j}\frac{\partial g^{mi}}{\Delta_j x^j}$$

$$= \frac{\partial g^{ki}}{\Delta_j x^j},$$

i. e.,

$$\frac{\partial g^{ki}}{\Delta_j x^j} = -g^{pk\sigma_j}\frac{\partial g_{pm}}{\Delta_j x^j}g^{mi}.$$

Now since

$$\frac{\partial g_{pm}}{\Delta_j x^j} = \Gamma_{jp,m} + \Gamma_{jm,p}, \tag{4.1}$$

we find

$$\frac{\partial g^{ki}}{\Delta_j x^j} = -g^{pk\sigma_j}g^{mi}\left(\Gamma_{jp,m} + \Gamma_{jm,p}\right)$$

$$= -g^{pk\sigma_j}g^{mi}\Gamma_{jp,m} - g^{pk\sigma_j}g^{mi}\Gamma_{jm,p}$$

$$= -g^{pk\sigma_j}\Gamma_{jp}^i - g^{pk\sigma_j}g^{mi}\Gamma_{jm,p}.$$

Next,

$$0 = \frac{\partial}{\Delta_j x^j}\left(g_{pm}g^{mi}\right)$$

$$= g^{mi\sigma_j}\frac{\partial g_{pm}}{\Delta_j x^j} + g_{pm}\frac{\partial g^{mi}}{\Delta_j x^j}.$$

Then

$$g_{pm}\frac{\partial g^{mi}}{\Delta_j x^j} = -g^{mi\sigma_j}\frac{\partial g_{pm}}{\Delta_j x^j},$$

which we multiply by g^{pk} and we find

$$-g^{pk}g^{mi\sigma_j}\frac{\partial g_{pm}}{\Delta_j x^j} = g^{pk}g_{pm}\frac{\partial g^{mi}}{\Delta_j x^j}$$

$$= \delta^k_m\frac{\partial g^{mi}}{\Delta_j x^j}$$

$$= \frac{\partial g^{ki}}{\Delta_j x^j}.$$

Now applying (4.1), we obtain

$$\frac{\partial g^{ki}}{\Delta_j x^j} = -g^{pk}g^{mi\sigma_j}(\Gamma_{jp,m} + \Gamma_{jm,p})$$

$$= -g^{pk}g^{mi\sigma_j}\Gamma_{jp,m} - g^{pk}g^{mi\sigma_j}\Gamma_{jm,p}$$

$$= -g^{pk}g^{mi\sigma_j}\Gamma_{jp,m} - g^{mi\sigma_j}\Gamma^k_{jm}. \qquad\square$$

6. Let G^{ij} be the cofactor of g_{ij} in $|g_{ij}|$. For $l \in \{2,3,\dots,N\}$, define the quantities H_{lj}, $j \in \{1,2,\dots,N\}$, as follows:

$$\det\begin{pmatrix} g^{\sigma_j}_{11} & g^{\sigma_j}_{12} & \cdots & g^{\sigma_j}_{1n} \\ g^{\sigma_j}_{21} & g^{\sigma_j}_{22} & \cdots & g^{\sigma_j}_{2n} \\ \vdots & \vdots & \vdots & \vdots \\ g^{\sigma_j}_{l-11} & g^{\sigma_j}_{l-12} & \cdots & g^{\sigma_j}_{l-1n} \\ \frac{\partial g_{l1}}{\Delta_j x^j} & \frac{\partial g_{l2}}{\Delta_j x^j} & \cdots & \frac{\partial g_{ln}}{\Delta_j x^j} \\ g_{l+11} & g_{l=12} & \cdots & g_{l+1n} \\ \vdots & \vdots & \vdots & \vdots \\ g_{n1} & g_{n2} & \cdots & g_{nn} \end{pmatrix} = \det\begin{pmatrix} g_{11} & g_{12} & \cdots & g_{1n} \\ g_{21} & g_{22} & \cdots & g_{2n} \\ \vdots & \vdots & \vdots & \vdots \\ g_{l-11} & g_{l-12} & \cdots & g_{l-1n} \\ \frac{\partial g_{l1}}{\Delta_j x^j} & \frac{\partial g_{l2}}{\Delta_j x^j} & \cdots & \frac{\partial g_{ln}}{\Delta_j x^j} \\ g_{l+11} & g_{l=12} & \cdots & g_{l+1n} \\ \vdots & \vdots & \vdots & \vdots \\ g_{n1} & g_{n2} & \cdots & g_{nn} \end{pmatrix} + \mu_j H_{lj}.$$

For $l = 2$, we have

$$H_{2j} = \det\begin{pmatrix} \frac{\partial g_{11}}{\Delta_j x^j} & \frac{\partial g_{12}}{\Delta_j x^j} & \cdots & \frac{\partial g_{1n}}{\Delta_j x^j} \\ \frac{\partial g_{21}}{\Delta_j x^j} & \frac{\partial g_{22}}{\Delta_j x^j} & \cdots & \frac{\partial g_{n2}}{\Delta_j x^j} \\ g_{31} & g_{32} & \cdots & g_{3n} \\ \vdots & \vdots & \vdots & \vdots \\ g_{n1} & g_{n2} & \cdots & g_{nn} \end{pmatrix}.$$

For $l = 3$, we have

$$H_{3j} = \det \begin{pmatrix} g_{11} & g_{12} & \cdots & g_{1n} \\ \frac{\partial g_{21}}{\Delta_j x^j} & \frac{\partial g_{22}}{\Delta_j x^j} & \cdots & \frac{\partial g_{2n}}{\Delta_j x^j} \\ \frac{\partial g_{31}}{\Delta_j x^j} & \frac{\partial g_{32}}{\Delta_j x^j} & \cdots & \frac{\partial g_{3n}}{\Delta_j x^j} \\ g_{41} & g_{42} & \cdots & g_{4n} \\ \vdots & \vdots & \vdots & \vdots \\ g_{n1} & g_{n2} & \cdots & g_{nn} \end{pmatrix} + \det \begin{pmatrix} \frac{\partial g_{11}}{\Delta_j x^j} & \frac{\partial g_{12}}{\Delta_j x^j} & \cdots & \frac{\partial g_{1n}}{\Delta_j x^j} \\ g_{21} & g_{22} & \cdots & g_{2n} \\ \frac{\partial g_{31}}{\Delta_j x^j} & \frac{\partial g_{32}}{\Delta_j x^j} & \cdots & \frac{\partial g_{3n}}{\Delta_j x^j} \\ g_{41} & g_{42} & \cdots & g_{4n} \\ \vdots & \vdots & \vdots & \vdots \\ g_{n1} & g_{n2} & \cdots & g_{nn} \end{pmatrix}$$

$$+ \mu_j \det \begin{pmatrix} \frac{\partial g_{11}}{\Delta_j x^j} & \frac{\partial g_{12}}{\Delta_j x^j} & \cdots & \frac{\partial g_{1n}}{\Delta_j x^j} \\ \frac{\partial g_{21}}{\Delta_j x^j} & \frac{\partial g_{22}}{\Delta_j x^j} & \cdots & \frac{\partial g_{2n}}{\Delta_j x^j} \\ \frac{\partial g_{31}}{\Delta_j x^j} & \frac{\partial g_{32}}{\Delta_j x^j} & \cdots & \frac{\partial g_{3n}}{\Delta_j x^j} \\ g_{41} & g_{42} & \cdots & g_{4n} \\ \vdots & \vdots & \vdots & \vdots \\ g_{n1} & g_{n2} & \cdots & g_{nn} \end{pmatrix}.$$

We have the following relation:

$$\Gamma_{ij}^i = \frac{1}{2g} \frac{\partial g}{\Delta_j x^j} - \frac{\mu_j}{2g} \sum_{l=2}^{N} H_{lj}.$$

In particular, if $1 \in \mathbb{T}_j$ and $\mathbb{T}_j$ contains at least two positive points,

$$g(\cdot, \cdot, \ldots, \mathbb{T}_j, \cdot, \ldots, \ldots) = \mathbb{T}_{(j)},$$

and

$$g(x^1, \ldots, x^{j-1}, \sigma_j(x^j), x^{j+1}, \ldots, x^n) = \sigma_{(j)}(g(x^1, \ldots, x^{j-1}, x^j, x^{j+1}, \ldots, x^n)),$$

then

$$\Gamma_{ij}^i = \frac{1}{2} \frac{\partial L_{\mathbb{T}_j}(g)}{\Delta_j x^j} - \frac{\mu_j}{2g} \sum_{l=2}^{N} H_{lj},$$

where $L_{\mathbb{T}_j}$ is the natural time scale logarithm defined by

$$L_{\mathbb{T}_j}(t) = \int_{1}^{t^j} \frac{1}{\tau} \Delta\tau, \quad t^j \in \mathbb{T}_j \cap (0, \infty).$$

Proof. Since $g = |g_{ij}|$, we get

$$g = \det \begin{pmatrix} g_{11} & g_{12} & \cdots & g_{1n} \\ g_{21} & g_{22} & \cdots & g_{2n} \\ \vdots & \vdots & \vdots & \vdots \\ g_{n1} & g_{n2} & \cdots & g_{nn} \end{pmatrix}.$$

Hence

$$\frac{\partial g}{\Delta_j x^j} = \det \begin{pmatrix} \frac{\partial g_{11}}{\Delta_j x^j} & \frac{\partial g_{12}}{\Delta_j x^j} & \cdots & \frac{\partial g_{1n}}{\Delta_j x^j} \\ g_{21} & g_{22} & \cdots & g_{2n} \\ g_{31} & g_{32} & \cdots & g_{3n} \\ \vdots & \vdots & \vdots & \vdots \\ g_{n1} & g_{n2} & \cdots & g_{nn} \end{pmatrix} + \det \begin{pmatrix} g_{11} & g_{12} & \cdots & g_{1n} \\ \frac{\partial g_{21}}{\Delta_j x^j} & \frac{\partial g_{22}}{\Delta_j x^j} & \cdots & \frac{\partial g_{2n}}{\Delta_j x^j} \\ g_{31} & g_{32} & \cdots & g_{3n} \\ \vdots & \vdots & \cdots & \vdots \\ g_{n1} & g_{n2} & \cdots & g_{nn} \end{pmatrix}$$

$$+ \cdots + \det \begin{pmatrix} g_{11} & g_{12} & \cdots & g_{1n} \\ g_{21} & g_{22} & \cdots & g_{2n} \\ \vdots & \vdots & \cdots & \vdots \\ g_{n-11} & g_{n-12} & \cdots & g_{n-1n} \\ \frac{\partial g_{n1}}{\Delta_j x^j} & \frac{\partial g_{n2}}{\Delta_j x^j} & \cdots & \frac{\partial g_{nn}}{\Delta_j x^j} \end{pmatrix} + \mu_j \sum_{l=2}^{N} H_{lj}$$

$$= G^{ik} \frac{\partial g_{ik}}{\Delta_j x^j} + \mu_j \sum_{l=2}^{N} H_{lj}$$

$$= g g^{ik} \frac{\partial g^{ik}}{\Delta_j x^j} + \mu_j \sum_{l=2}^{N} H_{lj}.$$

Now using that

$$\frac{\partial g_{ik}}{\Delta_j x^j} = \Gamma_{ij,k} + \Gamma_{kj,i},$$

we get

$$\frac{\partial g}{\Delta_j x^j} = g g^{ik}(\Gamma_{ij,k} + \Gamma_{kj,i}) + \mu_j \sum_{l=2}^{N} H_{lj}$$

$$= g(g^{ik}\Gamma_{ij,k} + g^{ik}\Gamma_{kj,i}) + \mu_j \sum_{l=2}^{N} H_{lj}$$

$$= g(\Gamma_{ij}^{i} + \Gamma_{kj}^{k}) + \mu_j \sum_{l=2}^{N} H_{lj}.$$

Replacing the dummy index k by i, we find

$$\frac{\partial g}{\Delta_j x^j} = g(\Gamma_{ij}^i + \Gamma_{ij}^i) + \mu_j \sum_{l=2}^{N} H_{lj}$$

$$= 2h\Gamma_{ij}^i + \mu_j \sum_{l=2}^{N} H_{lj},$$

whereupon

$$\Gamma_{ij}^i = \frac{1}{2g} \frac{\partial g}{\Delta_j x^j} - \frac{\mu_j}{2g} \sum_{l=2}^{N} H_{lj}.$$

In particular, if $1 \in \mathbb{T}_j$ and $\mathbb{T}_j$ contains at least two positive points,

$$g(\cdot, \cdot, \ldots, \mathbb{T}_j, \cdot, \ldots, \ldots) = \mathbb{T}_{(j)},$$

and

$$g(x^1, \ldots, x^{j-1}, \sigma_j(x^j), x^{j+1}, \ldots, x^n) = \sigma_{(j)}(g(x^1, \ldots, x^{j-1}, x^j, x^{j+1}, \ldots, x^n)),$$

then

$$\Gamma_{ij}^i = \frac{1}{2} \frac{\partial L_{\mathbb{T}_j}(g)}{\Delta_j x^j} - \frac{\mu_j}{2g} \sum_{l=2}^{N} H_{lj}.$$

This completes the proof. $\qquad\qquad\square$

Example 4.3. We will prove the relations

$$g_{\alpha\beta}^{\sigma_j} \frac{\partial \Gamma_{ik}^\beta}{\Delta_j x^j} = \frac{\partial \Gamma_{ik,\alpha}}{\Delta_j x^j} - \Gamma_{ik}^\beta (\Gamma_{\alpha j,\beta} + \Gamma_{\beta j,\alpha})$$

and

$$g_{\alpha\beta} \frac{\partial \Gamma_{ik}^\beta}{\Delta_j x^j} = \frac{\partial \Gamma_{ik,\alpha}}{\Delta_j x^j} - \Gamma_{ik}^{\beta\sigma_j} (\Gamma_{\alpha j,\beta} + \Gamma_{\beta j,\alpha}).$$

By the definition of the Christoffel symbols of the second kind we get

$$\Gamma_{ik}^\beta = g^{\alpha\beta} \Gamma_{ik,\alpha}.$$

Multiplying both sides of the last equation by $g_{\alpha\beta}$, we get

$$g_{\alpha\beta} \Gamma_{ik}^\beta = g_{\alpha\beta} g^{\alpha\beta} \Gamma_{ik,\alpha}$$
$$= \Gamma_{ik,\alpha}.$$

Now differentiating both sides of the last equation with respect to x^j, we find

$$\frac{\partial \Gamma_{ik,\alpha}}{\Delta_j x^j} = \frac{\partial}{\Delta_j x^j}\left(g_{\alpha\beta}\Gamma_{ik}^{\beta}\right)$$

$$= \frac{\partial g_{\alpha\beta}}{\Delta_j x^j}\Gamma_{ik}^{\beta} + g_{\alpha\beta}^{\sigma_j}\frac{\partial \Gamma_{ik}^{\beta}}{\Delta_j x^j}.$$

(4.2)

Now using that

$$\frac{\partial g_{\alpha\beta}}{\Delta_j x^j} = \Gamma_{\alpha j,\beta} + \Gamma_{\beta j,\alpha},$$

(4.3)

we find

$$\frac{\partial \Gamma_{ik,\alpha}}{\Delta_j x^j} = \Gamma_{ik}^{\beta}(\Gamma_{\alpha j,\beta} + \Gamma_{\beta j,\alpha}) + g_{\alpha\beta}^{\sigma_j}\frac{\partial \Gamma_{ik}^{\beta}}{\Delta_j x^j},$$

whereupon

$$\Gamma_{\beta j,\alpha}) + g_{\alpha\beta}^{\sigma_j}\frac{\partial \Gamma_{ik}^{\beta}}{\Delta_j x^j} = \frac{\partial \Gamma_{ik,\alpha}}{\Delta_j x^j} - \Gamma_{ik}^{\beta}(\Gamma_{\alpha j,\beta} + \Gamma_{\beta j,\alpha}).$$

Now applying (4.2) and (4.3), we arrive at

$$\frac{\partial \Gamma_{ik,\alpha}}{\Delta_j x^j} = \frac{\partial g_{\alpha\beta}}{\Delta_j x^j}\Gamma_{ik}^{\beta\sigma_j} + g_{\alpha\beta}\frac{\partial \Gamma_{ik}^{\beta}}{\Delta_j x^j}$$

$$= \Gamma_{ik}^{\beta\sigma_j}(\Gamma_{\alpha j,\beta} + \Gamma_{\beta j,\alpha}) + g_{\alpha\beta}\frac{\partial \Gamma_{ik}^{\beta}}{\Delta_j x^j},$$

from which we get

$$g_{\alpha\beta}\frac{\partial \Gamma_{ik}^{\beta}}{\Delta_j x^j} = \frac{\partial \Gamma_{ik,\alpha}}{\Delta_j x^j} - \Gamma_{ik}^{\beta\sigma_j}(\Gamma_{\alpha j,\beta} + \Gamma_{\beta j,\alpha}).$$

Example 4.4. We will evaluate the Christoffel symbols of both kinds for spaces where

$$g_{ij} = 0, \quad i \neq j.$$

We have the following cases.
1. $i = j = k$. Then

$$\Gamma_{ij,k} = \Gamma_{ii,i}$$

$$= \frac{1}{2}\left(\frac{\partial g_{ii}}{\Delta_i x^i} + \frac{\partial g_{ii}}{\Delta_i x^i} - \frac{\partial g_{ii}}{\Delta_i x^i}\right)$$

$$= \frac{1}{2}\frac{\partial g_{ii}}{\Delta_i x^i}.$$

2. $i = k \neq j$. Then

$$
\begin{aligned}
\Gamma_{ij,k} &= \Gamma_{ij,i} \\
&= \frac{1}{2}\left(\frac{\partial g_{ii}}{\Delta_j x^j} + \frac{\partial g_{ij}}{\Delta_i x^i} - \frac{\partial g_{ij}}{\Delta_i x^i}\right) \\
&= \frac{1}{2}\frac{\partial g_{ii}}{\Delta_j x^j}.
\end{aligned}
$$

3. $i = j \neq k$. Then

$$
\begin{aligned}
\Gamma_{ij,k} &= \Gamma_{ii,k} \\
&= \frac{1}{2}\left(\frac{\partial g_{ik}}{\Delta_i x^i} + \frac{\partial g_{ik}}{\Delta_i x^i} - \frac{\partial g_{ii}}{\Delta_k x^k}\right) \\
&= -\frac{1}{2}\frac{\partial g_{ii}}{\Delta_k x^k}.
\end{aligned}
$$

4. $i \neq j \neq k$. Then

$$
\begin{aligned}
\Gamma_{ij,k} &= \frac{1}{2}\left(\frac{\partial g_{ik}}{\Delta_j x^j} + \frac{\partial g_{jk}}{\Delta_i x^i} - \frac{\partial g_{ij}}{\Delta_k x^k}\right) \\
&= 0.
\end{aligned}
$$

Note that if $k \neq h$, then

$$
\begin{aligned}
\Gamma_{ij}^k &= g^{kh}\Gamma_{ij,h} \\
&= 0,
\end{aligned}
$$

and if $k == h$, then

$$
\begin{aligned}
\Gamma_{ij}^k &= g^{kk}\Gamma_{ij,k} \\
&= \frac{1}{g_{kk}}\Gamma_{ij,k}.
\end{aligned}
$$

Using the above cases for the Christoffel symbols of the first kind, we get the following cases for the Christoffel symbols of the second kind.

1. $i = k \neq j$. Then

$$
\begin{aligned}
\Gamma_{ij}^k &= \Gamma_{ij}^i \\
&= \frac{1}{g_{ii}}\Gamma_{ij,i} \\
&= \frac{1}{2g_{ii}}\frac{\partial g_{ii}}{\Delta_j x^j}.
\end{aligned}
$$

If

$$g_{ii}(\cdot,\,\cdot,\ldots,\cdot,\mathbb{T}_j,\cdot,\ldots,\cdot) = \mathbb{T}_{(j)}$$

and

$$g_{ii}(x^1,\ldots,x^{j-1},\sigma_j(x^j),x^{j+1},\ldots,x^N) = \sigma_{(j)}(g_{ii}(x^1,x^2,\ldots,x^N)),$$

then

$$\Gamma^i_{ij} = \frac{1}{2}\frac{\partial L_{\mathbb{T}_i}(g_{ii})}{\Delta_j x^j} - \frac{\mu_j}{g}\sum_{l=2}^N H_{li}.$$

2. $i = j = k$. Then

$$\begin{aligned}
\Gamma^k_{ij} &= \Gamma^i_{ii} \\
&= \frac{1}{g_{ii}}\Gamma_{ii,i} \\
&= \frac{1}{2g_{ii}}\frac{\partial g_{ii}}{\Delta_i x^i}.
\end{aligned}$$

If

$$g_{ii}(\cdot,\,\cdot,\ldots,\cdot,\mathbb{T}_i,\cdot,\ldots,\cdot) = \mathbb{T}_{(i)}$$

and

$$g_{ii}(x^1,\ldots,x^{i-1},\sigma_i(x^i),x^{i+1},\ldots,x^N) = \sigma_{(i)}(g_{ii}(x^1,x^2,\ldots,x^N)),$$

then

$$\Gamma^i_{ii} = \frac{1}{2}\frac{\partial L_{\mathbb{T}_i}(g_{ii})}{\Delta_i x^i} - \frac{\mu_i}{g}\sum_{l=2}^N H_{lj}.$$

3. $i = j \neq k$. Then

$$\begin{aligned}
\Gamma^k_{ij} &= \Gamma^k_{ii} \\
&= \frac{1}{g_{kk}}\Gamma_{ii,k} \\
&= -\frac{1}{2g_{kk}}\frac{\partial g_{ii}}{\Delta_k x^k}.
\end{aligned}$$

4. $i \neq j \neq k$. Then

$$\begin{aligned}
\Gamma^k_{ij} &= \frac{1}{g_{kk}}\Gamma_{ij,k} \\
&= 0.
\end{aligned}$$

Example 4.5. Let a^{ij} be the components of a symmetric tensor. We will show that

$$a^{jk}\Gamma_{ij}^k = \frac{1}{2}a^{jk}\frac{\partial g_{jk}}{\Delta_i x^i},$$

where g_{jk} have their usual meaning.

Indeed, using that

$$\frac{\partial g_{jk}}{\Delta_i x^i} = \Gamma_{ji,k} + \Gamma_{ki,j},$$

we get

$$\frac{1}{2}a^{jk}\frac{\partial g_{jk}}{\Delta_i x^i} = \frac{1}{2}a^{jk}(\Gamma_{ji,k} + \Gamma_{ki,j})$$
$$= \frac{1}{2}(a^{jk}\Gamma_{ji,k} + a^{jk}\Gamma_{ki,j}).$$

Now interchanging the dummy indices k and j in the second term of the last equality, we find

$$\frac{1}{2}a^{jk}\frac{\partial g_{jk}}{\Delta_i x^i} = \frac{1}{2}(a^{jk}\Gamma_{ji,k} + a^{jk}\Gamma_{ji,k})$$
$$= \frac{1}{2}(2a^{jk}\Gamma_{ji,k})$$
$$= a^{jk}\Gamma_{ji,k}.$$

Example 4.6. Let A^{ij} be a skew-symmetric tensor. We will prove that

$$A^{jk}\Gamma_{jk,i} = 0.$$

Since A^{jk} is a skew-symmetric tensor, we have

$$A^{jk} = -A^{kj}.$$

Now interchanging the dummy indices j and k, we find

$$A^{jk}\Gamma_{jk,i} = A^{kj}\Gamma_{kj,i}$$
$$= -A^{jk}\Gamma_{kj,i}$$
$$= -A^{jk}\Gamma_{jk,i},$$

whereupon

$$2A^{jk}\Gamma_{jk,i} = 0,$$

or

$$A^{jk}\Gamma_{jk,i} = 0.$$

Exercise 4.2. Let A^{pqr} be a skew-symmetric tensor. Prove that

$$A^{pqr}\Gamma^l_{pq} = 0. \tag{4.4}$$

Solution. Since A^{pqr} is a skew-symmetric tensor, we have

$$A^{pqr} = -A^{qpr}.$$

Now interchanging the dummy indices p and q in (4.4), we find

$$\begin{aligned}
A^{pqr}\Gamma^l_{pq} &= A^{qpr}\Gamma^l_{qp}\\
&= -A^{pqr}\Gamma^l_{qp}\\
&= -A^{pqr}\Gamma^l_{pq}.
\end{aligned}$$

Hence

$$2A^{pqr}\Gamma^l_{pq} = 0,$$

or

$$A^{pqr}\Gamma^l_{pq} = 0.$$

Definition 4.3. Define the σ-Christoffel symbols of the first kind as

$$\Gamma_{\sigma ij,k} = \frac{1}{2}\left(\left(\frac{\partial g_{jk}}{\Delta_i x^i}\right)^{\sigma_0\sigma_1\cdots\sigma_{i-1}} + \left(\frac{\partial g_{ik}}{\Delta_j x^j}\right)^{\sigma_0\sigma_1\cdots\sigma_{j-1}} - \left(\frac{\partial g_{ij}}{\Delta_k x^k}\right)^{\sigma_0\sigma_1\cdots\sigma_{k-1}}\right).$$

The σ-Christoffel symbols of the second kind are defined as

$$\Gamma^k_{\sigma ij} = g^{kl}\Gamma_{\sigma ij,l}.$$

Some of the properties of the σ-Christoffel symbols of the first and second kinds are as follows.

1. The σ-Christoffel symbols $\Gamma_{\sigma ij,k}$ and $\Gamma^k_{\sigma ij}$ are symmetric with respect to the indices i and j.

Proof. Since g_{ij} are symmetric, we have

$$\left(\frac{\partial g_{ij}}{\Delta_k x^k}\right)^{\sigma_0\sigma_1\cdots\sigma_{k-1}} - \left(\frac{\partial g_{ji}}{\Delta_k x^k}\right)^{\sigma_0\sigma_1\cdots\sigma_{k-1}},$$

and then

$$\Gamma_{\sigma ji,k} = \frac{1}{2}\left(\left(\frac{\partial g_{jk}}{\Delta_i x^i}\right)^{\sigma_0 \sigma_1 \cdots \sigma_i} + \left(\frac{\partial g_{ik}}{\Delta_j x^j}\right)^{\sigma_0 \sigma_1 \cdots \sigma_{j-1}} - \left(\frac{\partial g_{ji}}{\Delta_k x^k}\right)^{\sigma_0 \sigma_1 \cdots \sigma_{k-1}}\right)$$

$$= \frac{1}{2}\left(\left(\frac{\partial g_{jk}}{\Delta_i x^i}\right)^{\sigma_0 \sigma_1 \cdots \sigma_{i-1}} + \left(\frac{\partial g_{ik}}{\Delta_j x^j}\right)^{\sigma_0 \sigma_1 \cdots \sigma_{j-1}} - \left(\frac{\partial g_{ij}}{\Delta_k x^k}\right)^{\sigma_0 \sigma_1 \cdots \sigma_{k-1}}\right)$$

$$= \Gamma_{\sigma ij,k}.$$

Therefore

$$\Gamma_{\sigma ij}^k = g^{km}\Gamma_{\sigma ij,m}$$

$$= g^{km}\Gamma_{\sigma ji,m}$$

$$= \Gamma_{\sigma ji}^k.$$

This completes the proof. $\qquad\square$

2. The σ-Christoffel symbols vanish at a point if g_{ij} are constants.

Proof. Let g_{ij} vanish at a point P. Then

$$\frac{\partial g_{ik}}{\Delta_j x^j} = 0,$$

$$\frac{\partial g_{jk}}{\Delta_i x^i} = 0,$$

and

$$\frac{\partial g_{ij}}{\Delta_k x^k} = 0.$$

Hence

$$\left(\frac{\partial g_{ik}}{\Delta_j x^j}\right)^{\sigma_0 \sigma_1 \cdots \sigma_{j-1}} = 0,$$

$$\left(\frac{\partial g_{jk}}{\Delta_i x^i}\right)^{\sigma_0 \sigma_1 \cdots \sigma_{i-1}} = 0,$$

and

$$\left(\frac{\partial g_{ij}}{\Delta_k x^k}\right)^{\sigma_0 \sigma_1 \cdots \sigma_{k-1}} = 0,$$

and thus

$$\Gamma_{\sigma ij,k} = \frac{1}{2}\left(\left(\frac{\partial g_{jk}}{\Delta_i x^i}\right)^{\sigma_0\sigma_1\cdots\sigma_{i-1}} + \left(\frac{\partial g_{ik}}{\Delta_j x^j}\right)^{\sigma_0\sigma_1\cdots\sigma_{j-1}} - \left(\frac{\partial g_{ij}}{\Delta_k x^k}\right)^{\sigma_0\sigma_1\cdots\sigma_{k-1}}\right)$$
$$= 0.$$

This completes the proof. $\qquad\square$

3. We have

$$\Gamma_{\sigma ij,m} = g_{km}\Gamma^k_{\sigma ij}.$$

Proof. By the definition of $\Gamma^k_{\sigma ij}$ we have

$$\Gamma^k_{\sigma ij} = g^{km}\Gamma_{\sigma ij,m}.$$

Multiplying the last equation by g_{kp}, we obtain

$$\begin{aligned}
g_{kp}\Gamma^k_{\sigma ij} &= g_{kp}g^{km}\Gamma_{\sigma ij,m} \\
&= \delta^m_p \Gamma_{\sigma ij,m} \\
&= \Gamma_{\sigma ij,p}.
\end{aligned}$$

This completes the proof. $\qquad\square$

4. We have

$$\Gamma_{\sigma ij,k} + \Gamma_{\sigma jk,i} = \left(\frac{\partial g_{ik}}{\Delta_j x^j}\right)^{\sigma_0\sigma_1\cdots\sigma_{j-1}}.$$

Proof. By the definition of the σ-Christoffel indices we find

$$\Gamma_{\sigma ij,k} = \frac{1}{2}\left(\left(\frac{\partial g_{ik}}{\Delta_j x^j}\right)^{\sigma_0\sigma_1\cdots\sigma_{j-1}} + \left(\frac{\partial g_{jk}}{\Delta_i x^i}\right)^{\sigma_0\sigma_1\cdots\sigma_{i-1}} - \left(\frac{\partial g_{ij}}{\Delta_k x^k}\right)^{\sigma_0\sigma_1\cdots\sigma_{k-1}}\right),$$

$$\begin{aligned}
\Gamma_{\sigma jk,i} &= \frac{1}{2}\left(\left(\frac{\partial g_{ji}}{\Delta_k x^k}\right)^{\sigma_0\sigma_1\cdots\sigma_{k-1}} + \left(\frac{\partial g_{ki}}{\Delta_j x^j}\right)^{\sigma_0\sigma_1\cdots\sigma_{j-1}} - \left(\frac{\partial g_{jk}}{\Delta_i x^i}\right)^{\sigma_0\sigma_1\cdots\sigma_{i-1}}\right) \\
&= \frac{1}{2}\left(\left(\frac{\partial g_{ij}}{\Delta_k x^k}\right)^{\sigma_0\sigma_1\cdots\sigma_{k-1}} + \left(\frac{\partial g_{ik}}{\Delta_j x^j}\right)^{\sigma_0\sigma_1\cdots\sigma_{j-1}} - \left(\frac{\partial g_{jk}}{\Delta_i x^i}\right)^{\sigma_0\sigma_1\cdots\sigma_{i-1}}\right).
\end{aligned}$$

Add both equations, we find

$$\Gamma_{\sigma ij,k} + \Gamma_{\sigma jk,i} = \left(\frac{\partial g_{ik}}{\Delta_j x^j}\right)^{\sigma_0\sigma_1\cdots\sigma_{j-1}}.$$

This completes the proof. $\qquad\square$

Now we will deduct the transformation laws for the Christoffel symbols of the first and second kinds. Consider the transformation law for the covariant tensor g_{ij}:

$$\overline{g}_{lm} = \frac{\Delta_i x^i}{\Delta_{(l)} \overline{x}^l} \frac{\Delta_j x^j}{\Delta_{(m)} \overline{x}^m} g_{ij}.$$

Differentiating with respect to $\overline{x}^n$, we get

$$\frac{\partial \overline{g}_{lm}}{\Delta_{(n)} \overline{x}^n} = \frac{\Delta_i x^i}{\Delta_{(l)} \overline{x}^l} \frac{\Delta_j x^j}{\Delta_{(m)} \overline{x}^m} \left(\frac{\partial g_{ij}}{\Delta_k x^k} \right)^{\sigma_0 \sigma_1 \cdots \sigma_{k-1}} \frac{\Delta_k x^k}{\Delta_{(n)} \overline{x}^n} + \frac{\Delta_i^2 x^i}{\Delta_{(n)} \overline{x}^n \Delta_{(l)} \overline{x}^l} \frac{\Delta_j x^j}{\Delta_{(m)} \overline{x}^m} (g_{ij})^{\sigma_{(m)}}$$

$$+ \left(\frac{\Delta_i x^i}{\Delta_{(l)} \overline{x}^l} \right)^{\sigma_{(n)}} \frac{\Delta_j^2 x^j}{\Delta_{(n)} \overline{x}^n \Delta_{(m)} \overline{x}^m} (g_{ij})^{\sigma_{(n)}}.$$

Next, we have

$$\overline{g}_{mn} = \frac{\Delta_i x^i}{\Delta_{(m)} \overline{x}^m} \frac{\Delta_j x^j}{\Delta_{(n)} \overline{x}^n} g_{ij}.$$

Differentiating this with respect to $\overline{x}^l$, we find

$$\frac{\partial \overline{g}_{mn}}{\Delta_{(l)} \overline{x}^l} = \frac{\Delta_i x^i}{\Delta_{(m)} \overline{x}^m} \frac{\Delta_j x^j}{\Delta_{(n)} \overline{x}^n} \left(\frac{\partial g_{ij}}{\Delta_k x^k} \right)^{\sigma_0 \sigma_1 \cdots \sigma_{k-1}} \frac{\Delta_k x^k}{\Delta_{(l)} \overline{x}^l} + \frac{\Delta_i^2 x^i}{\Delta_{(l)} \overline{x}^l \Delta_{(m)} \overline{x}^m} \frac{\Delta_j x^j}{\Delta_{(n)} \overline{x}^n} (g_{ij})^{\sigma_{(l)}}$$

$$+ \left(\frac{\Delta_i x^i}{\Delta_{(m)} \overline{x}^m} \right)^{\sigma_{(l)}} \frac{\Delta_j^2 x^j}{\Delta_{(l)} \overline{x}^l \Delta_{(n)} \overline{x}^n} (g_{ij})^{\sigma_{(l)}}.$$

Now replacing the dummy indices i, j, and k by j, k, and i, respectively, we arrive at

$$\frac{\partial \overline{g}_{mn}}{\Delta_{(l)} \overline{x}^l} = \frac{\Delta_j x^j}{\Delta_{(m)} \overline{x}^m} \frac{\Delta_k x^k}{\Delta_{(n)} \overline{x}^n} \left(\frac{\partial g_{jk}}{\Delta_i x^i} \right)^{\sigma_0 \sigma_1 \cdots \sigma_{i-1}} \frac{\Delta_i x^i}{\Delta_{(l)} \overline{x}^l} + \frac{\Delta_j^2 x^j}{\Delta_{(l)} \overline{x}^l \Delta_{(m)} \overline{x}^m} \frac{\Delta_k x^k}{\Delta_{(n)} \overline{x}^n} (g_{jk})^{\sigma_{(l)}}$$

$$+ \left(\frac{\Delta_j x^j}{\Delta_{(m)} \overline{x}^m} \right)^{\sigma_{(l)}} \frac{\Delta_k^2 x^k}{\Delta_{(l)} \overline{x}^l \Delta_{(n)} \overline{x}^n} (g_{jk})^{\sigma_{(l)}}.$$

Next, we have

$$\overline{g}_{ln} = \frac{\Delta_i x^i}{\Delta_{(l)} \overline{x}^l} \frac{\Delta_j x^j}{\Delta_{(n)} \overline{x}^n} g_{ij}.$$

Differentiating this respect to $\overline{x}^m$, we obtain

$$\frac{\partial \overline{g}_{ln}}{\Delta_{(m)}\overline{x}^m} = \frac{\Delta_i x^i}{\Delta_{(l)}\overline{x}^l}\frac{\Delta_j x^j}{\Delta_{(n)}\overline{x}^n}\left(\frac{\partial g_{ij}}{\Delta_k x^k}\right)^{\sigma_0\sigma_1\cdots\sigma_{k-1}}\frac{\Delta_k x^k}{\Delta_{(m)}\overline{x}^m} + \frac{\Delta_i^2 x^i}{\Delta_{(m)}\overline{x}^m \Delta_{(l)}\overline{x}^l}\frac{\Delta_j x^j}{\Delta_{(n)}\overline{x}^n}\frac{\Delta_j x^j}{\Delta_{(n)}\overline{x}^n}(g_{ij})^{\sigma_{(m)}}$$

$$+ \left(\frac{\Delta_i x^i}{\Delta_{(l)}\overline{x}^l}\right)^{\sigma_{(m)}}\frac{\Delta_j^2 x^j}{\Delta_{(m)}\overline{x}^m \Delta_{(n)}\overline{x}^n}(g_{ij})^{\sigma_{(m)}}.$$

Interchanging the dummy indices j and k, we get

$$\frac{\partial \overline{g}_{ln}}{\Delta_{(m)}\overline{x}^m} = \frac{\Delta_i x^i}{\Delta_{(l)}\overline{x}^l}\frac{\Delta_k x^k}{\Delta_{(n)}\overline{x}^n}\left(\frac{\partial g_{ik}}{\Delta_j x^j}\right)^{\sigma_0\sigma_1\cdots\sigma_{j-1}}\frac{\Delta_j x^j}{\Delta_{(m)}\overline{x}^m} + \frac{\Delta_i^2 x^i}{\Delta_{(m)}\overline{x}^m \Delta_{(l)}\overline{x}^l}\frac{\Delta_k x^k}{\Delta_{(n)}\overline{x}^n}\frac{\Delta_k x^k}{\Delta_{(n)}\overline{x}^n}(g_{ik})^{\sigma_{(m)}}$$

$$+ \left(\frac{\Delta_i x^i}{\Delta_{(l)}\overline{x}^l}\right)^{\sigma_{(m)}}\frac{\Delta_k^2 x^k}{\Delta_{(m)}\overline{x}^m \Delta_{(n)}\overline{x}^n}(g_{ik})^{\sigma_{(m)}}.$$

Hence

$$\frac{\partial \overline{g}_{mn}}{\Delta_{(l)}\overline{x}^l} + \frac{\partial \overline{g}_{nl}}{\Delta_{(m)}\overline{x}^m} - \frac{\partial \overline{g}_{lm}}{\Delta_{(n)}\overline{x}^n}$$

$$= \frac{\Delta_i x^i}{\Delta_{(l)}\overline{x}^l}\frac{\Delta_j x^j}{\Delta_{(m)}\overline{x}^m}\left(\frac{\partial g_{ij}}{\Delta_k x^k}\right)^{\sigma_0\sigma_1\cdots\sigma_{k-1}}\frac{\Delta_k x^k}{\Delta_{(n)}\overline{x}^n} + \frac{\Delta_i^2 x^i}{\Delta_{(n)}\overline{x}^n \Delta_{(l)}\overline{x}^l}\frac{\Delta_j x^j}{\Delta_{(m)}\overline{x}^m}(g_{ij})^{\sigma_{(m)}}$$

$$+ \left(\frac{\Delta_i x^i}{\Delta_{(l)}\overline{x}^l}\right)^{\sigma_{(n)}}\frac{\Delta_j^2 x^j}{\Delta_{(n)}\overline{x}^n \Delta_{(m)}\overline{x}^m}(g_{ij})^{\sigma_{(n)}}$$

$$+ \frac{\Delta_j x^j}{\Delta_{(m)}\overline{x}^m}\frac{\Delta_k x^k}{\Delta_{(n)}\overline{x}^n}\left(\frac{\partial g_{jk}}{\Delta_i x^i}\right)^{\sigma_0\sigma_1\cdots\sigma_{i-1}}\frac{\Delta_i x^i}{\Delta_{(l)}\overline{x}^l} + \frac{\Delta_j^2 x^j}{\Delta_{(l)}\overline{x}^l \Delta_{(m)}\overline{x}^m}\frac{\Delta_k x^k}{\Delta_{(n)}\overline{x}^n}(g_{jk})^{\sigma_{(l)}}$$

$$+ \left(\frac{\Delta_j x^j}{\Delta_{(m)}\overline{x}^m}\right)^{\sigma_{(l)}}\frac{\Delta_k^2 x^k}{\Delta_{(l)}\overline{x}^l \Delta_{(n)}\overline{x}^n}(g_{jk})^{\sigma_{(l)}}$$

$$- \frac{\Delta_i x^i}{\Delta_{(l)}\overline{x}^l}\frac{\Delta_k x^k}{\Delta_{(n)}\overline{x}^n}\left(\frac{\partial g_{ik}}{\Delta_j x^j}\right)^{\sigma_0\sigma_1\cdots\sigma_{j-1}}\frac{\Delta_j x^j}{\Delta_{(m)}\overline{x}^m} - \frac{\Delta_i^2 x^i}{\Delta_{(m)}\overline{x}^m \Delta_{(l)}\overline{x}^l}\frac{\Delta_k x^k}{\Delta_{(n)}\overline{x}^n}\frac{\Delta_k x^k}{\Delta_{(n)}\overline{x}^n}(g_{ik})^{\sigma_{(m)}}$$

$$- \left(\frac{\Delta_i x^i}{\Delta_{(l)}\overline{x}^l}\right)^{\sigma_{(m)}}\frac{\Delta_k^2 x^k}{\Delta_{(m)}\overline{x}^m \Delta_{(n)}\overline{x}^n}(g_{ik})^{\sigma_{(m)}}$$

$$= \left(\left(\frac{\partial g_{jk}}{\Delta_i x^i}\right)^{\sigma_0\sigma_1\cdots\sigma_{i-1}} + \left(\frac{\partial g_{ik}}{\Delta_j x^j}\right)^{\sigma_0\sigma_1\cdots\sigma_{j-1}} - \left(\frac{\partial g_{ij}}{\Delta_k x^k}\right)^{\sigma_0\sigma_1\cdots\sigma_{k-1}}\right)\frac{\Delta_i x^i}{\Delta_{(l)}\overline{x}^l}\frac{\Delta_j x^j}{\Delta_{(m)}\overline{x}^m}\frac{\Delta_k x^k}{\Delta_{(n)}\overline{x}^n}$$

$$+ \left(\frac{\Delta_j^2 x^j}{\Delta_{(l)}\overline{x}^l \Delta_{(m)}\overline{x}^m}\frac{\Delta_k x^k}{\Delta_{(n)}\overline{x}^n} + \left(\frac{\Delta_j x^j}{\Delta_{(m)}\overline{x}^m}\right)^{\sigma_{(l)}}\frac{\Delta_k^2 x^k}{\Delta_{(l)}\overline{x}^l \Delta_{(n)}\overline{x}^n}\right)(g_{jk})^{\sigma_{(l)}}$$

$$+ \left(\frac{\Delta_i^2 x^i}{\Delta_{(m)}\overline{x}^m \Delta_{(l)}\overline{x}^l}\frac{\Delta_k x^k}{\Delta_{(n)}\overline{x}^n} + \left(\frac{\Delta_i x^i}{\Delta_{(l)}\overline{x}^l}\right)^{\sigma_{(m)}}\frac{\Delta_k^2 x^k}{\Delta_{(m)}\overline{x}^m \Delta_{(n)}\overline{x}^n}\right)(g_{ik})^{\sigma_{(m)}}$$

$$- \left(\frac{\Delta_i^2 x^i}{\Delta_{(n)}\overline{x}^n \Delta_{(l)}\overline{x}^l}\frac{\Delta_j x^j}{\Delta_{(m)}\overline{x}^m} + \left(\frac{\Delta_i x^i}{\Delta_{(l)}\overline{x}^l}\right)^{\sigma_{(n)}}\frac{\Delta_j^2 x^j}{\Delta_{(n)}\overline{x}^n \Delta_{(m)}\overline{x}^m}\right)(g_{ij})^{\sigma_{(n)}}.$$

Replacing k by i and k by j in the second and third terms in the last equality, we arrive at

$$
\frac{\partial \overline{g}_{mn}}{\Delta_{(l)}\overline{x}^l} + \frac{\partial \overline{g}_{nl}}{\Delta_{(m)}\overline{x}^m} - \frac{\partial \overline{g}_{lm}}{\Delta_{(n)}\overline{x}^n}
$$

$$
= \left(\left(\frac{\partial g_{jk}}{\Delta_i x^i}\right)^{\sigma_0\sigma_1\cdots\sigma_{i-1}} + \left(\frac{\partial g_{ik}}{\Delta_j x^j}\right)^{\sigma_0\sigma_1\cdots\sigma_{j-1}} - \left(\frac{\partial g_{ij}}{\Delta_k x^k}\right)^{\sigma_0\sigma_1\cdots\sigma_{k-1}}\right)\frac{\Delta_i x^i}{\Delta_{(l)}\overline{x}^l}\frac{\Delta_j x^j}{\Delta_{(m)}\overline{x}^m}\frac{\Delta_k x^k}{\Delta_{(n)}\overline{x}^n}
$$

$$
+ \left(\frac{\Delta_j^2 x^j}{\Delta_{(l)}\overline{x}^l\Delta_{(m)}\overline{x}^m}\frac{\Delta_i x^i}{\Delta_{(n)}\overline{x}^n} + \left(\frac{\Delta_j x^j}{\Delta_{(m)}\overline{x}^m}\right)^{\sigma_{(l)}}\frac{\Delta_i^2 x^i}{\Delta_{(l)}\overline{x}^l\Delta_{(n)}\overline{x}^n}\right)(g_{ji})^{\sigma_{(l)}} \cdot
$$

$$
+ \left(\frac{\Delta_i^2 x^i}{\Delta_{(m)}\overline{x}^m\Delta_{(l)}\overline{x}^l}\frac{\Delta_j x^j}{\Delta_{(n)}\overline{x}^n} + \left(\frac{\Delta_i x^i}{\Delta_{(l)}\overline{x}^l}\right)^{\sigma_{(m)}}\frac{\Delta_j^2 x^j}{\Delta_{(m)}\overline{x}^m\Delta_{(n)}\overline{x}^n}\right)(g_{ij})^{\sigma_{(m)}}
$$

$$
- \left(\frac{\Delta_i^2 x^i}{\Delta_{(n)}\overline{x}^n\Delta_{(l)}\overline{x}^l}\frac{\Delta_j x^j}{\Delta_{(m)}\overline{x}^m} + \left(\frac{\Delta_i x^i}{\Delta_{(l)}\overline{x}^l}\right)^{\sigma_{(n)}}\frac{\Delta_j^2 x^j}{\Delta_{(n)}\overline{x}^n\Delta_{(m)}\overline{x}^m}\right)(g_{ij})^{\sigma_{(n)}}
$$

$$
= \left(\left(\frac{\partial g_{jk}}{\Delta_i x^i}\right)^{\sigma_0\sigma_1\cdots\sigma_{i-1}} + \left(\frac{\partial g_{ik}}{\Delta_j x^j}\right)^{\sigma_0\sigma_1\cdots\sigma_{j-1}} - \left(\frac{\partial g_{ij}}{\Delta_k x^k}\right)^{\sigma_0\sigma_1\cdots\sigma_{k-1}}\right)\frac{\Delta_i x^i}{\Delta_{(l)}\overline{x}^l}\frac{\Delta_j x^j}{\Delta_{(m)}\overline{x}^m}\frac{\Delta_k x^k}{\Delta_{(n)}\overline{x}^n}
$$

$$
+ \frac{\Delta_j^2 x^j}{\Delta_{(l)}\overline{x}^l\Delta_{(m)}\overline{x}^m}\frac{\Delta_i x^i}{\Delta_{(n)}\overline{x}^n}\left((g_{ij})^{\sigma_{(l)}} + (g_{ij})^{\sigma_{(m)}}\right)
$$

$$
+ \frac{\Delta_i^2 x^i}{\Delta_{(l)}\overline{x}^l\Delta_{(n)}\overline{x}^n}\left(\left(\frac{\Delta_j x^j}{\Delta_{(m)}\overline{x}^m}\right)^{\sigma_{(l)}}(g_{ij})^{\sigma_{(l)}} - \frac{\Delta_j x^j}{\Delta_{(m)}\overline{x}^m}(g_{ij})^{\sigma_{(n)}}\right)
$$

$$
+ \frac{\Delta_j^2 x^j}{\Delta_{(m)}\overline{x}^m\Delta_{(n)}\overline{x}^n}\left(\left(\frac{\Delta_i x^i}{\Delta_{(l)}\overline{x}^l}\right)^{\sigma_{(m)}}(g_{ij})^{\sigma_{(m)}} - \left(\frac{\Delta_i x^i}{\Delta_{(l)}\overline{x}^l}\right)^{\sigma_{(n)}}(g_{ij})^{\sigma_{(n)}}\right).
$$

Therefore

$$
\overline{\Gamma}_{ml,n} = \frac{1}{2}\left(\left(\frac{\partial g_{jk}}{\Delta_i x^i}\right)^{\sigma_0\sigma_1\cdots\sigma_{i-1}} + \left(\frac{\partial g_{ik}}{\Delta_j x^j}\right)^{\sigma_0\sigma_1\cdots\sigma_{j-1}} - \left(\frac{\partial g_{ij}}{\Delta_k x^k}\right)^{\sigma_0\sigma_1\cdots\sigma_{k-1}}\right)\frac{\Delta_i x^i}{\Delta_{(l)}\overline{x}^l}\frac{\Delta_j x^j}{\Delta_{(m)}\overline{x}^m}\frac{\Delta_k x^k}{\Delta_{(n)}\overline{x}^n}
$$

$$
+ \frac{1}{2}\frac{\Delta_j^2 x^j}{\Delta_{(l)}\overline{x}^l\Delta_{(m)}\overline{x}^m}\frac{\Delta_i x^i}{\Delta_{(n)}\overline{x}^n}\left((g_{ij})^{\sigma_{(l)}} + (g_{ij})^{\sigma_{(m)}}\right)
$$

$$
+ \frac{1}{2}\frac{\Delta_i^2 x^i}{\Delta_{(l)}\overline{x}^l\Delta_{(n)}\overline{x}^n}\left(\left(\frac{\Delta_j x^j}{\Delta_{(m)}\overline{x}^m}\right)^{\sigma_{(l)}}(g_{ij})^{\sigma_{(l)}} - \frac{\Delta_j x^j}{\Delta_{(m)}\overline{x}^m}(g_{ij})^{\sigma_{(n)}}\right)
$$

$$
+ \frac{1}{2}\frac{\Delta_j^2 x^j}{\Delta_{(m)}\overline{x}^m\Delta_{(n)}\overline{x}^n}\left(\left(\frac{\Delta_i x^i}{\Delta_{(l)}\overline{x}^l}\right)^{\sigma_{(m)}}(g_{ij})^{\sigma_{(m)}} - \left(\frac{\Delta_i x^i}{\Delta_{(l)}\overline{x}^l}\right)^{\sigma_{(n)}}(g_{ij})^{\sigma_{(n)}}\right).
$$

Note that

$$
\begin{aligned}
(g_{ij})^{\sigma_{(l)}} + (g_{ij})^{\sigma_{(m)}} &= g_{ij} + \mu_{(l)} \frac{\partial g_{ij}}{\Delta_{(l)}\overline{x}^l} + g_{ij} + \mu_{(m)} \frac{\partial g_{ij}}{\Delta_{(m)}\overline{x}^m} \\
&= 2g_{ij} + \mu_{(l)} \frac{\partial g_{ij}}{\Delta_{(l)}\overline{x}^l} + \mu_{(m)} \frac{\partial g_{ij}}{\Delta_{(m)}\overline{x}^m} \\
&= 2g_{ij} + \mu_{(l)} \left(\frac{\partial g_{ij}}{\Delta_k x^k} \right)^{\sigma_0 \sigma_1 \cdots \sigma_{k-1}} \frac{\Delta_k x^k}{\Delta_{(l)}\overline{x}^l} + \mu_{(m)} \left(\frac{\partial g_{ij}}{\Delta_k x^k} \right)^{\sigma_0 \sigma_1 \cdots \sigma_{k-1}} \frac{\Delta_k x^k}{\Delta_{(m)}\overline{x}^m} \\
&= 2g_{ij} + \left(\frac{\partial g_{ij}}{\Delta_k x^k} \right)^{\sigma_0 \sigma_1 \cdots \sigma_{k-1}} \left(\mu_{(l)} \frac{\Delta_k x^k}{\Delta_{(l)}\overline{x}^l} + \mu_{(m)} \frac{\Delta_k x^k}{\Delta_{(m)}\overline{x}^m} \right),
\end{aligned}
$$

$$
\begin{aligned}
&\left(\frac{\Delta_j x^j}{\Delta_{(m)}\overline{x}^m} \right)^{\sigma_{(l)}} (g_{ij})^{\sigma_{(l)}} - \frac{\Delta_j x^j}{\Delta_{(m)}\overline{x}^m} (g_{ij})^{\sigma_{(k)}} \\
&= \left(\frac{\Delta_j x^j}{\Delta_{(m)}\overline{x}^m} + \mu_{(l)} \frac{\Delta_j^2 x^j}{\Delta_{(l)}\overline{x}^l \Delta_{(m)}\overline{x}^m} \right) \left(g_{ij} + \mu_{(l)} \left(\frac{\partial g_{ij}}{\Delta_k x^k} \right)^{\sigma_0 \sigma_1 \cdots \sigma_{k-1}} \frac{\Delta_k x^k}{\Delta_{(l)}\overline{x}^l} \right) \\
&\quad - \frac{\Delta_j x^j}{\Delta_{(m)}\overline{x}^m} \left(g_{ij} + \mu_{(n)} \left(\frac{\partial g_{ij}}{\Delta_k x^k} \right)^{\sigma_0 \sigma_1 \cdots \sigma_{k-1}} \frac{\Delta_k x^k}{\Delta_{(n)}\overline{x}^n} \right) \\
&= \mu_{(l)} \left(g_{ij} \frac{\Delta_j^2 x^j}{\Delta_{(l)}\overline{x}^l \Delta_{(m)}\overline{x}^m} + \frac{\Delta_j x^j}{\Delta_{(m)}\overline{x}^m} \left(\frac{\partial g_{ij}}{\Delta_k x^k} \right)^{\sigma_0 \sigma_1 \cdots \sigma_{k-1}} \frac{\Delta_k x^k}{\Delta_{(l)}\overline{x}^l} \right. \\
&\quad \left. + \mu_{(l)} \frac{\Delta^2 x^j}{\Delta_{(l)}\overline{x}^l \Delta_{(m)}\overline{x}^m} \left(\frac{\partial g_{ij}}{\Delta_k x^k} \right)^{\sigma_0 \sigma_1 \cdots \sigma_{k-1}} \frac{\Delta_k x^k}{\Delta_{(l)}\overline{x}^l} \right) \\
&\quad - \mu_{(n)} \left(\frac{\Delta_j x^j}{\Delta_{(m)}\overline{x}^m} \left(\frac{\partial g_{ij}}{\Delta_k x^k} \right)^{\sigma_0 \sigma_1 \cdots \sigma_{k-1}} \frac{\Delta_k x^k}{\Delta_{(n)}\overline{x}^n} \right) \\
&= \mu_{(l)} \left(g_{ij} \frac{\Delta_j^2 x^j}{\Delta_{(l)}\overline{x}^l \Delta_{(m)}\overline{x}^m} + \left(\frac{\Delta_j x^j}{\Delta_{(m)}\overline{x}^m} + \mu_{(l)} \frac{\Delta_j^2 x^j}{\Delta_{(l)}\overline{x}^l \Delta_{(m)}\overline{x}^m} \right) \left(\frac{\partial g_{ij}}{\Delta_k x^k} \right)^{\sigma_0 \sigma_1 \cdots \sigma_{k-1}} \frac{\Delta_k x^k}{\Delta_{(l)}\overline{x}^l} \right) \\
&\quad - \mu_{(n)} \frac{\Delta_j x^j}{\Delta_{(m)}\overline{x}^m} \left(\frac{\partial g_{ij}}{\Delta_k x^k} \right)^{\sigma_0 \sigma_1 \cdots \sigma_{k-1}} \frac{\Delta_k x^k}{\Delta_{(n)}\overline{x}^n} \\
&= \mu_{(l)} \left(g_{ij} \frac{\Delta_j^2 x^j}{\Delta_{(l)}\overline{x}^l \Delta_{(m)}\overline{x}^m} + \left(\frac{\Delta_j x^j}{\Delta_{(m)}\overline{x}^m} \right)^{\sigma_{(l)}} \left(\frac{\partial g_{ij}}{\Delta_k x^k} \right)^{\sigma_0 \sigma_1 \cdots \sigma_{k-1}} \frac{\Delta_k x^k}{\Delta_{(l)}\overline{x}^l} \right) \\
&\quad - \mu_{(n)} \frac{\Delta_j x^j}{\Delta_{(m)}\overline{x}^m} \left(\frac{\partial g_{ij}}{\Delta_k x^k} \right)^{\sigma_0 \sigma_1 \cdots \sigma_{k-1}} \frac{\Delta_k x^k}{\Delta_{(n)}\overline{x}^n},
\end{aligned}
$$

and

$$
\begin{aligned}
&\left(\frac{\Delta_i x^i}{\Delta_{(l)}\overline{x}^l} \right)^{\sigma_{(m)}} (g_{ij})^{\sigma_{(m)}} - \left(\frac{\Delta_i x^i}{\Delta_{(l)}\overline{x}^l} \right)^{\sigma_{(n)}} (g_{ij})^{\sigma_{(n)}} \\
&= \left(\frac{\Delta_i x^i}{\Delta_{(l)}\overline{x}^l} + \mu_{(m)} \frac{\Delta_i^2 x^i}{\Delta_{(m)}\overline{x}^m \Delta_{(l)}\overline{x}^l} \right) \left(g_{ij} + \mu_{(m)} \left(\frac{\partial g_{ij}}{\Delta_k x^k} \right)^{\sigma_0 \sigma_1 \cdots \sigma_{k-1}} \frac{\Delta_k x^k}{\Delta_{(m)}\overline{x}^m} \right)
\end{aligned}
$$

$$-\left(\frac{\Delta_i x^i}{\Delta_{(l)}\overline{x}^l} + \mu_{(n)}\frac{\Delta_i^2 x^i}{\Delta_{(n)}\overline{x}^n \Delta_{(l)}\overline{x}^l}\right)\left(g_{ij} + \mu_{(n)}\left(\frac{\partial g_{ij}}{\Delta_k x^k}\right)^{\sigma_0\sigma_1\cdots\sigma_{k-1}}\frac{\Delta_k x^k}{\Delta_{(n)}\overline{x}^n}\right)$$

$$= \mu_{(m)}\left(g_{ij}\frac{\Delta_i^2 x^i}{\Delta_{(m)}\overline{x}^m \Delta_{(l)}\overline{x}^l} + \frac{\Delta_i x^i}{\Delta_{(l)}\overline{x}^l}\left(\frac{\partial g_{ij}}{\Delta_k x^k}\right)^{\sigma_0\sigma_1\cdots\sigma_{k-1}}\frac{\Delta_k x^k}{\Delta_{(m)}\overline{x}^m}\right.$$

$$\left. + \mu_{(m)\frac{\Delta_i^2 x^i}{\Delta_{(m)}\overline{x}^m \Delta_{(l)}\overline{x}^l}}\left(\frac{\partial g_{ij}}{\Delta_k x^k}\right)^{\sigma_0\sigma_1\cdots\sigma_{k-1}}\frac{\Delta_k x^k}{\Delta_{(m)}\overline{x}^m}\right)$$

$$- \mu_{(n)}\left(g_{ij}\frac{\Delta_i^2 x^i}{\Delta_{(n)}\overline{x}^n \Delta_{(l)}\overline{x}^l} + \frac{\Delta_i x^i}{\Delta_{(l)}\overline{x}^l}\left(\frac{\partial g_{ij}}{\Delta_k x^k}\right)^{\sigma_0\sigma_1\cdots\sigma_{k-1}}\frac{\Delta_k x^k}{\Delta_{(n)}\overline{x}^n}\right.$$

$$\left. + \mu_{(n)\frac{\Delta_i^2 x^i}{\Delta_{(n)}\overline{x}^n \Delta_{(l)}\overline{x}^l}}\left(\frac{\partial g_{ij}}{\Delta_k x^k}\right)^{\sigma_0\sigma_1\cdots\sigma_{k-1}}\frac{\Delta_k x^k}{\Delta_{(n)}\overline{x}^n}\right)$$

$$= \mu_{(m)}\left(g_{ij}\frac{\Delta_i^2 x^i}{\Delta_{(m)}\overline{x}^m \Delta_{(l)}\overline{x}^l} + \left(\frac{\Delta_i x^i}{\Delta_{(l)}} + \mu_{(m)\frac{\Delta_i^2 x^i}{\Delta_{(m)}\overline{x}^m \Delta_{(l)}\overline{x}^l}}\right)\left(\frac{\partial g_{ij}}{\Delta_k x^k}\right)^{\sigma_0\sigma_1\cdots\sigma_{k-1}}\frac{\Delta_k x^k}{\Delta_{(m)}\overline{x}^m}\right)$$

$$- \mu_{(n)}\left(g_{ij}\frac{\Delta_i^2 x^i}{\Delta_{(n)}\overline{x}^n \Delta_{(l)}\overline{x}^l} + \left(\frac{\Delta_i x^i}{\Delta_{(l)}\overline{x}^l} + \mu_{(n)\frac{\Delta_i^2 x^i}{\Delta_{(n)}\overline{x}^n \Delta_{(l)}\overline{x}^l}}\right)\left(\frac{\partial g_{ij}}{\Delta_k x^k}\right)^{\sigma_0\sigma_1\cdots\sigma_{k-1}}\frac{\Delta_k x^k}{\Delta_{(n)}\overline{x}^n}\right)$$

$$= \mu_{(m)}\left(g_{ij}\frac{\Delta_i^2 x^i}{\Delta_{(m)}\overline{x}^m \Delta_{(l)}\overline{x}^l} + \left(\frac{\Delta_i x^i}{\Delta_{(l)}}\right)^{\sigma_{(m)}}\left(\frac{\partial g_{ij}}{\Delta_k x^k}\right)^{\sigma_0\sigma_1\cdots\sigma_{k-1}}\frac{\Delta_k x^k}{\Delta_{(m)}\overline{x}^m}\right)$$

$$- \mu_{(n)}\left(g_{ij}\frac{\Delta_i^2 x^i}{\Delta_{(n)}\overline{x}^n \Delta_{(l)}\overline{x}^l} + \left(\frac{\Delta_i x^i}{\Delta_{(l)}\overline{x}^l}\right)^{\sigma_{(n)}}\left(\frac{\partial g_{ij}}{\Delta_k x^k}\right)^{\sigma_0\sigma_1\cdots\sigma_{k-1}}\frac{\Delta_k x^k}{\Delta_{(n)}\overline{x}^n}\right).$$

Let

$$\widetilde{\Gamma}_{lm,n} = \frac{1}{2}\frac{\Delta_j^2 x^j}{\Delta_{(l)}\overline{x}^l \Delta_{(m)}\overline{x}^m}\frac{\Delta_i x^i}{\Delta_{(n)}\overline{x}^n}\left(\frac{\partial g_{ij}}{\Delta_k x^k}\right)^{\sigma_0\sigma_1\cdots\sigma_{k-1}}\left(\mu_{(l)}\frac{\Delta_k x^k}{\Delta_{(l)}\overline{x}^l} + \mu_{(m)}\frac{\Delta_k x^k}{\Delta_{(m)}\overline{x}^m}\right)$$

$$+ \frac{1}{2}\frac{\Delta_i^2 x^i}{\Delta_{(l)}\overline{x}^l \Delta_{(n)}\overline{x}^n}\left(\mu_{(l)}\left(g_{ij}\frac{\Delta_j^2 x^j}{\Delta_{(l)}\overline{x}^l \Delta_{(m)}\overline{x}^m} + \left(\frac{\Delta_j x^j}{\Delta_{(m)}\overline{x}^m}\right)^{\sigma_{(l)}}\left(\frac{\partial g_{ij}}{\Delta_k x^k}\right)^{\sigma_0\sigma_1\cdots\sigma_{k-1}}\frac{\Delta_k x^k}{\Delta_{(l)}\overline{x}^l}\right)\right.$$

$$\left. - \mu_{(n)}\frac{\Delta_j x^j}{\Delta_{(m)}\overline{x}^m}\left(\frac{\partial g_{ij}}{\Delta_k x^k}\right)^{\sigma_0\sigma_1\cdots\sigma_{k-1}}\frac{\Delta_k x^k}{\Delta_{(n)}\overline{x}^n}\right)$$

$$+ \frac{1}{2}\frac{\Delta_j^2 x^j}{\Delta_{(m)}\overline{x}^m \Delta_{(n)}\overline{x}^n}\left(\mu_{(m)}\left(g_{ij}\frac{\Delta_i^2 x^i}{\Delta_{(m)}\overline{x}^m \Delta_{(l)}\overline{x}^l} + \left(\frac{\Delta_i x^i}{\Delta_{(l)}}\right)^{\sigma_{(m)}}\left(\frac{\partial g_{ij}}{\Delta_k x^k}\right)^{\sigma_0\sigma_1\cdots\sigma_{k-1}}\frac{\Delta_k x^k}{\Delta_{(m)}\overline{x}^m}\right)\right.$$

$$\left. - \mu_{(n)}\left(g_{ij}\frac{\Delta_i^2 x^i}{\Delta_{(n)}\overline{x}^n \Delta_{(l)}\overline{x}^l} + \left(\frac{\Delta_i x^i}{\Delta_{(l)}\overline{x}^l}\right)^{\sigma_{(n)}}\left(\frac{\partial g_{ij}}{\Delta_k x^k}\right)^{\sigma_0\sigma_1\cdots\sigma_{k-1}}\frac{\Delta_k x^k}{\Delta_{(n)}\overline{x}^n}\right)\right).$$

Consequently,

$$\overline{\Gamma}_{ml,n} = \Gamma_{\sigma ijk}\frac{\Delta_i x^i}{\Delta_{(l)}\overline{x}^l}\frac{\Delta_j x^j}{\Delta_{(m)}\overline{x}^m}\frac{\Delta_k x^k}{\Delta_{(n)}\overline{x}^n} + g_{ij}\frac{\Delta_j^2 x^j}{\Delta_{(l)}\overline{x}^l \Delta_{(m)}\overline{x}^m}\frac{\Delta_i x^i}{\Delta_{(n)}\overline{x}^n} + \widetilde{\Gamma}_{lm,n}. \tag{4.5}$$

We see that the last expression does not correspond to some of the transformation laws for tensors of the third order. Thus the Christoffel symbols of the first kind are not tensors.

Now we will deduct the transformation law for Christoffel symbols of the second kind. Since g^{ij} is a contravariant tensor of the second kind, we have

$$\overline{g}^{np} = g^{rs}\frac{\Delta_{(n)}\overline{x}^n}{\Delta_r x^r}\frac{\Delta_{(p)}\overline{x}^p}{\Delta_s x^s}.$$
(4.6)

Inner multiplication of both sides of (4.5) by the corresponding sides of (4.6) gives

$$\overline{g}^{np}\overline{\Gamma}_{lmn} = g^{rs}\frac{\Delta_{(n)}\overline{x}^n}{\Delta_r x^r}\frac{\Delta_{(p)}\overline{x}^p}{\Delta_s x^s}$$

$$\times \left(\Gamma_{\sigma ijk}\frac{\Delta_i x^i}{\Delta_{(l)}\overline{x}^l}\frac{\Delta_j x^j}{\Delta_{(m)}\overline{x}^m}\frac{\Delta_k x^k}{\Delta_{(n)}\overline{x}^n} + g_{ij}\frac{\Delta_j^2 x^j}{\Delta_{(l)}\overline{x}^l\Delta_{(m)}\overline{x}^m}\frac{\Delta_i x^i}{\Delta_{(n)}\overline{x}^n} + \tilde{\Gamma}_{lm,n}\right)$$

$$= g^{rs}\Gamma_{\sigma ijk}\frac{\Delta_{(n)}\overline{x}^n}{\Delta_r x^r}\frac{\Delta_{(p)}\overline{x}^p}{\Delta_s x^s}\frac{\Delta_i x^i}{\Delta_{(l)}\overline{x}^l}\frac{\Delta_j x^j}{\Delta_{(m)}\overline{x}^m}\frac{\Delta_k x^k}{\Delta_{(n)}\overline{x}^n}$$

$$+ g^{rs}g_{ij}\frac{\Delta_{(n)}\overline{x}^n}{\Delta_r x^r}\frac{\Delta_{(p)}\overline{x}^p}{\Delta_s x^s}\frac{\Delta_j^2 x^j}{\Delta_{(l)}\overline{x}^l\Delta_{(m)}\overline{x}^m}\frac{\Delta_i x^i}{\Delta_{(n)}\overline{x}^n}$$

$$+ g^{rs}\tilde{\Gamma}_{lm,n}\frac{\Delta_{(n)}\overline{x}^n}{\Delta_r x^r}\frac{\Delta_{(p)}\overline{x}^p}{\Delta_s x^s}$$

$$= g^{rs}\Gamma_{\sigma ijk}\frac{\Delta_k x^k}{\Delta_r x^r}\frac{\Delta_{(p)}\overline{x}^p}{\Delta_s x^s}\frac{\Delta_i x^i}{\Delta_{(l)}\overline{x}^l}\frac{\Delta_j x^j}{\Delta_{(m)}\overline{x}^m}$$

$$+ g^{rs}g_{ij}\frac{\Delta_i x^i}{\Delta_r x^r}\frac{\Delta_{(p)}\overline{x}^p}{\Delta_s x^s}\frac{\Delta_j^2 x^j}{\Delta_{(l)}\overline{x}^l\Delta_{(m)}\overline{x}^m}$$

$$+ g^{rs}\tilde{\Gamma}_{lm,n}\frac{\Delta_{(n)}\overline{x}^n}{\Delta_r x^r}\frac{\Delta_{(p)}\overline{x}^p}{\Delta_s x^s}$$

$$= g^{ks}\Gamma_{\sigma ijk}\frac{\Delta_{(p)}\overline{x}^p}{\Delta_s x^s}\frac{\Delta_i x^i}{\Delta_{(l)}\overline{x}^l}\frac{\Delta_j x^j}{\Delta_{(m)}\overline{x}^m}$$

$$+ g^{rs}g_{rj}\frac{\Delta_{(p)}\overline{x}^p}{\Delta_s x^s}\frac{\Delta_j^2 x^j}{\Delta_{(l)}\overline{x}^l\Delta_{(m)}\overline{x}^m}$$

$$+ g^{rs}\tilde{\Gamma}_{lm,n}\frac{\Delta_{(n)}\overline{x}^n}{\Delta_r x^r}\frac{\Delta_{(p)}\overline{x}^p}{\Delta_s x^s}$$

$$= g^{ks}\Gamma_{\sigma ijk}\frac{\Delta_{(p)}\overline{x}^p}{\Delta_s x^s}\frac{\Delta_i x^i}{\Delta_{(l)}\overline{x}^l}\frac{\Delta_j x^j}{\Delta_{(m)}\overline{x}^m} + \delta_j^s\frac{\Delta_{(p)}\overline{x}^p}{\Delta_s x^s}\frac{\Delta_j^2 x^j}{\Delta_{(l)}\overline{x}^l\Delta_{(m)}\overline{x}^m}$$

$$+ g^{rs}\tilde{\Gamma}_{lm,n}\frac{\Delta_{(n)}\overline{x}^n}{\Delta_r x^r}\frac{\Delta_{(p)}\overline{x}^p}{\Delta_s x^s}$$

$$= g^{ks}\Gamma_{\sigma ijk}\frac{\Delta_{(p)}\overline{x}^p}{\Delta_s x^s}\frac{\Delta_i x^i}{\Delta_{(l)}\overline{x}^l}\frac{\Delta_j x^j}{\Delta_{(m)}\overline{x}^m} + \frac{\Delta_{(p)}\overline{x}^p}{\Delta_j x^j}\frac{\Delta_j^2 x^j}{\Delta_{(l)}\overline{x}^l\Delta_{(m)}\overline{x}^m}$$

$$+ g^{rs}\widetilde{\Gamma}_{lm,n}\frac{\Delta_{(n)}\overline{x}^n}{\Delta_r x^r}\frac{\Delta_{(p)}\overline{x}^p}{\Delta_s x^s}$$

$$= \Gamma_{\sigma ij}^s\frac{\Delta_{(p)}\overline{x}^p}{\Delta_s x^s}\frac{\Delta_i x^i}{\Delta_{(l)}\overline{x}^l}\frac{\Delta_j x^j}{\Delta_{(m)}\overline{x}^m} + \frac{\Delta_{(p)}\overline{x}^p}{\Delta_j x^j}\frac{\Delta_j^2 x^j}{\Delta_{(l)}\overline{x}^l\Delta_{(m)}\overline{x}^m}$$

$$+ g^{rs}\widetilde{\Gamma}_{lm,n}\frac{\Delta_{(n)}\overline{x}^n}{\Delta_r x^r}\frac{\Delta_{(p)}\overline{x}^p}{\Delta_s x^s}\,.$$

Therefore

$$\overline{\Gamma}_{lm}^p = \Gamma_{\sigma ij}^s\frac{\Delta_{(p)}\overline{x}^p}{\Delta_s x^s}\frac{\Delta_i x^i}{\Delta_{(l)}\overline{x}^l}\frac{\Delta_j x^j}{\Delta_{(m)}\overline{x}^m} + \frac{\Delta_{(p)}\overline{x}^p}{\Delta_j x^j}\frac{\Delta_j^2 x^j}{\Delta_{(l)}\overline{x}^l\Delta_{(m)}\overline{x}^m}$$

$$+ g^{rs}\widetilde{\Gamma}_{lm,n}\frac{\Delta_{(n)}\overline{x}^n}{\Delta_r x^r}\frac{\Delta_{(p)}\overline{x}^p}{\Delta_s x^s}\,.$$

Hence

$$\overline{\Gamma}_{lm}^p - g^{rs}\widetilde{\Gamma}_{lm,n}\frac{\Delta_{(n)}\overline{x}^n}{\Delta_r x^r}\frac{\Delta_{(p)}\overline{x}^p}{\Delta_s x^s} = \Gamma_{\sigma ij}^s\frac{\Delta_{(p)}\overline{x}^p}{\Delta_s x^s}\frac{\Delta_i x^i}{\Delta_{(l)}\overline{x}^l}\frac{\Delta_j x^j}{\Delta_{(m)}\overline{x}^m} + \frac{\Delta_{(p)}\overline{x}^p}{\Delta_j x^j}\frac{\Delta_j^2 x^j}{\Delta_{(l)}\overline{x}^l\Delta_{(m)}\overline{x}^m}\,.$$

Inner multiplication of the last equation by $\frac{\Delta_r x^r}{\Delta_{(p)}\overline{x}^p}$ gives

$$\frac{\Delta_r x^r}{\Delta_{(p)}\overline{x}^p}\left(\overline{\Gamma}_{lm}^p - g^{rs}\widetilde{\Gamma}_{lm,n}\frac{\Delta_{(n)}\overline{x}^n}{\Delta_r x^r}\frac{\Delta_{(p)}\overline{x}^p}{\Delta_s x^s}\right)$$

$$= \Gamma_{\sigma ij}^s\frac{\Delta_r x^r}{\Delta_{(p)}\overline{x}^p}\frac{\Delta_{(p)}\overline{x}^p}{\Delta_s x^s}\frac{\Delta_i x^i}{\Delta_{(l)}\overline{x}^l}\frac{\Delta_j x^j}{\Delta_{(m)}\overline{x}^m} + \frac{\Delta_r x^r}{\Delta_{(p)}\overline{x}^p}\frac{\Delta_{(p)}\overline{x}^p}{\Delta_j x^j}\frac{\Delta_j^2 x^j}{\Delta_{(l)}\overline{x}^l\Delta_{(m)}\overline{x}^m}$$

$$= \Gamma_{\sigma ij}^s\delta_s^r\frac{\Delta_i x^i}{\Delta_{(l)}\overline{x}^l}\frac{\Delta_j x^j}{\Delta_{(m)}\overline{x}^m} + \delta_j^r\frac{\Delta_j^2 x^j}{\Delta_{(l)}\overline{x}^l\Delta_{(m)}\overline{x}^m}$$

$$= \Gamma_{\sigma ij}^r\frac{\Delta_i x^i}{\Delta_{(l)}\overline{x}^l}\frac{\Delta_r x^r}{\Delta_{(m)}\overline{x}^m} + \frac{\Delta_j^2 x^j}{\Delta_{(l)}\overline{x}^l\Delta_{(m)}\overline{x}^m}\,,$$

whereupon

$$\frac{\Delta_j^2 x^j}{\Delta_{(l)}\overline{x}^l\Delta_{(m)}\overline{x}^m} = \frac{\Delta_r x^r}{\Delta_{(p)}\overline{x}^p}\left(\overline{\Gamma}_{lm}^p - g^{rs}\widetilde{\Gamma}_{lm,n}\frac{\Delta_{(n)}\overline{x}^n}{\Delta_r x^r}\frac{\Delta_{(p)}\overline{x}^p}{\Delta_s x^s}\right) - \Gamma_{\sigma ij}^r\frac{\Delta_i x^i}{\Delta_{(l)}\overline{x}^l}\frac{\Delta_r x^r}{\Delta_{(m)}\overline{x}^m}\,. \tag{4.7}$$

4.2 Transformation laws for Christoffel symbols

In this section, we deduct the transformation laws for Christoffel symbols of the first and second kinds. We know that

$$\overline{g}_{lm} = \frac{\Delta_i x^i}{\Delta_{(l)}\overline{x}^l}\,\frac{\Delta_j x^j}{\Delta_{(m)}\overline{x}^m}\,g_{ij}.$$

Differentiating this expression with respect to $\overline{x}^n$, we find

$$
\begin{aligned}
\frac{\partial \overline{g}_{lm}}{\Delta_{(n)}\overline{x}^n} &= \frac{\Delta_i x^i}{\Delta_{(l)}\overline{x}^l}\,\frac{\Delta_j x^j}{\Delta_{(m)}\overline{x}^m}\,\frac{\partial g_{ij}}{\Delta_{(n)}\overline{x}^n} \\[4pt]
&\quad + \frac{\Delta_i^2 x^i}{\Delta_{(n)}\overline{x}^n \Delta_{(l)}\overline{x}^l}\,\frac{\Delta_j x^j}{\Delta_{(m)}\overline{x}^m}\,g_{ij}^{\sigma_{(n)}} \\[4pt]
&\quad + \left(\frac{\Delta_i x^i}{\Delta_{(l)}\overline{x}^l}\right)^{\sigma_{(n)}}\frac{\Delta_j^2 x^j}{\Delta_{(n)}\overline{x}^n \Delta_{(m)}\overline{x}^m}\,f_{ij}^{\sigma_{(n)}} \\[8pt]
&= \frac{\Delta_i x^i}{\Delta_{(l)}\overline{x}^l}\,\frac{\Delta_j x^j}{\Delta_{(m)}\overline{x}^m}\,\frac{\partial g_{ij}^{\sigma_0\sigma_1\cdots\sigma_{k-1}}}{\Delta_k x^k}\,\frac{\Delta_k x^k}{\Delta_{(n)}\overline{x}^n} \\[4pt]
&\quad + \frac{\Delta_i^2 x^i}{\Delta_{(n)}\overline{x}^n \Delta_{(l)}\overline{x}^l}\,\frac{\Delta_j x^j}{\Delta_{(m)}\overline{x}^m}\,g_{ij}^{\sigma_{(n)}} \\[4pt]
&\quad + \left(\frac{\Delta_i x^i}{\Delta_{(l)}\overline{x}^l}\right)^{\sigma_{(n)}}\frac{\Delta_j^2 x^j}{\Delta_{(n)}\overline{x}^n \Delta_{(m)}\overline{x}^m}\,f_{ij}^{\sigma_{(n)}} \\[8pt]
&= \frac{\Delta_i x^i}{\Delta_{(l)}\overline{x}^l}\,\frac{\Delta_j x^j}{\Delta_{(m)}\overline{x}^m}\,\frac{\Delta_k x^k}{\Delta_{(n)}\overline{x}^n}\,\frac{\partial g_{ij}}{\Delta_k x^k} \\[4pt]
&\quad + \frac{\Delta_i x^i}{\Delta_{(l)}\overline{x}^l}\,\frac{\Delta_j x^j}{\Delta_{(m)}\overline{x}^m}\left(\sum_{r=1}^{k-1}\mu_{k-r}\,\frac{\partial^2 g_{ij}^{\sigma_0\sigma_1\cdots\sigma_{r-1}}}{\Delta_{k-r} x^{k-r}\Delta_k x^k}\,\frac{\Delta_k x^k}{\Delta_{(n)}\overline{x}^n}\right) \\[4pt]
&\quad + \frac{\Delta_i^2 x^i}{\Delta_{(n)}\overline{x}^n \Delta_{(l)}\overline{x}^l}\,\frac{\Delta_j x^j}{\Delta_{(m)}\overline{x}^m}\,g_{ij} \\[4pt]
&\quad + \mu_{(n)}\frac{\Delta_i^2 x^i}{\Delta_{(n)}\overline{x}^n \Delta_{(l)}\overline{x}^l}\,\frac{\Delta_j x^j}{\Delta_{(m)}\overline{x}^m}\,\frac{\partial g_{ij}^{\sigma_0\sigma_1\cdots\sigma_{k-1}}}{\Delta_k x^k}\,\frac{\Delta_k x^k}{\Delta_{(n)}\overline{x}^n} \\[4pt]
&\quad + \left(\frac{\Delta_i x^i}{\Delta_{(l)}\overline{x}^l} + \mu_{(n)}\frac{\Delta_i^2 x^i}{(\Delta_{(n)}\overline{x}^n)^2}\right)\frac{\Delta_j^2 x^j}{\Delta_{(n)}\overline{x}^n \Delta_{(m)}\overline{x}^m}\left(g_{ij} + \mu_{(n)}\frac{\partial g_{ij}}{\Delta_{(n)}\overline{x}^n}\right) \\[8pt]
&= \frac{\Delta_i x^i}{\Delta_{(l)}\overline{x}^l}\,\frac{\Delta_j x^j}{\Delta_{(m)}\overline{x}^m}\,\frac{\Delta_k x^k}{\Delta_{(n)}\overline{x}^n}\,\frac{\partial g_{ij}}{\Delta_k x^k} \\[4pt]
&\quad + \frac{\Delta_i^2 x^i}{\Delta_{(n)}\overline{x}^n \Delta_{(l)}\overline{x}^l}\,\frac{\Delta_j x^j}{\Delta_{(m)}\overline{x}^m}\,g_{ij}
\end{aligned}
$$

$$+ \frac{\Delta_i x^i}{\Delta_{(l)} \overline{x}^l} \frac{\Delta_j^2 x^j}{\Delta_{(n)} \overline{x}^n \Delta_{(m)} \overline{x}^m} g_{ij}$$

$$+ \frac{\Delta_i x^i}{\Delta_{(l)} \overline{x}^l} \frac{\Delta_j x^j}{\Delta_{(m)} \overline{x}^m} \left(\sum_{r=1}^{k-1} \mu_{k-r} \frac{\partial^2 g_{ij}^{\sigma_0 \sigma_1 \cdots \sigma_{k-r-1}}}{\Delta_{k-r} x^{k-r} \Delta_k x^k} \frac{\Delta_k x^k}{\Delta_{(r)} \overline{x}^r} \right)$$

$$+ \mu_{(n)} \frac{\Delta_i^2 x^i}{\Delta_{(n)} \overline{x}^n \Delta_{(l)} \overline{x}^l} \frac{\Delta_j x^j}{\Delta_{(m)} \overline{x}^m} \frac{\partial g_{ij}^{\sigma_0 \sigma_1 \cdots \sigma_{k-1}}}{\Delta_k x^k} \frac{\Delta_k x^k}{\Delta_{(n)} \overline{x}^n}$$

$$+ \mu_{(n)} \frac{\Delta_i x^i}{\Delta_{(l)} \overline{x}^l} \frac{\Delta_j^2 x^j}{\Delta_{(n)} \overline{x}^n \Delta_{(m)} \overline{x}^m} \frac{\partial g_{ij}}{\Delta_{(n)} \overline{x}^n}$$

$$+ \mu_{(n)} \frac{\Delta_i^2 x^i}{(\Delta_{(n)} \overline{x}^n)^2} \frac{\Delta_j^2 x^j}{\Delta_{(n)} \overline{x}^n \Delta_{(m)} \overline{x}^m} \left(g_{ij} + \mu_{(n)} \frac{\partial g_{ij}}{\Delta_{(n)} \overline{x}^n} \right).$$

Let

$$G_{lmn} = \frac{\Delta_i x^i}{\Delta_{(l)} \overline{x}^l} \frac{\Delta_j x^j}{\Delta_{(m)} \overline{x}^m} \left(\sum_{r=1}^{k-1} \mu_{k-r} \frac{\partial^2 g_{ij}^{\sigma_0 \sigma_1 \cdots \sigma_{k-r-1}}}{\Delta_{k-r} x^{k-r} \Delta_k x^k} \frac{\Delta_k x^k}{\Delta_{(r)} \overline{x}^r} \right)$$

$$+ \mu_{(n)} \frac{\Delta_i^2 x^i}{\Delta_{(n)} \overline{x}^n \Delta_{(l)} \overline{x}^l} \frac{\Delta_j x^j}{\Delta_{(m)} \overline{x}^m} \frac{\partial g_{ij}^{\sigma_0 \sigma_1 \cdots \sigma_{k-1}}}{\Delta_k x^k} \frac{\Delta_k x^k}{\Delta_{(n)} \overline{x}^n}$$

$$+ \mu_{(n)} \frac{\Delta_i x^i}{\Delta_{(l)} \overline{x}^l} \frac{\Delta_j^2 x^j}{\Delta_{(n)} \overline{x}^n \Delta_{(m)} \overline{x}^m} \frac{\partial g_{ij}}{\Delta_{(n)} \overline{x}^n}$$

$$+ \mu_{(n)} \frac{\Delta_i^2 x^i}{(\Delta_{(n)} \overline{x}^n)^2} \frac{\Delta_j^2 x^j}{\Delta_{(n)} \overline{x}^n \Delta_{(m)} \overline{x}^m} \left(g_{ij} + \mu_{(n)} \frac{\partial g_{ij}}{\Delta_{(n)} \overline{x}^n} \right).$$

Then

$$\frac{\partial \overline{g}_{lm}}{\Delta_{(n)} \overline{x}^n} = \frac{\Delta_i x^i}{\Delta_{(l)} \overline{x}^l} \frac{\Delta_j x^j}{\Delta_{(m)} \overline{x}^m} \frac{\Delta_k x^k}{\Delta_{(n)} \overline{x}^n} \frac{\partial g_{ij}}{\Delta_k x^k}$$

$$+ \frac{\Delta_i^2 x^i}{\Delta_{(n)} \overline{x}^n \Delta_{(l)} \overline{x}^l} \frac{\Delta_j x^j}{\Delta_{(m)} \overline{x}^m} g_{ij}$$

$$+ \frac{\Delta_i x^i}{\Delta_{(l)} \overline{x}^l} \frac{\Delta_j^2 x^j}{\Delta_{(n)} \overline{x}^n \Delta_{(m)} \overline{x}^m} g_{ij} + G_{lmn}.$$

Now interchanging l with m, m with n, and n with l, we obtain

$$\frac{\partial \overline{g}_{mn}}{\Delta_{(l)} \overline{x}^l} = \frac{\Delta_i x^i}{\Delta_{(m)} \overline{x}^m} \frac{\Delta_j x^j}{\Delta_{(n)} \overline{x}^n} \frac{\Delta_k x^k}{\Delta_{(l)} \overline{x}^l} \frac{\partial g_{ij}}{\Delta_k x^k}$$

$$+ \frac{\Delta_i^2 x^i}{\Delta_{(l)} \overline{x}^l \Delta_{(m)} \overline{x}^m} \frac{\Delta_j x^j}{\Delta_{(n)} \overline{x}^n} g_{ij}$$

$$+ \frac{\Delta_i x^i}{\Delta_{(m)} \overline{x}^m} \frac{\Delta_j^2 x^j}{\Delta_{(l)} \overline{x}^l \Delta_{(n)} \overline{x}^n} g_{ij} + G_{mnl}.$$

Replacing the dummy indices i, j, k with j, k, i, respectively, we find

$$\frac{\partial \overline{g}_{mn}}{\Delta_{(l)} \overline{x}^l} = \frac{\Delta_i x^i}{\Delta_{(l)} \overline{x}^l} \frac{\Delta_j x^j}{\Delta_{(m)} \overline{x}^m} \frac{\Delta_k x^k}{\Delta_{(n)} \overline{x}^n} \frac{\partial g_{jk}}{\Delta_i x^i}$$

$$+ \frac{\Delta_j^2 x^j}{\Delta_{(l)} \overline{x}^l \Delta_{(m)} \overline{x}^m} \frac{\Delta_k x^k}{\Delta_{(n)} \overline{x}^n} g_{jk}$$

$$+ \frac{\Delta_j x^j}{\Delta_{(m)} \overline{x}^m} \frac{\Delta_k^2 x^k}{\Delta_{(l)} \overline{x}^l \Delta_{(n)} \overline{x}^n} g_{jk} + G_{mnl}.$$

Next,

$$\frac{\partial \overline{g}_{ln}}{\Delta_{(m)} \overline{x}^m} = \frac{\Delta_i x^i}{\Delta_{(l)} \overline{x}^l} \frac{\Delta_j x^j}{\Delta_{(n)} \overline{x}^n} \frac{\Delta_k x^k}{\Delta_{(m)} \overline{x}^m} \frac{\partial g_{ij}}{\Delta_k x^k}$$

$$+ \frac{\Delta_i^2 x^i}{\Delta_{(l)} \overline{x}^l \Delta_{(m)} \overline{x}^m} \frac{\Delta_j x^j}{\Delta_{(n)} \overline{x}^n} g_{ij}$$

$$+ \frac{\Delta_i x^i}{\Delta_{(l)} \overline{x}^l} \frac{\Delta_j^2 x^j}{\Delta_{(n)} \overline{x}^n \Delta_{(m)} \overline{x}^m} g_{ij} + G_{lnm}.$$

Now interchanging the dummy indices k and j, we get

$$\frac{\partial \overline{g}_{ln}}{\Delta_{(m)} \overline{x}^m} = \frac{\Delta_i x^i}{\Delta_{(l)} \overline{x}^l} \frac{\Delta_j x^j}{\Delta_{(m)} \overline{x}^m} \frac{\Delta_k x^k}{\Delta_{(n)} \overline{x}^n} \frac{\partial g_{ik}}{\Delta_j x^j}$$

$$+ \frac{\Delta_i^2 x^i}{\Delta_{(l)} \overline{x}^l \Delta_{(m)} \overline{x}^m} \frac{\Delta_k x^k}{\Delta_{(n)} \overline{x}^n} g_{ik}$$

$$+ \frac{\Delta_i x^i}{\Delta_{(l)} \overline{x}^l} \frac{\Delta_k^2 x^k}{\Delta_{(n)} \overline{x}^n \Delta_{(m)} \overline{x}^m} g_{ij} + G_{lnm}.$$

Hence

$$\frac{\partial \overline{g}_{mn}}{\Delta_{(l)} \overline{x}^l} + \frac{\partial \overline{g}_{nl}}{\Delta_{(m)} \overline{x}^m} - \frac{\partial \overline{g}_{lm}}{\Delta_{(n)} \overline{x}^n}$$

$$= \frac{\Delta_i x^i}{\Delta_{(l)} \overline{x}^l} \frac{\Delta_j x^j}{\Delta_{(m)} \overline{x}^m} \frac{\Delta_k x^k}{\Delta_{(n)} \overline{x}^n} \frac{\partial g_{jk}}{\Delta_i x^i}$$

$$+ \frac{\Delta_j^2 x^j}{\Delta_{(l)} \overline{x}^l \Delta_{(m)} \overline{x}^m} \frac{\Delta_k x^k}{\Delta_{(n)} \overline{x}^n} g_{jk}$$

$$+ \frac{\Delta_j x^j}{\Delta_{(m)}\overline{x}^m} \frac{\Delta_k^2 x^k}{\Delta_{(l)}\overline{x}^l \Delta_{(n)}\overline{x}^n} g_{jk} + G_{mnl}$$

$$+ \frac{\Delta_i x^i}{\Delta_{(l)}\overline{x}^l} \frac{\Delta_j x^j}{\Delta_{(m)}\overline{x}^m} \frac{\Delta_k x^k}{\Delta_{(n)}\overline{x}^n} \frac{\partial g_{ik}}{\Delta_j x^j}$$

$$+ \frac{\Delta_i^2 x^i}{\Delta_{(l)}\overline{x}^l \Delta_{(m)}\overline{x}^m} \frac{\Delta_k x^k}{\Delta_{(n)}\overline{x}^n} g_{ik}$$

$$+ \frac{\Delta_i x^i}{\Delta_{(l)}\overline{x}^l} \frac{\Delta_k^2 x^k}{\Delta_{(n)}\overline{x}^n \Delta_{(m)}\overline{x}^m} g_{ij} + G_{lnm}$$

$$- \frac{\Delta_i x^i}{\Delta_{(l)}\overline{x}^l} \frac{\Delta_j x^j}{\Delta_{(m)}\overline{x}^m} \frac{\Delta_k x^k}{\Delta_{(n)}\overline{x}^n} \frac{\partial g_{ij}}{\Delta_k x^k}$$

$$- \frac{\Delta_i^2 x^i}{\Delta_{(n)}\overline{x}^n \Delta_{(l)}\overline{x}^l} \frac{\Delta_j x^j}{\Delta_{(m)}\overline{x}^m} g_{ij}$$

$$- \frac{\Delta_i x^i}{\Delta_{(l)}\overline{x}^l} \frac{\Delta_j^2 x^j}{\Delta_{(n)}\overline{x}^n \Delta_{(m)}\overline{x}^m} g_{ij} - G_{lmn}$$

$$= \left(\frac{\partial g_{jk}}{\Delta_i x^i} + \frac{\partial g_{ik}}{\Delta_j x^j} - \frac{\partial g_{ij}}{\Delta_k x^k} \right) \frac{\Delta_i x^i}{\Delta_{(l)}\overline{x}^l} \frac{\Delta_j x^j}{\Delta_{(m)}\overline{x}^m} \frac{\Delta_k x^k}{\Delta_{(n)}\overline{x}^n}$$

$$+ \left(\frac{\Delta_j^2 x^j}{\Delta_{(l)}\overline{x}^l \Delta_{(m)}\overline{x}^m} \frac{\Delta_k x^k}{\Delta_{(n)}\overline{x}^n} g_{jk} + \frac{\Delta_i^2 x^i}{\Delta_{(l)}\overline{x}^l \Delta_{(m)}\overline{x}^m} \frac{\Delta_k x^k}{\Delta_{(n)}\overline{x}^n} g_{ik} \right)$$

$$+ \left(\frac{\Delta_j x^j}{\Delta_{(m)}\overline{x}^m} \frac{\Delta_k^2 x^k}{\Delta_{(l)}\overline{x}^l \Delta_{(n)}\overline{x}^n} g_{jk} - \frac{\Delta_i^2 x^i}{\Delta_{(n)}\overline{x}^n \Delta_{(l)}\overline{x}^l} \frac{\Delta_j x^j}{\Delta_{(m)}\overline{x}^m} g_{ij} \right)$$

$$+ \left(\frac{\Delta_k^2 x^k}{\Delta_{(n)}\overline{x}^n \Delta_{(m)}\overline{x}^m} \frac{\Delta_i x^i}{\Delta_{(l)}\overline{x}^l} g_{ik} - \frac{\Delta_i x^i}{\Delta_{(l)}\overline{x}^l} \frac{\Delta_j^2 x^j}{\Delta_{(n)}\overline{x}^n \Delta_{(m)}\overline{x}^m} g_{ij} \right)$$

$$+ G_{mnl} + G_{lnm} - G_{lmn}$$

$$= 2\Gamma_{ij,k} \frac{\Delta_i x^i}{\Delta_{(l)}\overline{x}^l} \frac{\Delta_j x^j}{\Delta_{(m)}\overline{x}^m} \frac{\Delta_k x^k}{\Delta_{(n)}\overline{x}^n}$$

$$+ \left(\frac{\Delta_j^2 x^j}{\Delta_{(l)}\overline{x}^l \Delta_{(m)}\overline{x}^m} \frac{\Delta_k x^k}{\Delta_{(n)}\overline{x}^n} g_{jk} + \frac{\Delta_j^2 x^j}{\Delta_{(l)}\overline{x}^l \Delta_{(m)}\overline{x}^m} \frac{\Delta_k x^k}{\Delta_{(n)}\overline{x}^n} g_{jk} \right)$$

$$+ \left(\frac{\Delta_j x^j}{\Delta_{(m)}\overline{x}^m} \frac{\Delta_k^2 x^k}{\Delta_{(l)}\overline{x}^l \Delta_{(n)}\overline{x}^n} g_{jk} - \frac{\Delta_k^2 x^k}{\Delta_{(n)}\overline{x}^n \Delta_{(l)}\overline{x}^l} \frac{\Delta_j x^j}{\Delta_{(m)}\overline{x}^m} g_{kj} \right)$$

$$+ \left(\frac{\Delta_k^2 x^k}{\Delta_{(n)}\overline{x}^n \Delta_{(m)}\overline{x}^m} \frac{\Delta_i x^i}{\Delta_{(l)}\overline{x}^l} g_{ik} - \frac{\Delta_i x^i}{\Delta_{(l)}\overline{x}^l} \frac{\Delta_k^2 x^k}{\Delta_{(n)}\overline{x}^n \Delta_{(m)}\overline{x}^m} g_{ik} \right)$$

$$+ G_{mnl} + G_{lnm} - G_{lmn}$$

$$= 2\Gamma_{ij,k}\frac{\Delta_i x^i}{\Delta_{(l)}\overline{x}^l}\frac{\Delta_j x^j}{\Delta_{(m)}\overline{x}^m}\frac{\Delta_k x^k}{\Delta_{(n)}\overline{x}^n} + 2\frac{\Delta_j^2 x^j}{\Delta_{(l)}\overline{x}^l\Delta_{(m)}\overline{x}^m}\frac{\Delta_k x^k}{\Delta_{(n)}\overline{x}^n}g_{jk}$$

$$+ 2\frac{\Delta_j^2 x^j}{\Delta_{(l)}\overline{x}^l\Delta_{(m)}\overline{x}^m}\frac{\Delta_k x^k}{\Delta_{(n)}\overline{x}^n}g_{jk}$$

$$+ (G_{mnl} + G_{lnm} - G_{lmn}).$$

Therefore

$$\overline{\Gamma}_{lm,n} = \frac{1}{2}\left(\frac{\partial\overline{g}_{mn}}{\Delta_{(l)}\overline{x}^l} + \frac{\partial\overline{g}_{nl}}{\Delta_{(m)}\overline{x}^m} - \frac{\partial\overline{g}_{lm}}{\Delta_{(n)}\overline{x}^n}\right)$$

$$= \Gamma_{ij,k}\frac{\Delta_i x^i}{\Delta_{(l)}\overline{x}^l}\frac{\Delta_j x^j}{\Delta_{(m)}\overline{x}^m}\frac{\Delta_k x^k}{\Delta_{(n)}\overline{x}^n} + \frac{\Delta_j^2 x^j}{\Delta_{(l)}\overline{x}^l\Delta_{(m)}\overline{x}^m}\frac{\Delta_k x^k}{\Delta_{(n)}\overline{x}^n}g_{jk}$$

$$+ \frac{\Delta_j^2 x^j}{\Delta_{(l)}\overline{x}^l\Delta_{(m)}\overline{x}^m}\frac{\Delta_k x^k}{\Delta_{(n)}\overline{x}^n}g_{jk}$$

$$+ \frac{1}{2}(G_{mnl} + G_{lnm} - G_{lmn}). \tag{4.8}$$

Now we will deduct the transformation law for the Christoffel symbols of the second kinds. We have

$$\overline{g}^{np} = g^{rs}\frac{\Delta_{(n)}\overline{x}^{n\sigma_0\sigma_1\cdots\sigma_{r-1}}}{\Delta_r x^r}\frac{\Delta_{(p)}\overline{x}^{p\sigma_0\sigma_1\cdots\sigma_{s-1}}}{\Delta_s x^s}. \tag{4.9}$$

Inner multiplication of both sides of (4.8) by the corresponding side of (4.9) gives

$$\overline{g}^{np}\overline{\Gamma}_{lm,n} = g^{rs}\frac{\Delta_{(n)}\overline{x}^{n\sigma_0\sigma_1\cdots\sigma_{r-1}}}{\Delta_r x^r}\frac{\Delta_{(p)}\overline{x}^{p\sigma_0\sigma_1\cdots\sigma_{s-1}}}{\Delta_s x^s}$$

$$+ g^{rs}\frac{\Delta_{(n)}\overline{x}^{n\sigma_0\sigma_1\cdots\sigma_{r-1}}}{\Delta_r x^r}\frac{\Delta_{(p)}\overline{x}^{p\sigma_0\sigma_1\cdots\sigma_{s-1}}}{\Delta_s x^s}\frac{\Delta_j^2 x^j}{\Delta_{(l)}\overline{x}^l\Delta_{(m)}\overline{x}^m}\frac{\Delta_k x^k}{\Delta_{(n)}\overline{x}^n}g_{jk}$$

$$+ \frac{1}{2}g^{rs}\frac{\Delta_{(n)}\overline{x}^{n\sigma_0\sigma_1\cdots\sigma_{r-1}}}{\Delta_r x^r}\frac{\Delta_{(p)}\overline{x}^{p\sigma_0\sigma_1\cdots\sigma_{s-1}}}{\Delta_s x^s}(G_{mnl} + G_{lnm} - G_{lmn})$$

$$= g^{ks}\Gamma_{ij,k}\frac{\Delta_{(p)}\overline{x}^{p\sigma_0\sigma_1\cdots\sigma_{s-1}}}{\Delta_s x^s}\frac{\Delta_i x^i}{\Delta_{(l)}\overline{x}^l}\frac{\Delta_j x^j}{\Delta_{(m)}\overline{x}^m}$$

$$+ g^{rs}g_{jr}\frac{\Delta_{(p)}\overline{x}^{p\sigma_0\sigma_1\cdots\sigma_{s-1}}}{\Delta_s x^s}\frac{\Delta_j^2 x^j}{\Delta_{(l)}\overline{x}^l\Delta_{(m)}\overline{x}^m}$$

$$+ \frac{1}{2}g^{rs}\frac{\Delta_{(n)}\overline{x}^{n\sigma_0\sigma_1\cdots\sigma_{r-1}}}{\Delta_r x^r}\frac{\Delta_{(p)}\overline{x}^{p\sigma_0\sigma_1\cdots\sigma_{s-1}}}{\Delta_s x^s}(G_{mnl} + G_{lnm} - G_{lmn})$$

$$= g^{ks}\Gamma_{ij,k}\frac{\Delta_{(p)}\overline{x}^{p\sigma_0\sigma_1\cdots\sigma_{s-1}}}{\Delta_s x^s}\frac{\Delta_i x^i}{\Delta_{(l)}\overline{x}^l}\frac{\Delta_j x^j}{\Delta_{(m)}\overline{x}^m}$$

$$
\begin{aligned}
&+ \delta_j^s \frac{\Delta_{(p)}\overline{x}^{p\sigma_0\sigma_1\cdots\sigma_{s-1}}}{\Delta_s x^s} \frac{\Delta_j^2 x^j}{\Delta_{(l)}\overline{x}^l \Delta_{(m)}\overline{x}^m} \\
&+ \frac{1}{2} g^{rs} \frac{\Delta_{(n)}\overline{x}^{n\sigma_0\sigma_1\cdots\sigma_{r-1}}}{\Delta_r x^r} \frac{\Delta_{(p)}\overline{x}^{p\sigma_0\sigma_1\cdots\sigma_{s-1}}}{\Delta_s x^s} (G_{mnl} + G_{lnm} - G_{lmn}) \\
&= g^{ks}\Gamma_{ij,k} \frac{\Delta_{(p)}\overline{x}^{p\sigma_0\sigma_1\cdots\sigma_{s-1}}}{\Delta_s x^s} \frac{\Delta_i x^i}{\Delta_{(l)}\overline{x}^l} \frac{\Delta_j x^j}{\Delta_{(m)}\overline{x}^m} \\
&+ \frac{\Delta_{(p)}\overline{x}^{p\sigma_0\sigma_1\cdots\sigma_{j-1}}}{\Delta_j x^j} \frac{\Delta_j^2 x^j}{\Delta_{(l)}\overline{x}^l \Delta_{(m)}\overline{x}^m} \\
&+ \frac{1}{2} g^{rs} \frac{\Delta_{(n)}\overline{x}^{n\sigma_0\sigma_1\cdots\sigma_{r-1}}}{\Delta_r x^r} \frac{\Delta_{(p)}\overline{x}^{p\sigma_0\sigma_1\cdots\sigma_{s-1}}}{\Delta_s x^s} (G_{mnl} + G_{lnm} - G_{lmn})
\end{aligned}
$$

or

$$
\begin{aligned}
\overline{\Gamma}^p_{lm} = {}& g^{ks}\Gamma_{ij,k} \frac{\Delta_{(p)}\overline{x}^{p\sigma_0\sigma_1\cdots\sigma_{s-1}}}{\Delta_s x^s} \frac{\Delta_i x^i}{\Delta_{(l)}\overline{x}^l} \frac{\Delta_j x^j}{\Delta_{(m)}\overline{x}^m} \\
&+ \frac{\Delta_{(p)}\overline{x}^{p\sigma_0\sigma_1\cdots\sigma_{j-1}}}{\Delta_j x^j} \frac{\Delta_j^2 x^j}{\Delta_{(l)}\overline{x}^l \Delta_{(m)}\overline{x}^m} \\
&+ \frac{1}{2} g^{rs} \frac{\Delta_{(n)}\overline{x}^{n\sigma_0\sigma_1\cdots\sigma_{r-1}}}{\Delta_r x^r} \frac{\Delta_{(p)}\overline{x}^{p\sigma_0\sigma_1\cdots\sigma_{s-1}}}{\Delta_s x^s} (G_{mnl} + G_{lnm} - G_{lmn}).
\end{aligned}
$$

Let

$$
\widetilde{G}_{plm} = \frac{1}{2} g^{rs} \frac{\Delta_{(n)}\overline{x}^{n\sigma_0\sigma_1\cdots\sigma_{r-1}}}{\Delta_r x^r} \frac{\Delta_{(p)}\overline{x}^{p\sigma_0\sigma_1\cdots\sigma_{s-1}}}{\Delta_s x^s} (G_{mnl} + G_{lnm} - G_{lmn}).
$$

Then

$$
\begin{aligned}
\overline{\Gamma}^p_{lm} = {}& g^{ks}\Gamma_{ij,k} \frac{\Delta_{(p)}\overline{x}^{p\sigma_0\sigma_1\cdots\sigma_{s-1}}}{\Delta_s x^s} \frac{\Delta_i x^i}{\Delta_{(l)}\overline{x}^l} \frac{\Delta_j x^j}{\Delta_{(m)}\overline{x}^m} \\
&+ \frac{\Delta_{(p)}\overline{x}^{p\sigma_0\sigma_1\cdots\sigma_{j-1}}}{\Delta_j x^j} \frac{\Delta_j^2 x^j}{\Delta_{(l)}\overline{x}^l \Delta_{(m)}\overline{x}^m} + \widetilde{G}_{plm}.
\end{aligned}
$$

By the last equation we find

$$
\begin{aligned}
\overline{\Gamma}^p_{lm} - \widetilde{G}_{plm} = {}& g^{ks}\Gamma_{ij,k} \frac{\Delta_{(p)}\overline{x}^{p\sigma_0\sigma_1\cdots\sigma_{s-1}}}{\Delta_s x^s} \frac{\Delta_i x^i}{\Delta_{(l)}\overline{x}^l} \frac{\Delta_j x^j}{\Delta_{(m)}\overline{x}^m} \\
&+ \frac{\Delta_{(p)}\overline{x}^{p\sigma_0\sigma_1\cdots\sigma_{j-1}}}{\Delta_j x^j} \frac{\Delta_j^2 x^j}{\Delta_{(l)}\overline{x}^l \Delta_{(m)}\overline{x}^m},
\end{aligned}
$$

which we multiply by $\dfrac{\Delta_r x^r}{\Delta_{(p)}\overline{x}^p}$ and find

$$\frac{\Delta_r x^r}{\Delta_{(p)}\overline{x}^p}(\overline{\Gamma}^p_{lm} - \widetilde{G}_{plm}) = g^{ks}\Gamma_{ij,k}\frac{\Delta_{(p)}\overline{x}^{p\sigma_0\sigma_1\cdots\sigma_{s-1}}}{\Delta_s x^s}\frac{\Delta_i x^i}{\Delta_{(l)}\overline{x}^l}\frac{\Delta_j x^j}{\Delta_{(m)}\overline{x}^m}\frac{\Delta_r x^r}{\Delta_{(p)}\overline{x}^p}$$

$$+ \frac{\Delta_r x^r}{\Delta_{(p)}\overline{x}^p}\frac{\Delta_{(p)}\overline{x}^{p\sigma_0\sigma_1\cdots\sigma_{j-1}}}{\Delta_j x^j}\frac{\Delta_j^2 x^j}{\Delta_{(l)}\overline{x}^l\Delta_{(m)}\overline{x}^m}$$

$$= g^{kr}\Gamma_{ij,k}\frac{\Delta_i x^i}{\Delta_{(l)}\overline{x}^l}\frac{\Delta_j x^j}{\Delta_{(m)}\overline{x}^m} + \frac{\Delta_r^2 x^r}{\Delta_{(l)}\overline{x}^l\Delta_{(m)}\overline{x}^m},$$

whereupon

$$\frac{\Delta_r^2 x^r}{\Delta_{(l)}\overline{x}^l\Delta_{(m)}\overline{x}^m} = \frac{\Delta_r x^r}{\Delta_{(p)}\overline{x}^p}(\overline{\Gamma}^p_{lm} - \widetilde{G}_{plm}) - \Gamma^r_{ij}\frac{\Delta_i x^i}{\Delta_{(l)}\overline{x}^l}\frac{\Delta_j x^j}{\Delta_{(m)}\overline{x}^m}.$$

4.3 Covariant differentiation

In this section, we introduce the concept of covariant differentiation tensors of arbitrary order.

Definition 4.4. The covariant derivative of a tensor $A^{i_1 i_2 \cdots i_p}_{j_1 j_2 \cdots j_q}$ is defined as

$$A^{i_1 i_2 \cdots i_p}_{j_1 j_2 \cdots j_q,k} = \frac{\partial A^{i_1 i_2 \cdots i_p}_{j_1 j_2 \cdots j_q}}{\Delta_k x^k} + A^{a i_2 \cdots i_p}_{j_1 j_2 \cdots j_q}\Gamma^{i_1}_{ak}$$

$$+ A^{i_1 a i_3 \cdots i_p}_{j_1 j_2 j_3 \cdots j_q,k}\Gamma^{i_2}_{ak} + \cdots + A^{i_1 i_2 \cdots i_{p-1} a}_{j_1 j_2 \cdots j_{q-1} j_q}\Gamma^{i_p}_{ak}$$

$$- A^{i_1 i_2 \cdots i_p}_{\beta j_2 \cdots j_q}\Gamma^{\beta}_{j_1 k} - A^{i_1 i_2 \cdots i_p}_{j_1 \beta j_3 \cdots j_q}\Gamma^{\beta}_{j_2 k} - \cdots - A^{i_1 i_2 \cdots i_p}_{j_1 j_2 \cdots j_{q-1}\beta}\Gamma^{\beta}_{j_q k}.$$

Definition 4.5. Let ϕ be an invariant. Then its covariant derivative is defined by

$$\phi_{,j} = \frac{\partial \phi}{\Delta_j x^j}.$$

Example 4.7. Consider a covariant vector A_i. The transformation law for covariant vectors gives

$$\overline{A}_i = \frac{\Delta_k x^k}{\Delta_{(i)}\overline{x}^i}A_k.$$

Differentiating the last equation with respect to $\overline{x}^j$, we get

$$\frac{\partial \overline{A}_i}{\Delta_{(j)}\overline{x}^j} = \frac{\Delta_k^2 x^k}{\Delta_{(j)}\overline{x}^j \Delta_{(i)}\overline{x}^i} A_k^{\sigma_{(j)}} + \frac{\partial A_k}{\Delta_{(j)}\overline{x}^j} \frac{\Delta_k x^k}{\Delta_{(i)}\overline{x}^i}$$

$$= \frac{\Delta_k^2 x^k}{\Delta_{(j)}\overline{x}^j \Delta_{(i)}\overline{x}^i} A_k^{\sigma_{(j)}} + \left(\frac{\partial A_k}{\Delta_l x^l}\right)^{\sigma_0 \sigma_1 \cdots \sigma_{l-1}} \frac{\Delta_l x^l}{\Delta_{(j)}\overline{x}^j} \frac{\Delta_k x^k}{\Delta_{(i)}\overline{x}^i}.$$

Now applying (4.7), we find

$$\frac{\partial \overline{A}_i}{\Delta_{(j)}\overline{x}^j} = \left(\frac{\partial A_k}{\Delta_l x^l}\right)^{\sigma_0 \sigma_1 \cdots \sigma_{l-1}} \frac{\Delta_l x^l}{\Delta_{(j)}\overline{x}^j} \frac{\Delta_k x^k}{\Delta_{(i)}\overline{x}^i}$$

$$+ \frac{\Delta_k x^k}{\Delta_{(p)}\overline{x}^p} \left(\overline{\Gamma}_{ij}^p - g^{ks}\widetilde{\Gamma}_{ij,q} \frac{\Delta_{(q)}\overline{x}^q \Delta_{(p)}\overline{x}^p}{\Delta_k x^k \Delta_s x^s}\right) A_k^{\sigma_{(j)}}$$

$$- \Gamma_{\sigma mn}^k A_k^{\sigma_{(j)}} \frac{\Delta_m x^m \Delta_n x^n}{\Delta_{(i)}\overline{x}^i \Delta_{(j)}\overline{x}^j},$$

whereupon

$$\frac{\partial \overline{A}_i}{\Delta_{(j)}\overline{x}^j} - \frac{\Delta_k x^k}{\Delta_{(p)}\overline{x}^p} \left(\overline{\Gamma}_{ij}^p - g^{ks}\widetilde{\Gamma}_{ij,q} \frac{\Delta_{(q)}\overline{x}^q \Delta_{(p)}\overline{x}^p}{\Delta_k x^k \Delta_s x^s}\right) A_k^{\sigma_{(j)}}$$

$$= \left(\frac{\partial A_k}{\Delta_l x^l}\right)^{\sigma_0 \sigma_1 \cdots \sigma_{l-1}} \frac{\Delta_l x^l}{\Delta_{(j)}\overline{x}^j} \frac{\Delta_k x^k}{\Delta_{(i)}\overline{x}^i} - \Gamma_{\sigma mn}^k A_k^{\sigma_{(j)}} \frac{\Delta_m x^m \Delta_n x^n}{\Delta_{(i)}\overline{x}^i \Delta_{(j)}\overline{x}^j}.$$

Replacing the dummy indices k and l in the first term of the right-hand side by t and s, respectively, and the dummy indices k, m, n in the second term of the right-hand side by r, t, s, respectively, we get

$$\frac{\partial \overline{A}_i}{\Delta_{(j)}\overline{x}^j} - \frac{\Delta_k x^k}{\Delta_{(p)}\overline{x}^p} \left(\overline{\Gamma}_{ij}^p - g^{ks}\widetilde{\Gamma}_{ij,q} \frac{\Delta_{(q)}\overline{x}^q \Delta_{(p)}\overline{x}^p}{\Delta_k x^k \Delta_s x^s}\right) A_k^{\sigma_{(j)}}$$

$$= \left(\frac{\partial A_t}{\Delta_s x^s}\right)^{\sigma_0 \sigma_1 \cdots \sigma_{s-1}} \frac{\Delta_t x^t}{\Delta_{(j)}\overline{x}^j} \frac{\Delta_s x^s}{\Delta_{(i)}\overline{x}^i} - \Gamma_{\sigma ts}^r A_r^{\sigma_{(j)}} \frac{\Delta_t x^t}{\Delta_{(i)}\overline{x}^i} \frac{\Delta_s x^s}{\Delta_{(j)}\overline{x}^j}$$

$$= \left(\left(\frac{\partial A_t}{\Delta_s x^s}\right)^{\sigma_0 \sigma_1 \cdots \sigma_{s-1}} - \Gamma_{\sigma ts}^r A_r^{\sigma_{(j)}}\right) \frac{\Delta_t x^t}{\Delta_{(i)}\overline{x}^i} \frac{\Delta_s x^s}{\Delta_{(j)}\overline{x}^j}$$

$$= \left(\left(\frac{\partial A_t}{\Delta_s x^s}\right)^{\sigma_0 \sigma_1 \cdots \sigma_{s-1}} - \Gamma_{\sigma ts}^r A_r - \mu_{(j)}\Gamma_{\sigma ts}^r \frac{\partial A_r}{\Delta_{(j)}\overline{x}^j}\right) \frac{\Delta_t x^t}{\Delta_{(i)}\overline{x}^i} \frac{\Delta_s x^s}{\Delta_{(j)}\overline{x}^j}.$$

Hence

$$\frac{\partial \overline{A}_i}{\Delta_{(j)}\overline{x}^j} - \frac{\Delta_k x^k}{\Delta_{(p)}\overline{x}^p} \left(\overline{\Gamma}_{ij}^p - g^{ks}\widetilde{\Gamma}_{ij,q} \frac{\Delta_{(q)}\overline{x}^q \Delta_{(p)}\overline{x}^p}{\Delta_k x^k \Delta_s x^s}\right) A_k^{\sigma_{(j)}} + \mu_{(j)}\Gamma_{\sigma ts}^r \frac{\partial A_r}{\Delta_{(j)}\overline{x}^j} \frac{\Delta_t x^t}{\Delta_{(i)}\overline{x}^i} \frac{\Delta_s x^s}{\Delta_{(j)}\overline{x}^j}$$

$$= \left(\left(\frac{\partial A_t}{\Delta_s x^s}\right)^{\sigma_0 \sigma_1 \cdots \sigma_{s-1}} - \Gamma_{\sigma ts}^r A_r\right) \frac{\Delta_t x^t}{\Delta_{(i)}\overline{x}^i} \frac{\Delta_s x^s}{\Delta_{(j)}\overline{x}^j}.$$

Note that

$$\left(\frac{\partial A_t}{\Delta_s x^s}\right)^{\sigma_0\sigma_1\cdots\sigma_{s-1}} = \frac{\partial}{\Delta_s x^s}\left(A_t^{\sigma_0\sigma_1\cdots\sigma_{s-1}}\right)$$

$$= \frac{\partial}{\Delta_s x^s}\left(A_t^{\sigma_0\sigma_1\cdots\sigma_{s-2}} + \mu_{s-1}\frac{\partial A_t^{\sigma_0\sigma_1\cdots\sigma_{s-2}}}{\Delta_{s-1}x^{s-1}}\right)$$

$$= \frac{\partial}{\Delta_s x^s}\left(A_t^{\sigma_0\sigma_1\cdots\sigma_{s-3}} + \mu_{s-2}\frac{\partial A_t^{\sigma_0\sigma_1\cdots\sigma_{s-3}}}{\Delta_{s-2}x^{s-2}} + \mu_{s-1}\frac{\partial A_t^{\sigma_0\sigma_1\cdots\sigma_{s-2}}}{\Delta_{s-1}x^{s-1}}\right)$$

$$\vdots$$

$$= \frac{\partial}{\Delta_s x^s}\left(A_t + \mu_1\frac{\partial A_t}{\Delta_1 x^1} + \mu_2\frac{\partial A_t^{\sigma_0\sigma_1}}{\Delta_2 x^2} + \cdots + \mu_{s-1}\frac{\partial A_t^{\sigma_0\sigma_1\cdots\sigma_{s-2}}}{\Delta_{s-1}x^{s-1}}\right)$$

$$= \frac{\partial A_t}{\Delta_s x^s} + \frac{\partial}{\Delta_s x^s}\left(\mu_1\frac{\partial A_t}{\Delta_1 x^1} + \mu_2\frac{\partial A_t^{\sigma_0\sigma_1}}{\Delta_2 x^2} + \cdots + \mu_{s-1}\frac{\partial A_t^{\sigma_0\sigma_1\cdots\sigma_{s-2}}}{\Delta_{s-1}x^{s-1}}\right).$$

Let

$$\tilde{A}_{t,s} = \frac{\partial}{\Delta_s x^s}\left(\mu_1\frac{\partial A_t}{\Delta_1 x^1} + \mu_2\frac{\partial A_t^{\sigma_0\sigma_1}}{\Delta_2 x^2} + \cdots + \mu_{s-1}\frac{\partial A_t^{\sigma_0\sigma_1\cdots\sigma_{s-2}}}{\Delta_{s-1}x^{s-1}}\right).$$

Then

$$\left(\frac{\partial A_t}{\Delta_s x^s}\right)^{\sigma_0\sigma_1\cdots\sigma_{s-1}} = \frac{\partial A_t}{\Delta_s x^s} + \tilde{A}_{t,s}.$$

Therefore

$$\frac{\partial \bar{A}_i}{\Delta_{(j)}\bar{x}^j} - \frac{\Delta_k x^k}{\Delta_{(p)}\bar{x}^p}\left(\bar{\Gamma}_{ij}^p - g^{ks}\tilde{\Gamma}_{ij,q}\frac{\Delta_{(q)}\bar{x}^q}{\Delta_k x^k}\frac{\Delta_{(p)}\bar{x}^p}{\Delta_s x^s}\right)A_k^{\sigma_{(j)}}$$

$$+ \mu_{(j)}\Gamma_{\sigma ts}^r\frac{\partial A_r}{\Delta_{(j)}\bar{x}^j}\frac{\Delta_t x^t}{\Delta_{(i)}\bar{x}^i}\frac{\Delta_s x^s}{\Delta_{(j)}\bar{x}^j} - \tilde{A}_{t,s}\frac{\Delta_t x^t}{\Delta_{(i)}\bar{x}^i}\frac{\Delta_s x^s}{\Delta_{(j)}\bar{x}^j}$$

$$= \left(\frac{\partial A_t}{\Delta_s x^s} - \Gamma_{\sigma ts}^r A_r\right)\frac{\Delta_t x^t}{\Delta_{(i)}\bar{x}^i}\frac{\Delta_s x^s}{\Delta_{(j)}\bar{x}^j}.$$

Observe that

$$\Gamma_{\sigma ts}^r = \frac{1}{2}\left(\left(\frac{\partial g_{rs}}{\Delta_t x^t}\right)^{\sigma_0\sigma_1\cdots\sigma_{t-1}} + \left(\frac{\partial g_{rt}}{\Delta_s x^s}\right)^{\sigma_0\sigma_1\cdots\sigma_{s-1}} - \left(\frac{\partial g_{ts}}{\Delta_r x^r}\right)^{\sigma_0\sigma_1\cdots\sigma_{r-1}}\right)$$

$$= \frac{1}{2}\left(\frac{\partial g_{rs}}{\Delta_t x^t} + \sum_{l=1}^{t-1}\mu_l\left(\frac{\partial g_{rs}}{\Delta_t x^t}\right)^{\sigma_0\sigma_1\cdots\sigma_{l-1}} + \frac{\partial g_{rt}}{\Delta_s x^s} + \sum_{l=1}^{s-1}\mu_l\left(\frac{\partial g_{rt}}{\Delta_s x^s}\right)^{\sigma_0\sigma_1\cdots\sigma_{l-1}}\right)$$

$$
-\frac{\partial g_{ts}}{\Delta_r x^r} - \sum_{l=1}^{r-1} \mu_l \left(\frac{\partial g_{ts}}{\Delta_r x^r} \right)^{\sigma_0 \sigma_1 \cdots \sigma_{l-1}}
$$

$$
= \Gamma_{rt}^s + \frac{1}{2} \left(\sum_{l=1}^{t-1} \mu_l \left(\frac{\partial g_{rs}}{\Delta_t x^t} \right)^{\sigma_0 \sigma_1 \cdots \sigma_{l-1}} + \sum_{l=1}^{s-1} \mu_l \left(\frac{\partial g_{rt}}{\Delta_s x^s} \right)^{\sigma_0 \sigma_1 \cdots \sigma_{l-1}} - \sum_{l=1}^{r-1} \mu_l \left(\frac{\partial g_{ts}}{\Delta_r x^r} \right)^{\sigma_0 \sigma_1 \cdots \sigma_{l-1}} \right).
$$

Consequently,

$$
\frac{\partial \overline{A}_i}{\Delta_{(j)} \overline{x}^j} - \frac{\Delta_k x^k}{\Delta_{(p)} \overline{x}^p} \left(\overline{\Gamma}_{ij}^p - g^{ks} \widetilde{\Gamma}_{ij,q} \frac{\Delta_{(q)} \overline{x}^q}{\Delta_k x^k} \frac{\Delta_{(p)} \overline{x}^p}{\Delta_s x^s} \right) A_k^{\sigma_{(j)}}
$$

$$
+ \mu_{(j)} \Gamma_{\sigma ts}^r \frac{\partial A_r}{\Delta_{(j)} \overline{x}^j} \frac{\Delta_t x^t}{\Delta_{(i)} \overline{x}^i} \frac{\Delta_s x^s}{\Delta_{(j)} \overline{x}^j} - \widetilde{A}_{t,s} \frac{\Delta_t x^t}{\Delta_{(i)} \overline{x}^i} \frac{\Delta_s x^s}{\Delta_{(j)} \overline{x}^j}
$$

$$
+ \frac{1}{2} \left(\sum_{l=1}^{t-1} \mu_l \left(\frac{\partial g_{rs}}{\Delta_t x^t} \right)^{\sigma_0 \sigma_1 \cdots \sigma_{l-1}} + \sum_{l=1}^{s-1} \mu_l \left(\frac{\partial g_{rt}}{\Delta_s x^s} \right)^{\sigma_0 \sigma_1 \cdots \sigma_{l-1}} - \sum_{l=1}^{r-1} \mu_l \left(\frac{\partial g_{ts}}{\Delta_r x^r} \right)^{\sigma_0 \sigma_1 \cdots \sigma_{l-1}} \right)
$$

$$
\times A_r \frac{\Delta_t x^t}{\Delta_{(i)} \overline{x}^i} \frac{\Delta_s x^s}{\Delta_{(j)} \overline{x}^j}
$$

$$
= \left(\frac{\partial A_t}{\Delta_s x^s} - \Gamma_{ts}^r A_r \right) \frac{\Delta_t x^t}{\Delta_{(i)} \overline{x}^i} \frac{\Delta_s x^s}{\Delta_{(j)} \overline{x}^j}.
$$

Example 4.8. We will show that if the covariant derivative of a covariant vector is symmetric, then the vector is a gradient. Indeed, we have

$$
\frac{\partial A_t}{\Delta_s x^s} - \Gamma_{ts}^r A_r = \frac{\partial A_s}{\Delta_t x^t} - \Gamma_{st}^r A_r
$$

$$
= \frac{\partial A_s}{\Delta_t x^t} - \Gamma_{ts}^r A_r.
$$

Then

$$
\frac{\partial A_t}{\Delta_s x^s} = \frac{\partial A_s}{\Delta_t x^t},
$$

$$
\Delta A_t = \frac{\partial}{\Delta_t x^t} (A_s \Delta_s x^s),
$$

and

$$
A_t = \int \frac{\partial}{\Delta_t x^t} (A_s \Delta_s x^s)
$$

$$
= \frac{\partial}{\Delta_t x^t} \left(\int A_s \Delta_s x^s \right).
$$

Since

$$
\int A_s \Delta_s x^s
$$

is a scalar quantity, let it be ϕ. Then

$$A_t = \frac{\partial \phi}{\Delta_t x^t}.$$

Example 4.9. Now we consider the contravariant vector A^i. The transformation law for contravariant vectors gives

$$A^k = \frac{\Delta_k x^k}{\Delta_{(i)}\overline{x}^i}\overline{A}^i.$$

Differentiating the last equation with respect to x^j, we get

$$\frac{\partial A^k}{\Delta_j x^j} = \frac{\partial}{\Delta_j x^j}\left(\frac{\Delta_k x^k}{\Delta_{(i)}\overline{x}^i}\overline{A}^i\right)$$

$$= \frac{\partial}{\Delta_j x^j}\left(\frac{\Delta_k x^k}{\Delta_{(i)}\overline{x}^i}\right)\overline{A}^{i\sigma_j} + \frac{\Delta_k x^k}{\Delta_{(i)}\overline{x}^i}\frac{\partial \overline{A}^j}{\Delta_j x^j}$$

$$= \left(\frac{\Delta_k^2 x^k}{\Delta_{(p)}\overline{x}^p \Delta_{(i)}\overline{x}^i}\right)^{\sigma_{(0)}\sigma_{(1)}\cdots\sigma_{(p-1)}}\frac{\Delta_{(p)}\overline{x}^p}{\Delta_j x^j}\overline{A}^{i\sigma_j} + \frac{\Delta_k x^k}{\Delta_{(i)}\overline{x}^i}\frac{\partial \overline{A}^j}{\Delta_j x^j}.$$

Applying (4.7), we obtain

$$\frac{\partial A^k}{\Delta_j x^j} = \frac{\Delta_k x^k}{\Delta_{(i)}\overline{x}^i}\left(\frac{\partial \overline{A}^i}{\Delta_{(p)}\overline{x}^p}\right)^{\sigma_{(0)}\sigma_{(1)}\cdots\sigma_{(p-1)}}\frac{\Delta_{(p)}\overline{x}^p}{\Delta_j x^j}$$

$$+ \left(\frac{\Delta_k x^k}{\Delta_{(m)}\overline{x}^m}\left(\overline{\Gamma}^m_{pi}g^{ks}\widetilde{\Gamma}_{pi,n}\frac{\Delta_{(n)}\overline{x}^n}{\Delta_k x^k}\frac{\Delta_{(m)}\overline{x}^m}{\Delta_s x^s}\right) - \Gamma^k_{\sigma ts}\frac{\Delta_t x^t}{\Delta_{(i)}\overline{x}^i}\frac{\Delta_s x^s}{\Delta_{(p)}\overline{x}^p}\right)^{\sigma_{(0)}\sigma_{(1)}\cdots\sigma_{(p-1)}}\overline{A}^{i\sigma_j}\frac{\Delta_{(p)}\overline{x}^p}{\Delta_j x^j}$$

$$= \frac{\Delta_k x^k}{\Delta_{(i)}\overline{x}^i}\left(\frac{\partial \overline{A}^i}{\Delta_{(p)}\overline{x}^p}\right)^{\sigma_{(0)}\sigma_{(1)}\cdots\sigma_{(p-1)}}\frac{\Delta_{(p)}\overline{x}^p}{\Delta_j x^j}$$

$$+ \overline{A}^{i\sigma_j}\frac{\Delta_{(p)}\overline{x}^p}{\Delta_j x^j}\left(\frac{\Delta_k x^k}{\Delta_{(m)}\overline{x}^m}\left(\overline{\Gamma}^m_{pi} - g^{ks}\widetilde{\Gamma}_{pi,n}\frac{\Delta_{(n)}\overline{x}^n}{\Delta_k x^k}\frac{\Delta_{(m)}\overline{x}^m}{\Delta_s x^s}\right)\right)^{\sigma_{(0)}\sigma_{(1)}\cdots\sigma_{(p-1)}}$$

$$- \overline{A}^{i\sigma_j}\frac{\Delta_{(p)}}{\Delta_j x^j}\left(\Gamma^k_{\sigma ts}\frac{\Delta_t x^t}{\Lambda_{(i)}\overline{x}^i}\frac{\Delta_s x^s}{\Delta_{(p)}\overline{x}^p}\right)^{\sigma_{(0)}\sigma_{(1)}\cdots\sigma_{(p-1)}}$$

$$= \frac{\Delta_k x^k}{\Delta_{(i)}\overline{x}^i}\left(\frac{\partial \overline{A}^i}{\Delta_{(p)}\overline{x}^p}\right)^{\sigma_{(0)}\sigma_{(1)}\cdots\sigma_{(p-1)}}\frac{\Delta_{(p)}\overline{x}^p}{\Delta_j x^j}$$

$$+ \overline{A}^{i\sigma_j}\frac{\Delta_{(p)}\overline{x}^p}{\Delta_j x^j}\left(\frac{\Delta_k x^k}{\Delta_{(m)}\overline{x}^m}\left(\overline{\Gamma}^m_{pi} - g^{ks}\widetilde{\Gamma}_{pi,n}\frac{\Delta_{(n)}\overline{x}^n}{\Delta_k x^k}\frac{\Delta_{(m)}\overline{x}^m}{\Delta_s x^s}\right)\right)^{\sigma_{(0)}\sigma_{(1)}\cdots\sigma_{(p-1)}}$$

$$- \left(\overline{A}^i + \mu_j\frac{\partial \overline{A}^i}{\Delta_j x^j}\right)\frac{\Delta_{(p)}\overline{x}^p}{\Delta_j x^j}\left(\Gamma^k_{\sigma ts}\frac{\Delta_t x^t}{\Delta_{(i)}\overline{x}^i}\frac{\Delta_s x^s}{\Delta_{(p)}\overline{x}^p} + \sum_{l=1}^{p-2}\mu_l\frac{\partial A_t^{\sigma_{(0)}\sigma_{(1)}\cdots\sigma_{(l-1)}}}{\Delta_l x^l}\right)$$

$$= \frac{\Delta_k x^k}{\Delta_{(i)} \overline{x}^i} \left(\frac{\partial \overline{A}^i}{\Delta_{(p)} \overline{x}^p} \right)^{\sigma_{(0)} \sigma_{(1)} \cdots \sigma_{(p-1)}} \frac{\Delta_{(p)} \overline{x}^p}{\Delta_j x^j}$$

$$+ \overline{A}^{i\sigma_j} \frac{\Delta_{(p)} \overline{x}^p}{\Delta_j x^j} \left(\frac{\Delta_k x^k}{\Delta_{(m)} \overline{x}^m} \left(\overline{\Gamma}^m_{pi} - g^{ks} \widetilde{\Gamma}_{pi,n} \frac{\Delta_{(n)} \overline{x}^n}{\Delta_k x^k} \frac{\Delta_{(m)} \overline{x}^m}{\Delta_s x^s} \right) \right)^{\sigma_{(0)} \sigma_{(1)} \cdots \sigma_{(p-1)}}$$

$$- \mu_j \frac{\partial \overline{A}^i}{\Delta_j x^j} \frac{\Delta_{(p)} \overline{x}^p}{\Delta_j x^j} \left(\Gamma^k_{\sigma ts} \frac{\Delta_t x^t}{\Delta_{(i)} \overline{x}^i} \frac{\Delta_s x^s}{\Delta_{(p)} \overline{x}^p} + \sum_{l=1}^{p-2} \mu_l \frac{\partial A_t^{\sigma_{(0)} \sigma_{(1)} \cdots \sigma_{(l-1)}}}{\Delta_l x^l} \right)$$

$$- \overline{A}^i \frac{\partial \overline{A}^i}{\Delta_j x^j} \frac{\Delta_{(p)} \overline{x}^p}{\Delta_j x^j} \left(\sum_{l=1}^{p-2} \mu_l \frac{\partial A_t^{\sigma_{(0)} \sigma_{(1)} \cdots \sigma_{(l-1)}}}{\Delta_l x^l} \right)$$

$$= \frac{\Delta_k x^k}{\Delta_{(i)} \overline{x}^i} \left(\frac{\partial \overline{A}^i}{\Delta_{(p)} \overline{x}^p} \right)^{\sigma_{(0)} \sigma_{(1)} \cdots \sigma_{(p-1)}} \frac{\Delta_{(p)} \overline{x}^p}{\Delta_j x^j}$$

$$+ \overline{A}^{i\sigma_j} \frac{\Delta_{(p)} \overline{x}^p}{\Delta_j x^j} \left(\frac{\Delta_k x^k}{\Delta_{(m)} \overline{x}^m} \left(\overline{\Gamma}^m_{pi} - g^{ks} \widetilde{\Gamma}_{pi,n} \frac{\Delta_{(n)} \overline{x}^n}{\Delta_k x^k} \frac{\Delta_{(m)} \overline{x}^m}{\Delta_s x^s} \right) \right)^{\sigma_{(0)} \sigma_{(1)} \cdots \sigma_{(p-1)}}$$

$$- \mu_j \frac{\partial \overline{A}^i}{\Delta_j x^j} \frac{\Delta_{(p)} \overline{x}^p}{\Delta_j x^j} \left(\Gamma^k_{\sigma ts} \frac{\Delta_t x^t}{\Delta_{(i)} \overline{x}^i} \frac{\Delta_s x^s}{\Delta_{(p)} \overline{x}^p} + \sum_{l=1}^{p-2} \mu_l \frac{\partial A_t^{\sigma_{(0)} \sigma_{(1)} \cdots \sigma_{(l-1)}}}{\Delta_l x^l} \right)$$

$$- \overline{A}^i \Gamma^k_{\sigma ts} \frac{\Delta_s x^s}{\Delta_j x^j} \frac{\Delta_t x^t}{\Delta_{(i)} \overline{x}^i}$$

$$= \frac{\Delta_k x^k}{\Delta_{(i)} \overline{x}^i} \left(\frac{\partial \overline{A}^i}{\Delta_{(p)} \overline{x}^p} \right)^{\sigma_{(0)} \sigma_{(1)} \cdots \sigma_{(p-1)}} \frac{\Delta_{(p)} \overline{x}^p}{\Delta_j x^j}$$

$$+ \overline{A}^{i\sigma_j} \frac{\Delta_{(p)} \overline{x}^p}{\Delta_j x^j} \left(\frac{\Delta_k x^k}{\Delta_{(m)} \overline{x}^m} \left(\overline{\Gamma}^m_{pi} - g^{ks} \widetilde{\Gamma}_{pi,n} \frac{\Delta_{(n)} \overline{x}^n}{\Delta_k x^k} \frac{\Delta_{(m)} \overline{x}^m}{\Delta_s x^s} \right) \right)^{\sigma_{(0)} \sigma_{(1)} \cdots \sigma_{(p-1)}}$$

$$- \mu_j \frac{\partial \overline{A}^i}{\Delta_j x^j} \frac{\Delta_{(p)} \overline{x}^p}{\Delta_j x^j} \left(\Gamma^k_{\sigma ts} \frac{\Delta_t x^t}{\Delta_{(i)} \overline{x}^i} \frac{\Delta_s x^s}{\Delta_{(p)} \overline{x}^p} + \sum_{l=1}^{p-2} \mu_l \frac{\partial A_t^{\sigma_{(0)} \sigma_{(1)} \cdots \sigma_{(l-1)}}}{\Delta_l x^l} \right)$$

$$- A^t \Gamma^k_{\sigma tj}$$

$$= \frac{\Delta_k x^k}{\Delta_{(i)} \overline{x}^i} \left(\frac{\partial \overline{A}^i}{\Delta_{(p)} \overline{x}^p} \right)^{\sigma_{(0)} \sigma_{(1)} \cdots \sigma_{(p-1)}} \frac{\Delta_{(p)} \overline{x}^p}{\Delta_j x^j}$$

$$+ \overline{A}^{i\sigma_j} \frac{\Delta_{(p)} \overline{x}^p}{\Delta_j x^j} \left(\frac{\Delta_k x^k}{\Delta_{(m)} \overline{x}^m} \left(\overline{\Gamma}^m_{pi} - g^{ks} \widetilde{\Gamma}_{pi,n} \frac{\Delta_{(n)} \overline{x}^n}{\Delta_k x^k} \frac{\Delta_{(m)} \overline{x}^m}{\Delta_s x^s} \right) \right)^{\sigma_{(0)} \sigma_{(1)} \cdots \sigma_{(p-1)}}$$

$$- \mu_j \frac{\partial \overline{A}^i}{\Delta_j x^j} \frac{\Delta_{(p)} \overline{x}^p}{\Delta_j x^j} \left(\Gamma^k_{\sigma ts} \frac{\Delta_t x^t}{\Delta_{(i)} \overline{x}^i} \frac{\Delta_s x^s}{\Delta_{(p)} \overline{x}^p} + \sum_{l=1}^{p-2} \mu_l \frac{\partial A_t^{\sigma_{(0)} \sigma_{(1)} \cdots \sigma_{(l-1)}}}{\Delta_l x^l} \right) - A^t \Gamma^k_{tj}$$

$$- \frac{A^t}{2} \left(\sum_{l=1}^{t-1} \mu_l \left(\frac{\partial g_{kj}}{\Delta_t x^t} \right)^{\sigma_0 \sigma_1 \cdots \sigma_{l-1}} + \sum_{l=1}^{j-1} \mu_l \left(\frac{\partial g_{kt}}{\Delta_j x^j} \right)^{\sigma_0 \sigma_1 \cdots \sigma_{l-1}} - \sum_{l=1}^{k-1} \mu_l \left(\frac{\partial g_{tj}}{\Delta_k x^k} \right)^{\sigma_0 \sigma_1 \cdots \sigma_{l-1}} \right).$$

From this we obtain

$$\frac{\partial A^k}{\Delta_j x^j} + A^t \Gamma_{tj}^k$$

$$= \frac{\Delta_k x^k}{\Delta_{(i)} \overline{x}^i} \left(\frac{\partial \overline{A}^i}{\Delta_{(p)} \overline{x}^p} \right)^{\sigma_{(0)} \sigma_{(1)} \cdots \sigma_{(p-1)}} \frac{\Delta_{(p)} \overline{x}^p}{\Delta_j x^j}$$

$$+ \overline{A}^{i\sigma_j} \frac{\Delta_{(p)} \overline{x}^p}{\Delta_j x^j} \left(\frac{\Delta_k x^k}{\Delta_{(m)} \overline{x}^m} \left(\overline{\Gamma}_{pi}^m - g^{ks} \widetilde{\Gamma}_{pi,n} \frac{\Delta_{(n)} \overline{x}^n}{\Delta_k x^k} \frac{\Delta_{(m)} \overline{x}^m}{\Delta_s x^s} \right) \right)^{\sigma_{(0)} \sigma_{(1)} \cdots \sigma_{(p-1)}}$$

$$- \mu_j \frac{\partial \overline{A}^i}{\Delta_j x^j} \frac{\Delta_{(p)} \overline{x}^p}{\Delta_j x^j} \left(\Gamma_{\sigma ts}^k \frac{\Delta_t x^t}{\Delta_{(i)} \overline{x}^i} \frac{\Delta_s x^s}{\Delta_{(p)} \overline{x}^p} + \sum_{l=1}^{p-2} \mu_l \frac{\partial A_t^{\sigma_{(0)} \sigma_{(1)} \cdots \sigma_{(l-1)}}}{\Delta_l x^l} \right)$$

$$- A^t \Gamma_{tj}^k - \frac{A^t}{2} \left(\sum_{l=1}^{t-1} \mu_l \left(\frac{\partial g_{kj}}{\Delta_t x^t} \right)^{\sigma_0 \sigma_1 \cdots \sigma_{l-1}} + \sum_{l=1}^{j-1} \mu_l \left(\frac{\partial g_{kt}}{\Delta_j x^j} \right)^{\sigma_0 \sigma_1 \cdots \sigma_{l-1}} - \sum_{l=1}^{k-1} \mu_l \left(\frac{\partial g_{tj}}{\Delta_k x^k} \right)^{\sigma_0 \sigma_1 \cdots \sigma_{l-1}} \right).$$

Remark 4.1. By the above examples we see that in the case of an arbitrary time scale, the covariant derivative of a tensor is not a tensor.

Theorem 4.1 (Ricci theorem). *We have*

$$g_{ij,k} = 0 \quad and \quad \delta_{k,j}^i = 0$$

for all i, j, and k.

Proof. By the definition of the covariant derivative we find

$$g_{ij,k} = \frac{\partial g_{ij}}{\Delta_k x^k} - g_{aj} \Gamma_{ik}^\alpha - g_{ia} \Gamma_{jk}^\alpha$$

$$= \frac{\partial g_{ij}}{\Delta_k x^k} - \Gamma_{ik,j} - \Gamma_{jk,i}$$

$$= \frac{\partial g_{ij}}{\Delta_k x^k} - \frac{1}{2} \left(\frac{\partial g_{ij}}{\Delta_k x^k} + \frac{\partial g_{kj}}{\Delta_i x^i} - \frac{\partial g_{ik}}{\Delta_j x^j} \right)$$

$$- \frac{1}{2} \left(\frac{\partial g_{ij}}{\Delta_k x^k} + \frac{\partial g_{ki}}{\Delta_j x^j} - \frac{\partial g_{jk}}{\Delta_i x^i} \right)$$

$$= \frac{\partial g_{ij}}{\Delta_k x^k} - \frac{1}{2} \frac{\partial g_{ij}}{\Delta_k x^k} - \frac{1}{2} \frac{\partial g_{kj}}{\Delta_i x^i} + \frac{1}{2} \frac{\partial g_{ik}}{\Delta_j x^j}$$

$$- \frac{1}{2} \frac{\partial g_{ij}}{\Delta_k x^k} - \frac{1}{2} \frac{\partial g_{ki}}{\Delta_j x^j} + \frac{1}{2} \frac{\partial g_{jk}}{\Delta_i x^i}$$

$$= \frac{\partial g_{ij}}{\Delta_k x^k} - \frac{\partial g_{ij}}{\Delta_k x^k}$$

$$= 0$$

and

$$\delta^i_{j,k} = \frac{\partial \delta^i_j}{\Delta_k x^k} + \delta^a_j \Gamma^i_{ak} - \delta^i_\beta \Gamma^\beta_{jk}$$

$$= \Gamma^i_{jk} - \Gamma^i_{jk}$$

$$= 0.$$

This completes the proof. $\square$

Exercise 4.3. Is it true that

$$g^{ij}_{,k} = 0?$$

Answer 4.2. In the general case, it is not true.

Example 4.10. We will write down an expression for A^i_j and will prove that $A^i_{,i}$ is an invariant.

By the transformation law for a mixed tensor we find

$$A^i_j = \frac{\Delta_{(l)} \overline{x}^{l\sigma_0 \sigma_1 \cdots \sigma_{j-1}}}{\Delta_j x^j} \frac{\Delta_i x^i}{\Delta_{(m)} \overline{x}^m} \overline{A}^m_l.$$

Hence

$$A^i_{,i} = \frac{\Delta_{(l)} \overline{x}^{l\sigma_0 \sigma_1 \cdots \sigma_{i-1}}}{\Delta_i x^i} \frac{\Delta_i x^i}{\Delta_{(m)} \overline{x}^m} \overline{A}^m_l$$

$$= \frac{\Delta_{(l)} \overline{x}^l}{\Delta_{(m)} \overline{x}^m} \overline{A}^m_l$$

$$= \delta^l_m \overline{A}^m_l$$

$$= \overline{A}^m_m.$$

Example 4.11. Suppose that P is a given point and

$$\frac{\partial g_{ij}}{\Delta_k x^k} = 0 \quad \text{at } P$$

for all $i, j,$ and k. We will prove that the components of the covariant derivative at that point are the same as those of ordinary derivatives. Indeed, we have

$$\Gamma^a_{il} = 0 \quad \text{at } P$$

for all $i, l,$ and α. Hence

$$g_{ij,l} = \frac{\partial g_{ij}}{\Delta_l x^l} - \Gamma^a_{il} g_{aj} - \Gamma^a_{jl} g_{ai}$$

$$= \frac{\partial g_{ij}}{\Delta_l x^l}.$$

Example 4.12. If A^i is a contravariant vector, then we will prove that

$$A^i_{,i} = \frac{\partial A^i}{\Delta_i x^i} + \frac{1}{2g}\frac{\partial g}{\Delta_i x^i}A_i - \frac{\mu_i}{g}A_i\sum_{l=2}^{N}H_{li}. \tag{4.10}$$

Indeed, we have

$$A^i_{,i} = \frac{\partial A^i}{\Delta_i x^i} + \Gamma^i_{ai}A^a$$

$$= \frac{\partial A^i}{\Delta_i x^i} + \left(\frac{1}{2g}\frac{\partial g}{\Delta_a x^a} - \frac{\mu_a}{g}\sum_{l=2}^{N}H_{al}\right)A_a.$$

Now replacing the dummy index a by i, we obtain (4.10).

Example 4.13. Let A^{ij} be a skew-symmetric tensor. We will show that

$$A^{ij}_{,i} = \frac{\partial A^{ij}}{\Delta_i x^i} + A^{ij}\left(\frac{1}{2g}\frac{\partial g}{\Delta_i x^i} - \frac{\mu_i}{g}\sum_{l=2}^{N}H_{li}\right).$$

Since A^{ij} is skew-symmetric and Γ^k_{ij} is symmetric with respect to the lower indices, we find

$$A^{jk}\Gamma^i_{jk} = A^{kj}\Gamma^i_{kj}$$
$$= -A^{jk}\Gamma^i_{kj}$$
$$= -A^{jk}\Gamma^i_{jk},$$

whereupon

$$A^{jk}\Gamma^i_{jk} = 0.$$

Hence, applying the definition of a covariant derivative, we arrive at

$$A^{ij}_{,i} = \frac{\partial A^{ij}}{\Delta_i x^i} + A^{ai}\Gamma^i_{ai} + A^{ia}\Gamma^j_{ai}$$

$$= \frac{\partial A^{ij}}{\Delta_i x^i} + A^{aj}\Gamma^i_{ia} + A^{ia}\Gamma^j_{ia}$$

$$= \frac{\partial A^{ij}}{\Delta_i x^i} + A^{aj}\Gamma^i_{ia}$$

$$= \frac{\partial A^{ij}}{\Delta_i x^i} + A^{aj}\left(\frac{1}{2g}\frac{\partial g}{\Delta_a x^a} - \frac{\mu_a}{g}\sum_{l=2}^{N}H_{la}\right)$$

$$= \frac{\partial A^{ij}}{\Delta_i x^i} + A^{ij}\left(\frac{1}{2g}\frac{\partial g}{\Delta_i x^i} - \frac{\mu_i}{g}\sum_{l=2}^{N}H_{li}\right),$$

where we have replaced the dummy index a by i in the second term.

Example 4.14. We will find the covariant derivative of

$$C^{jlm}_{kn} = A^j_k B^{lm}_n$$

with respect to x^q. Since C^{jlm}_{kn} is a mixed tensor, contravariant of the third order and covariant of the second order, applying the definition of a covariant derivative, we find

$$\begin{aligned}
C^{jlm}_{kn,q} &= \left(A^j_k B^{lm}_n\right)_q \\
&= \frac{\partial\left(A^j_k B^{lm}_n\right)}{\Delta_q x^q} + C^{amn}_{kn}\Gamma^j_{aq} + C^{jam}_{kn}\Gamma^l_{aq} \\
&\quad + C^{jla}_{kn}\Gamma^m_{aq} - C^{jlm}\beta n\Gamma^\beta_{kq} - C^{jlm}_{k\beta}\Gamma^\beta_{nq} \\
&= \frac{\partial A^j_k}{\Delta_q x^q}B^{lm}_n + A^{j\sigma_q}_k\frac{\partial B^{lm}_n}{\Delta_q x^q} \\
&\quad + A^a_k B^{lm}_n\Gamma^j_{aq} + A^j_k B^{am}_n\Gamma^l_{aq} + A^j_k B^{la}_n\Gamma^m_{aq} \\
&\quad - A^j_\beta B^{lm}_n\Gamma^\beta_{kq} - A^j_k B^{lm}_\beta\Gamma^\beta_{nq} \\
&= \left(\frac{\partial A^j_k}{\Delta_q x^q} + A^l_k\Gamma^j_{aq} - A^j_\beta\Gamma^\beta_{kq}\right)B^{lm}_n + \left(A^j_k + \mu_q\frac{\partial A^j_k}{\Delta_q x^q}\right)\frac{\partial B^{lm}_n}{\Delta_q x^q} \\
&\quad + A^j_k B^{am}_n\Gamma^l_{aq} + A^j_k B^{la}_n\Gamma^m_{aq} - A^j_k B^{lm}_\beta\Gamma^\beta_{nq} \\
&= A^j_{k,q}B^{lm}_n + A^j_k\left(\frac{\partial B^{lm}_n}{\Delta_q x^q} + B^{am}_n\Gamma^l_{aq} + B^{la}_n\Gamma^m_{aq} - B^{lm}_\beta\Gamma^\beta_{nq}\right) \\
&\quad + \mu_q\frac{\partial A^j_k}{\Delta_q x^q}\frac{\partial B^{lm}_n}{\Delta_q x^q} \\
&= A^j_{k,q}B^{lm}_n + A^j_k B^{lm}_{n,q} + A^j_k B^{lm}_\beta\Gamma^\beta_{nq}.
\end{aligned}$$

Example 4.15. Let A^{ijk} be a skew-symmetric tensor. We will show that

$$\frac{\partial A^{ijl}}{\Delta_l x^l} + A^{ijl}\left(\frac{1}{2g}\frac{\partial g}{\Delta_l x^l} - \frac{\mu_l}{g}\sum_{k=2}^N H_{kl}\right) \tag{4.11}$$

is a tensor. By the definition of a covariant derivative we get

$$\begin{aligned}
A^{ijl}_{,l} &= \frac{\partial A^{ijl}}{\Delta_l x^l} + A^{ajl}\Gamma^i_{al} + A^{ial}\Gamma^j_{al} + A^{ijl}\Gamma^l_{al} \\
&= \frac{\partial A^{ijl}}{\Delta_l x^l} - A^{jal}\Gamma^i_{al} + A^{ial}\Gamma^j_{al} + A^{ija}\Gamma^l_{al}.
\end{aligned}$$

Since A^{ial} is a skew-symmetric tensor, we find

$$A^{ial}\Gamma^{j}_{al} = A^{ial}\Gamma^{j}_{al}$$
$$= -A^{ial}\Gamma^{j}_{la}$$
$$= -A^{ial}\Gamma^{j}_{al},$$

whereupon

$$A^{ial}\Gamma^{j}_{al} = 0.$$

As above, we can conclude that

$$A^{jal}\Gamma^{i}_{al} = 0.$$

Therefore

$$A^{ijl}_{,l} = \frac{\partial A^{ijl}}{\Delta_l x^l} + A^{ija}\Gamma^{l}_{al}$$

$$= \frac{\partial A^{ijl}}{\Delta_l x^l} + A^{ija}\left(\frac{1}{2g}\frac{\partial g}{\Delta_a x^a} - \frac{\mu_a}{g}\sum_{k=2}^{N} H_{ka} \right)$$

$$= \frac{\partial A^{ijl}}{\Delta_l x^l} + A^{ijl}\left(\frac{1}{2g}\frac{\partial g}{\Delta_l x^l} - \frac{\mu_l}{g}\sum_{k=2}^{N} H_{kl} \right).$$

Since the left-hand side of the last equation is a tensor, we conclude that (4.11) is a tensor.

Example 4.16. Let A_{ij} be a symmetric tensor. We will show that $A_{ij,k}$ is a symmetric tensor with respect to the indices i and j.

By the definition of a covariant derivative we get

$$A_{ij,k} = \frac{\partial A_{ij}}{\Delta_k x^k} - \Gamma^{a}_{ik}A_{aj} - \Gamma^{a}_{jk}A_{ia}$$

$$= \frac{\partial A_{ji}}{\Delta_k x^k} - \Gamma^{a}_{ik}A_{ja} - \Gamma^{a}_{jk}A_{ai}$$

$$= A_{ji,k}.$$

Example 4.17. Let A^{ij} be a symmetric tensor. Then

$$A^{j}_{i,k} = \frac{\partial A^{j}_{i}}{\Delta_k x^k} + \Gamma^{j}_{ak}A^{a}_{i} - \Gamma^{a}_{ik}A^{j}_{a}$$

and

$$A^{j}_{i,j} = \frac{\partial A^{j}_{i}}{\Delta_j x^j} + \Gamma^{j}_{ak}A^{a}_{i} - \Gamma^{a}_{ij}A^{j}_{a}$$

$$= \frac{\partial A^{j}_{i}}{\Delta_j x^j} + \left(\frac{1}{2g}\frac{\partial g}{\Delta_a x^a} - \frac{\mu_a}{g}\sum_{l=2}^{N} H_{la} \right)A^{a}_{i}$$

$$-g^{la}\Gamma_{ij,l}A^j_\alpha$$

$$= \frac{\partial A^j_i}{\Delta_j x^j} + \left(\frac{1}{2g}\frac{\partial g}{\Delta_\alpha x^\alpha} - \frac{\mu_\alpha}{g}\sum_{l=2}^N H_{l\alpha}\right)A^\alpha_i - \Gamma_{ij,l}A^{lj}$$

$$= \frac{\partial A^j_i}{\Delta_j x^j} + \left(\frac{1}{2g}\frac{\partial g}{\Delta_\alpha x^\alpha} - \frac{\mu_\alpha}{g}\sum_{l=2}^N H_{l\alpha}\right)A^\alpha_i - \Gamma_{ij,k}A^{jk}.$$

Since A^{jk} is symmetric, we have

$$A^{jk} = A^{kj}$$

and

$$\Gamma_{ij,k}A^{jk} = \frac{1}{2}A^{jk}\left(\frac{\partial g_{ik}}{\Delta_j x^j} + \frac{\partial g_{jk}}{\Delta_i x^i} - \frac{\partial g_{ij}}{\Delta_k x^k}\right)$$

$$= \frac{1}{2}\left(A^{jk}\frac{\partial g_{ik}}{\Delta_j x^j} + A^{jk}\frac{\partial g_{jk}}{\Delta_i x^i} - A^{jk}\frac{\partial g_{ij}}{\Delta_k x^k}\right)$$

$$= \frac{1}{2}\left(A^{kj}\frac{\partial g_{ij}}{\Delta_k x^k} + A^{jk}\frac{\partial g_{jk}}{\Delta_i x^i} - A^{kj}\frac{\partial g_{ij}}{\Delta_k x^k}\right)$$

$$= \frac{1}{2}A^{jk}\frac{\partial g_{jk}}{\Delta_i x^i}.$$

Consequently,

$$A^j_{i,j} = \frac{\partial A^j_i}{\Delta_j x^j} + \left(\frac{1}{2g}\frac{\partial g}{\Delta_\alpha x^\alpha} - \frac{\mu_\alpha}{g}\sum_{l=2}^N H_{l\alpha}\right)A^\alpha_i - \frac{1}{2}A^{jk}\frac{\partial g_{jk}}{\Delta_i x^i}.$$

Exercise 4.4. Prove that:

1.

$$A^{jk}_{a,q} = \frac{\partial A^{jk}_l}{\Delta_q x^q} - \Gamma^a_{lq}A^{jk}_a + \Gamma^l_{qa}A^{ak}_l + \Gamma^k_{qa}A^{ja}_i.$$

2.

$$A^{jk}_{lm,q} = \frac{\partial A^{jk}_{lm}}{\Delta_q x^q} - \Gamma^a_{lq}A^{jk}_{lm} - \Gamma^a_{mq}A^{jk}_{la} + \Gamma^j_{qa}A_{lm}A^{ak}$$
$$+ \Gamma^k_{qa}A^{ja}_{lm}.$$

3.

$$A^{jkl}m,q = \frac{\partial A^{jkl}_m}{\Delta_q x^q} - \Gamma^a_{mq}A^{jkl}_a + \Gamma^j_{qa}A^{akl}_m + \Gamma^k_{qa}A^{jkl}_m$$
$$+ \Gamma^l_{qa}A^{jka}_m.$$

4.

$$A^{jk}_{lmn,q} = \frac{\partial A^{jk}_{lmn}}{\Delta_q x^q} - \Gamma^{\alpha}_{lq} A^{jk}_{\alpha mn} - \Gamma^{\alpha}_{mq} A^{jk}_{\alpha ln} - \Gamma^{\alpha}_{nq} A^{jk}_{lm\alpha}$$
$$+ \Gamma^{j}_{q\alpha} A^{\alpha k}_{lmn} + \Gamma^{k}_{q\alpha} A^{j\alpha}_{lmn}.$$

Hint 4.3. Use the definition of a covariant derivative.

4.4 Divergence

Let A^i be a contravariant vector. Then its covariant derivative is given by

$$A^i_{,j} = \frac{\partial A^j}{\Delta_j x^j} + \Gamma^i_{\alpha j} A^{\alpha},$$

and in particular, we get

$$A^i_{,i} = \frac{\partial A^i}{\Delta_i x^i} + \Gamma^i_{\alpha i} A^{\alpha}$$
$$= \frac{\partial A^i}{\Delta_i x^i} + \left(\frac{1}{2g} \frac{\partial g}{\Delta_\alpha x^\alpha} - \frac{\mu_\alpha}{g} \sum_{l=2}^{N} H_{l\alpha} \right) A^{\alpha}.$$

Note that $A^i_{,i}$ is a tensor that is contravariant and covariant of the zero order. Thus $A^i_{,i}$ is an invariant.

Definition 4.6. The divergence of the contravariant vector A^i defined as

$$\operatorname{div} A^i = A^i_{,i}.$$

Theorem 4.4. *Let A^i and B^i be two contravariant vectors. Then*

$$\operatorname{div}(A^i + B^i) = \operatorname{div} A^i + \operatorname{div} B^i.$$

Proof. Note that $A^i + B^i$ is a contravariant vector. Then

$$\operatorname{div}(A^i + B^i) = \frac{\partial(A^i + B^i)}{\Delta_i x^i} + \left(\frac{1}{2g} \frac{\partial g}{\Delta_\alpha x^\alpha} - \frac{\mu_\alpha}{g} \sum_{l=2}^{N} H_{l\alpha} \right)(A^\alpha + B^\alpha)$$
$$= \frac{\partial A^i}{\Delta_i x^i} + \frac{\partial B^i}{\Delta_i x^i} + \left(\frac{1}{2g} \frac{\partial g}{\Delta_\alpha x^\alpha} - \frac{\mu_\alpha}{g} \sum_{l=2}^{N} H_{l\alpha} \right) A^\alpha$$
$$+ \left(\frac{1}{2g} \frac{\partial g}{\Delta_\alpha x^\alpha} - \frac{\mu_\alpha}{g} \sum_{l=2}^{N} H_{l\alpha} \right) B^\alpha$$

$$= \left(\frac{\partial A^i}{\Delta_i x^i} + \left(\frac{1}{2g} \frac{\partial g}{\Delta_\alpha x^\alpha} - \frac{\mu_\alpha}{g} \sum_{l=2}^{N} H_{l\alpha} \right) A^\alpha \right)$$

$$+ \left(\frac{\partial B^i}{\Delta_i x^i} + \left(\frac{1}{2g} \frac{\partial g}{\Delta_\alpha x^\alpha} - \frac{\mu_\alpha}{g} \sum_{l=2}^{N} H_{l\alpha} \right) B^\alpha \right)$$

$$= A^i_{,i} + B^i_{,i}$$

$$= \operatorname{div} A^i + \operatorname{div} B^i.$$

This completes the proof. $\qquad\qquad\square$

Definition 4.7. Let A_i be a covariant vector. Then $g^{jk} A_{j,k}$ is an invariant and is called the divergence of A_i:

$$\operatorname{div} A_i = g^{jk} A_{j,k}.$$

We have

$$\operatorname{div} A_i = g^{jk} A_{j,k}$$

$$= (g^{jk} A_j)_{,k} - \mu_k \frac{\partial g_{jk}}{\Delta_k x^k} \frac{\partial A_j}{\Delta_k x^k}$$

$$= A^k_{,k} - \mu_k \frac{\partial g_{jk}}{\Delta_k x^k} \frac{\partial A_j}{\Delta_k x^k}$$

$$= \operatorname{div} A^k - \mu_k \frac{\partial g_{jk}}{\Delta_k x^k} \frac{\partial A_j}{\Delta_k x^k}.$$

Thus, in the general case, we have that

$$\operatorname{div} A^i \neq \operatorname{div} A_i.$$

Example 4.18. We will show that

$$\operatorname{div} A^{ij} = A^{ij}_{,j}$$

$$= \frac{\partial A^{ij}}{\Delta_j x^j} + \Gamma^i_{\alpha j} A^{\alpha j} + \left(\frac{1}{2g} \frac{\partial g}{\Delta_\alpha x^\alpha} - \frac{\mu_\alpha}{g} \sum_{l=2}^{N} H_{l\alpha} \right) A^{i\alpha}.$$

Indeed, we have

$$\operatorname{div} A^{ij} = A^{ij}_{,j}$$

$$= \frac{\partial A^{ij}}{\Delta_j x^j} + \Gamma^i_{\alpha j} A^{\alpha j} + \Gamma^j_{\alpha j} A^{i\alpha}$$

$$= \frac{\partial A^{ij}}{\Delta_j x^j} + \Gamma^i_{\alpha j} A^{\alpha j} + \left(\frac{1}{2g} \frac{\partial g}{\Delta_\alpha x^\alpha} - \frac{\mu_\alpha}{g} \sum_{l=2}^{N} H_{l\alpha} \right) A^{i\alpha}.$$

If A^{ij} is skew-symmetric, then

$$A^{ij} = -A^{ji},$$

and

$$A^{aj}\Gamma^i_{aj} = A^{ja}\Gamma^i_{ja}$$
$$= -A^{aj}\Gamma^i_{ja}$$
$$= -A^{aj}\Gamma^i_{aj},$$

whereupon

$$A^{aj}\Gamma^i_{aj} = 0,$$

and

$$\operatorname{div} A^{ij} = \frac{\partial A^{ij}}{\varDelta_j x^j} + \left(\frac{1}{2g}\frac{\partial g}{\varDelta_a x^a} - \frac{\mu_a}{g}\sum_{l=2}^{N} H_{la} \right) A^{ia}.$$

Example 4.19. We will express the divergence of a vector A^i in terms of its physical components for cylindrical coordinates

$$\varDelta s^2 = \left(\varDelta_1 x^1\right)^2 + \left(x^1\right)^2\left(\varDelta_2 x^2\right)^2 + \left(\varDelta_3 x^3\right)^2.$$

Here

$$g_{11} = 1,$$
$$g_{22} = \left(x^1\right)^2,$$
$$g_{33} = 1,$$
$$g_{ij} = 0, \quad i,j \in \{1,2,3\}, \ i \neq j.$$

Then

$$g = \det\begin{pmatrix} g_{11} & g_{12} & g_{13} \\ g_{21} & g_{22} & g_{23} \\ g_{31} & g_{32} & g_{33} \end{pmatrix}$$
$$= \begin{pmatrix} 1 & 0 & 0 \\ 0 & (x^1)^2 & 0 \\ 0 & 0 & 1 \end{pmatrix}$$
$$= \left(x^1\right)^2.$$

Hence

$$\frac{\partial g}{\Delta_1 x^1} = \sigma_1(x^1) + x^1,$$

$$\frac{\partial g}{\Delta_2 x^2} = 0,$$

$$\frac{\partial g}{\Delta_3 x^3} = 0,$$

$$\frac{\partial g_{22}}{\Delta_1 x^1} = \sigma_1(x^1) + x^1,$$

$$\frac{\partial g_{22}}{\Delta_2 x^2} = 0,$$

$$\frac{\partial g_{22}}{\Delta_3 x^3} = 0,$$

and

$$\frac{\partial g_{ij}}{\Delta_k x^k} = 0, \quad i, j, k \in \{1, 2, 3\} \text{ with } i \neq 2 \text{ if } i = j.$$

Now we will find H_{lj}, $l \in \{2, 3\}$, $j \in \{1, 2, 3\}$. We have

$$
H_{21} = \det \begin{pmatrix} \frac{\partial g_{11}}{\Delta_1 x^1} & \frac{\partial g_{12}}{\Delta_1 x^1} & \frac{\partial g_{13}}{\Delta_1 x^1} \\ \frac{\partial g_{21}}{\Delta_1 x^1} & \frac{\partial g_{22}}{\Delta_1 x^1} & \frac{\partial g_{23}}{\Delta_1 x^1} \\ g_{31} & g_{32} & g_{33} \end{pmatrix}
$$

$$
= \det \begin{pmatrix} 0 & 0 & 0 \\ 0 & \sigma_1(x^1) + x^1 & 0 \\ 0 & 0 & 1 \end{pmatrix}
$$

$$= 1,$$

$$
H_{22} = \det \begin{pmatrix} \frac{\partial g_{11}}{\Delta_2 x^2} & \frac{\partial g_{12}}{\Delta_2 x^2} & \frac{\partial g_{13}}{\Delta_2 x^2} \\ \frac{\partial g_{21}}{\Delta_2 x^2} & \frac{\partial g_{22}}{\Delta_2 x^2} & \frac{\partial g_{23}}{\Delta_2 x^2} \\ g_{31} & g_{32} & g_{33} \end{pmatrix}
$$

$$
= \det \begin{pmatrix} 0 & 0 & 0 \\ 0 & 0 & 0 \\ 0 & 0 & 1 \end{pmatrix}
$$

$$= 0,$$

$$
H_{23} = \det \begin{pmatrix} \frac{\partial g_{11}}{\Delta_3 x^3} & \frac{\partial g_{12}}{\Delta_3 x^3} & \frac{\partial g_{13}}{\Delta_3 x^3} \\ \frac{\partial g_{21}}{\Delta_3 x^3} & \frac{\partial g_{22}}{\Delta_3 x^3} & \frac{\partial g_{23}}{\Delta_3 x^3} \\ g_{31} & g_{32} & g_{33} \end{pmatrix}
$$

$$
= \begin{pmatrix} 0 & 0 & 0 \\ 0 & 0 & 0 \\ 0 & 0 & 1 \end{pmatrix}
$$

$$= 0,$$

$$H_{31} = \det \begin{pmatrix} g_{11} & g_{12} & g_{13} \\ \frac{\partial g_{21}}{\Delta_1 x^1} & \frac{\partial g_{22}}{\Delta_1 x^1} & \frac{\partial g_{23}}{\Delta_1 x^1} \\ \frac{\partial g_{31}}{\Delta_1 x^1} & \frac{\partial g_{32}}{\Delta_1 x^1} & \frac{\partial g_{33}}{\Delta_1 x^1} \end{pmatrix} + \det \begin{pmatrix} \frac{\partial g_{11}}{\Delta_1 x^1} & \frac{\partial g_{12}}{\Delta_1 x^1} & \frac{\partial g_{13}}{\Delta_1 x^1} \\ g_{21} & g_{22} & g_{23} \\ \frac{\partial g_{31}}{\Delta_1 x^1} & \frac{\partial g_{32}}{\Delta_1 x^1} & \frac{\partial g_{33}}{\Delta_1 x^1} \end{pmatrix}$$

$$+ \det \begin{pmatrix} \frac{\partial g_{11}}{\Delta_1 x^1} & \frac{\partial g_{12}}{\Delta_1 x^1} & \frac{\partial g_{13}}{\Delta_1 x^1} \\ \frac{\partial g_{21}}{\Delta_1 x^1} & \frac{\partial g_{22}}{\Delta_1 x^1} & \frac{\partial g_{23}}{\Delta_1 x^1} \\ \frac{\partial g_{31}}{\Delta_1 x^1} & \frac{\partial g_{32}}{\Delta_1 x^1} & \frac{\partial g_{33}}{\Delta_1 x^1} \end{pmatrix}$$

$$= \det \begin{pmatrix} 1 & 0 & 0 \\ 0 & \sigma_1(x^1) + x^1 & 0 \\ 0 & 0 & 0 \end{pmatrix} + \det \begin{pmatrix} 0 & 0 & 0 \\ 0 & (x^1)^2 & 0 \\ 0 & 0 & 0 \end{pmatrix}$$

$$+ \det \begin{pmatrix} 0 & 0 & 0 \\ 0 & \sigma_1(x^1) + x^1 & 0 \\ 0 & 0 & 0 \end{pmatrix}$$

$$= 0,$$

$$H_{32} = \det \begin{pmatrix} g_{11} & g_{12} & g_{13} \\ \frac{\partial g_{21}}{\Delta_2 x^2} & \frac{\partial g_{22}}{\Delta_2 x^2} & \frac{\partial g_{23}}{\Delta_2 x^2} \\ \frac{\partial g_{31}}{\Delta_2 x^2} & \frac{\partial g_{32}}{\Delta_2 x^2} & \frac{\partial g_{33}}{\Delta_2 x^2} \end{pmatrix} + \det \begin{pmatrix} \frac{\partial g_{11}}{\Delta_2 x^2} & \frac{\partial g_{12}}{\Delta_2 x^2} & \frac{\partial g_{13}}{\Delta_2 x^2} \\ g_{21} & g_{22} & g_{23} \\ \frac{\partial g_{31}}{\Delta_2 x^2} & \frac{\partial g_{32}}{\Delta_2 x^2} & \frac{\partial g_{33}}{\Delta_2 x^2} \end{pmatrix}$$

$$+ \det \begin{pmatrix} \frac{\partial g_{11}}{\Delta_2 x^2} & \frac{\partial g_{12}}{\Delta_2 x^2} & \frac{\partial g_{13}}{\Delta_2 x^2} \\ \frac{\partial g_{21}}{\Delta_2 x^2} & \frac{\partial g_{22}}{\Delta_2 x^2} & \frac{\partial g_{23}}{\Delta_2 x^2} \\ \frac{\partial g_{31}}{\Delta_2 x^2} & \frac{\partial g_{32}}{\Delta_2 x^2} & \frac{\partial g_{33}}{\Delta_2 x^2} \end{pmatrix}$$

$$= \det \begin{pmatrix} 1 & 0 & 0 \\ 0 & 0 & 0 \\ 0 & 0 & 0 \end{pmatrix} + \det \begin{pmatrix} 0 & 0 & 0 \\ 0 & (x^1)^2 & 0 \\ 0 & 0 & 0 \end{pmatrix}$$

$$+ \det \begin{pmatrix} 0 & 0 & 0 \\ 0 & 0 & 0 \\ 0 & 0 & 0 \end{pmatrix}$$

$$= 0,$$

$$H_{33} = \det \begin{pmatrix} g_{11} & g_{12} & g_{13} \\ \frac{\partial g_{21}}{\Delta_3 x^3} & \frac{\partial g_{22}}{\Delta_3 x^3} & \frac{\partial g_{23}}{\Delta_3 x^3} \\ \frac{\partial g_{31}}{\Delta_3 x^3} & \frac{\partial g_{32}}{\Delta_3 x^3} & \frac{\partial g_{33}}{\Delta_3 x^3} \end{pmatrix} + \det \begin{pmatrix} \frac{\partial g_{11}}{\Delta_3 x^3} & \frac{\partial g_{12}}{\Delta_3 x^3} & \frac{\partial g_{13}}{\Delta_3 x^3} \\ g_{21} & g_{22} & g_{23} \\ \frac{\partial g_{31}}{\Delta_3 x^3} & \frac{\partial g_{32}}{\Delta_3 x^3} & \frac{\partial g_{33}}{\Delta_3 x^3} \end{pmatrix}$$

$$+ \det \begin{pmatrix} \frac{\partial g_{11}}{\Delta_3 x^3} & \frac{\partial g_{12}}{\Delta_3 x^3} & \frac{\partial g_{13}}{\Delta_3 x^3} \\ \frac{\partial g_{21}}{\Delta_3 x^3} & \frac{\partial g_{22}}{\Delta_3 x^3} & \frac{\partial g_{23}}{\Delta_3 x^3} \\ \frac{\partial g_{31}}{\Delta_3 x^3} & \frac{\partial g_{32}}{\Delta_3 x^3} & \frac{\partial g_{33}}{\Delta_3 x^3} \end{pmatrix}$$

$$= \det \begin{pmatrix} 1 & 0 & 0 \\ 0 & 0 & 0 \\ 0 & 0 & 0 \end{pmatrix} + \det \begin{pmatrix} 0 & 0 & 0 \\ 0 & (x^1)^2 & 0 \\ 0 & 0 & 0 \end{pmatrix}$$

$$+ \det \begin{pmatrix} 0 & 0 & 0 \\ 0 & 0 & 0 \\ 0 & 0 & 0 \end{pmatrix}$$

$$= 0.$$

Let the physical components be denoted by A_1, A_2, and A_3. Then

$$A_1 = \sqrt{g_{11}}A^1$$
$$= A^1,$$
$$A_2 = \sqrt{g_{22}}A^2$$
$$= x^1 A^2,$$
$$A_3 = \sqrt{g_{33}}A^3$$
$$= A^3,$$

from which

$$A^1 = A_1,$$
$$A^2 = \frac{1}{x^1}A_2,$$
$$A^3 = A_3.$$

Therefore

$$\mathrm{div}\, A^i = \frac{\partial A^1}{\varDelta_1 x^1} + \frac{\partial A^2}{\varDelta_2 x^2} + \frac{\partial A^3}{\varDelta_3 x^3}$$

$$+ \left(\frac{1}{2g}\frac{\partial g}{\varDelta_1 x^1} - \frac{\mu_1}{g}(H_{21} + H_{31}) \right) A^1$$

$$+ \left(\frac{1}{2g}\frac{\partial g}{\varDelta_2 x^2} - \frac{\mu_2}{g}(H_{22} + H_{32}) \right) A^2$$

$$+ \left(\frac{1}{2g}\frac{\partial g}{\varDelta_3 x^3} - \frac{\mu_3}{g}(H_{23} + H_{33}) \right) A^3$$

$$= \frac{\partial A_1}{\varDelta_1 x^1} + \frac{1}{x^1}\frac{\partial A_2}{\varDelta_2 x^2} + \frac{\partial A_3}{\varDelta_3 x^3} + \frac{1}{2(x^1)^2}(\sigma_1(x^1) + x^1)A_1.$$

Example 4.20. We will express the divergence of a vector A^i in terms of its physical components for cylindrical coordinates

$$\varDelta s^2 = (\varDelta_1 x^1)^2 + (x^1)^2(\varDelta_2 x^2)^2 + (x^1)^2(\sin_1(x^2, x_0^2))^2(\varDelta_3 x^3)^2.$$

Here

$$g_{11} = 1,$$
$$g_{22} = (x^1)^2,$$
$$g_{33} = (x^1)^2(\sin_1(x^2, x_0^2))^2,$$
$$g_{ij} = 0, \quad i, j \in \{1, 2, 3\}, \ i \neq j.$$

Then

$$g = \det \begin{pmatrix} g_{11} & g_{12} & g_{13} \\ g_{21} & g_{22} & g_{23} \\ g_{31} & g_{32} & g_{33} \end{pmatrix}$$
$$= \begin{pmatrix} 1 & 0 & 0 \\ 0 & (x^1)^2 & 0 \\ 0 & 0 & (x^1)^2(\sin_1(x^2, x_0^2))^2 \end{pmatrix}$$
$$= (x^1)^4(\sin_1(x^2, x_0^2))^2.$$

Hence

$$\frac{\partial g}{\Delta_1 x^1} = ((\sigma_1(x^1))^3 + x^1(\sigma_1(x^1))^2 + (x^1)^2\sigma_1(x^1) + (x^1)^3)(\sin_1(x^2, x_0^2))^2,$$

$$\frac{\partial g}{\Delta_2 x^2} = (x^1)^4 \cos_1(x^2, x_0^2)(\sin_1(x^2, x_0^2) + \sin_1(\sigma_2(x^2), x_0^2)),$$

$$\frac{\partial g}{\Delta_3 x^3} = 0,$$

$$\frac{\partial g_{22}}{\Delta_1 x^1} = \sigma_1(x^1) + x^1,$$

$$\frac{\partial g_{22}}{\Delta_2 x^2} = 0,$$

$$\frac{\partial g_{22}}{\Delta_3 x^3} = 0,$$

$$\frac{\partial g_{33}}{\Delta_1 x^1} = (\sigma_1(x^1) + x^1)(\sin_1(x^2, x_0^2))^2,$$

$$\frac{\partial g_{33}}{\Delta_2 x^2} = (x^1)^2 \cos_1(x^2, x_0^2)(\sin_1(x^2, x_0^2) + \sin_1(\sigma_2(x^2), x_0^2)),$$

$$\frac{\partial g_{33}}{\Delta_3 x^3} = 0,$$

and

$$\frac{\partial g_{ij}}{\Delta_k x^k} = 0, \quad i, j, k \in \{1, 2, 3\} \text{ with } i \neq 2, 3 \text{ if } i = j.$$

Now we will find H_{lj}, $l \in \{2, 3\}$, $j \in \{1, 2, 3\}$. We have

$$H_{21} = \det \begin{pmatrix} \frac{\partial g_{11}}{\Delta_1 x^1} & \frac{\partial g_{12}}{\Delta_1 x^1} & \frac{\partial g_{13}}{\Delta_1 x^1} \\ \frac{\partial g_{21}}{\Delta_1 x^1} & \frac{\partial g_{22}}{\Delta_1 x^1} & \frac{\partial g_{23}}{\Delta_1 x^1} \\ g_{31} & g_{32} & g_{33} \end{pmatrix}$$

$$= \det \begin{pmatrix} 0 & 0 & 0 \\ 0 & \sigma_1(x^1) + x^1 & 0 \\ 0 & 0 & (x^1)^2(\sin_1(x^2, x_0^2))^2 \end{pmatrix}$$

$$= 1,$$

$$H_{22} = \det \begin{pmatrix} \frac{\partial g_{11}}{\Delta_2 x^2} & \frac{\partial g_{12}}{\Delta_2 x^2} & \frac{\partial g_{13}}{\Delta_2 x^2} \\ \frac{\partial g_{21}}{\Delta_2 x^2} & \frac{\partial g_{22}}{\Delta_2 x^2} & \frac{\partial g_{23}}{\Delta_2 x^2} \\ g_{31} & g_{32} & g_{33} \end{pmatrix}$$

$$= \det \begin{pmatrix} 0 & 0 & 0 \\ 0 & 0 & 0 \\ 0 & 0 & (x^1)^2(\sin_1(x^2, x_0^2))^2 \end{pmatrix}$$

$$= 0,$$

$$H_{23} = \det \begin{pmatrix} \frac{\partial g_{11}}{\Delta_3 x^3} & \frac{\partial g_{12}}{\Delta_3 x^3} & \frac{\partial g_{13}}{\Delta_3 x^3} \\ \frac{\partial g_{21}}{\Delta_3 x^3} & \frac{\partial g_{22}}{\Delta_3 x^3} & \frac{\partial g_{23}}{\Delta_3 x^3} \\ g_{31} & g_{32} & g_{33} \end{pmatrix}$$

$$= \begin{pmatrix} 0 & 0 & 0 \\ 0 & 0 & 0 \\ 0 & 0 & (x^1)^2(\sin_1(x^2, x_0^2))^2 \end{pmatrix}$$

$$= 0,$$

$$H_{31} = \det \begin{pmatrix} g_{11} & g_{12} & g_{13} \\ \frac{\partial g_{21}}{\Delta_1 x^1} & \frac{\partial g_{22}}{\Delta_1 x^1} & \frac{\partial g_{23}}{\Delta_1 x^1} \\ \frac{\partial g_{31}}{\Delta_1 x^1} & \frac{\partial g_{32}}{\Delta_1 x^1} & \frac{\partial g_{33}}{\Delta_1 x^1} \end{pmatrix} + \det \begin{pmatrix} \frac{\partial g_{11}}{\Delta_1 x^1} & \frac{\partial g_{12}}{\Delta_1 x^1} & \frac{\partial g_{13}}{\Delta_1 x^1} \\ g_{21} & g_{22} & g_{23} \\ \frac{\partial g_{31}}{\Delta_1 x^1} & \frac{\partial g_{32}}{\Delta_1 x^1} & \frac{\partial g_{33}}{\Delta_1 x^1} \end{pmatrix}$$

$$+ \det \begin{pmatrix} \frac{\partial g_{11}}{\Delta_1 x^1} & \frac{\partial g_{12}}{\Delta_1 x^1} & \frac{\partial g_{13}}{\Delta_1 x^1} \\ \frac{\partial g_{21}}{\Delta_1 x^1} & \frac{\partial g_{22}}{\Delta_1 x^1} & \frac{\partial g_{23}}{\Delta_1 x^1} \\ \frac{\partial g_{31}}{\Delta_1 x^1} & \frac{\partial g_{32}}{\Delta_1 x^1} & \frac{\partial g_{33}}{\Delta_1 x^1} \end{pmatrix}$$

$$= \det \begin{pmatrix} 1 & 0 & 0 \\ 0 & \sigma_1(x^1) + x^1 & 0 \\ 0 & 0 & (\sigma_1(x^1) + x^1)(\sin_1(x^2, x_0^2))^2 \end{pmatrix}$$

$$+ \det \begin{pmatrix} 0 & 0 & 0 \\ 0 & (x^1)^2 & 0 \\ 0 & 0 & (\sigma_1(x^1) + x^1)(\sin_1(x^2, x_0^2))^2 \end{pmatrix}$$

$$+ \det \begin{pmatrix} 0 & 0 & 0 \\ 0 & \sigma_1(x^1) + x^1 & 0 \\ 0 & 0 & (\sigma_1(x^1) + x^1)(\sin_1(x^2, x_0^2))^2 \end{pmatrix}$$

$$= 0,$$

$$H_{32} = \det \begin{pmatrix} g_{11} & g_{12} & g_{13} \\ \frac{\partial g_{21}}{\Delta_2 x^2} & \frac{\partial g_{22}}{\Delta_2 x^2} & \frac{\partial g_{23}}{\Delta_2 x^2} \\ \frac{\partial g_{31}}{\Delta_2 x^2} & \frac{\partial g_{32}}{\Delta_2 x^2} & \frac{\partial g_{33}}{\Delta_2 x^2} \end{pmatrix} + \det \begin{pmatrix} \frac{\partial g_{11}}{\Delta_2 x^2} & \frac{\partial g_{12}}{\Delta_2 x^2} & \frac{\partial g_{13}}{\Delta_2 x^2} \\ g_{21} & g_{22} & g_{23} \\ \frac{\partial g_{31}}{\Delta_2 x^2} & \frac{\partial g_{32}}{\Delta_2 x^2} & \frac{\partial g_{33}}{\Delta_2 x^2} \end{pmatrix}$$

$$+ \det \begin{pmatrix} \frac{\partial g_{11}}{\Delta_2 x^2} & \frac{\partial g_{12}}{\Delta_2 x^2} & \frac{\partial g_{13}}{\Delta_2 x^2} \\ \frac{\partial g_{21}}{\Delta_2 x^2} & \frac{\partial g_{22}}{\Delta_2 x^2} & \frac{\partial g_{23}}{\Delta_2 x^2} \\ \frac{\partial g_{31}}{\Delta_2 x^2} & \frac{\partial g_{32}}{\Delta_2 x^2} & \frac{\partial g_{33}}{\Delta_2 x^2} \end{pmatrix}$$

$$= \det \begin{pmatrix} 1 & 0 & 0 \\ 0 & 0 & 0 \\ 0 & 0 & (x^1)^2 \cos_1(x^2, x_0^2)(\sin_1(x^2, x_0^2) + \sin_1(\sigma_2(x^2), x_0^2)) \end{pmatrix}$$

$$+ \det \begin{pmatrix} 0 & 0 & 0 \\ 0 & (x^1)^2 & 0 \\ 0 & 0 & (x^1)^2 \cos_1(x^2, x_0^2)(\sin_1(x^2, x_0^2) + \sin_1(\sigma_2(x^2), x_0^2)) \end{pmatrix}$$

$$+ \det \begin{pmatrix} 0 & 0 & 0 \\ 0 & 0 & 0 \\ 0 & 0 & (x^1)^2 \cos_1(x^2, x_0^2)(\sin_1(x^2, x_0^2) + \sin_1(\sigma_2(x^2), x_0^2)) \end{pmatrix}$$

$$= 0,$$

$$H_{33} = \det \begin{pmatrix} g_{11} & g_{12} & g_{13} \\ \frac{\partial g_{21}}{\Delta_3 x^3} & \frac{\partial g_{22}}{\Delta_3 x^3} & \frac{\partial g_{23}}{\Delta_3 x^3} \\ \frac{\partial g_{31}}{\Delta_3 x^3} & \frac{\partial g_{32}}{\Delta_3 x^3} & \frac{\partial g_{33}}{\Delta_3 x^3} \end{pmatrix} + \det \begin{pmatrix} \frac{\partial g_{11}}{\Delta_3 x^3} & \frac{\partial g_{12}}{\Delta_3 x^3} & \frac{\partial g_{13}}{\Delta_3 x^3} \\ g_{21} & g_{22} & g_{23} \\ \frac{\partial g_{31}}{\Delta_3 x^3} & \frac{\partial g_{32}}{\Delta_3 x^3} & \frac{\partial g_{33}}{\Delta_3 x^3} \end{pmatrix}$$

$$+ \det \begin{pmatrix} \frac{\partial g_{11}}{\Delta_3 x^3} & \frac{\partial g_{12}}{\Delta_3 x^3} & \frac{\partial g_{13}}{\Delta_3 x^3} \\ \frac{\partial g_{21}}{\Delta_3 x^3} & \frac{\partial g_{22}}{\Delta_3 x^3} & \frac{\partial g_{23}}{\Delta_3 x^3} \\ \frac{\partial g_{31}}{\Delta_3 x^3} & \frac{\partial g_{32}}{\Delta_3 x^3} & \frac{\partial g_{33}}{\Delta_3 x^3} \end{pmatrix}$$

$$= \det \begin{pmatrix} 1 & 0 & 0 \\ 0 & 0 & 0 \\ 0 & 0 & 0 \end{pmatrix} + \det \begin{pmatrix} 0 & 0 & 0 \\ 0 & (x^1)^2 & 0 \\ 0 & 0 & 0 \end{pmatrix}$$

$$+ \det \begin{pmatrix} 0 & 0 & 0 \\ 0 & 0 & 0 \\ 0 & 0 & 0 \end{pmatrix}$$

$$= 0.$$

Let the physical components be denoted by A_1, A_2, and A_3. Then

$$A_1 = \sqrt{g_{11}}A^1$$
$$= A^1,$$
$$A_2 = \sqrt{g_{22}}A^2$$
$$= x^1 A^2,$$
$$A_3 = \sqrt{g_{33}}A^3$$
$$= x^1 \sin_1(x^2, x_0^2)A^3,$$

from which

$$A^1 = A_1,$$
$$A^2 = \frac{1}{x^1}A_2,$$
$$A^3 = \frac{1}{x^1 \sin_1(x^2, x_0^2)}A_3.$$

Therefore

$$\operatorname{div} A^i = \frac{\partial A^1}{\Delta_1 x^1} + \frac{\partial A^2}{\Delta_2 x^2} + \frac{\partial A^3}{\Delta_3 x^3}$$
$$+ \left(\frac{1}{2g} \frac{\partial g}{\Delta_1 x^1} - \frac{\mu_1}{g}(H_{21} + H_{31}) \right) A^1$$
$$+ \left(\frac{1}{2g} \frac{\partial g}{\Delta_2 x^2} - \frac{\mu_2}{g}(H_{22} + H_{32}) \right) A^2$$
$$+ \left(\frac{1}{2g} \frac{\partial g}{\Delta_3 x^3} - \frac{\mu_3}{g}(H_{23} + H_{33}) \right) A^3$$
$$= \frac{\partial A_1}{\Delta_1 x^1} + \frac{1}{x^1} \frac{\partial A_2}{\Delta_2 x^2} + \frac{\partial A_3}{\Delta_3 x^3}$$
$$+ \left(\frac{((\sigma_1(x^1))^3 + (\sigma_1(x^1))^2 x^1 + \sigma_1(x^1)(x^1)^2 + (x^1)^3)(\sin_1(x^2, x_0^2))^2}{2(x^1)^4(\sin_1(x^2, x_0^2))^2} \right.$$
$$\left. - \frac{\mu_1(\sigma_1(x^1) + x^1)^2(\sin_1(x^2, x_0^2))^2}{(x^1)^4(\sin_1(x^2, x_0^2))^2} \right) A_1$$
$$+ \frac{1}{2(x^1)^4(\sin_1(x^2, x_0^2))^2}(x^1)^4 \cos_1(x^2, x_0^2)(\sin_1(\sigma_2(x^2), x_0^2) + \sin_1(x^2, x_0^2))\frac{1}{x^1}A_2$$
$$= \frac{\partial A_1}{\Delta_1 x^1} + \frac{1}{x^1} \frac{\partial A_2}{\Delta_2 x^2} + \frac{\partial A_3}{\Delta_3 x^3}$$
$$+ \frac{(\sigma_1(x^1))^3 + (\sigma_1(x^1))^2 x^1 + \sigma_1(x^1)(x^1)^2 + (x^1)^3 - 2\mu_1(\sigma_1(x^1) + x^1)^2}{2(x^1)^4}A_1$$
$$+ \frac{\cos_1(x^2, x_0^2)(\sin_1(\sigma_2(x^2), x_0^2) + \sin_1(x^2, x_0^2))}{2x^1(\sin_1(x^2, x_0^2))^2}A_2.$$

Example 4.21. Consider V_2 with the line element

$$\Delta s^2 = \left(\Delta_1 x^1\right)^2 + \left(x^1\right)^2 \left(\Delta_2 x^2\right)^2.$$

We will find the divergence of the covariant vector with components

$$x^1 \cos_2(x^2, x_0^2) \quad \text{and} \quad -\left(x^1\right)^2 \sin_2(x^2, x_0^2).$$

Here

$$\begin{aligned}
g_{11} &= 1, \\
g_{12} &= 0, \\
g_{21} &= 0, \\
g_{22} &= \left(x^1\right)^2, \\
A_1 &= x^1 \cos_2(x^2, x_0^2), \\
A_2 &= -\left(x^1\right)^2 \sin_2(x^2, x_0^2).
\end{aligned}$$

Then

$$\begin{aligned}
g &= \det \begin{pmatrix} g_{11} & g_{12} \\ g_{21} & g_{22} \end{pmatrix} \\
&= \det \begin{pmatrix} 1 & 0 \\ 0 & (x^1)^2 \end{pmatrix} \\
&= \left(x^1\right)^2.
\end{aligned}$$

Hence

$$\begin{aligned}
\frac{\partial g}{\Delta_1 x^1} &= \sigma_1(x^1) + x^1, \\
\frac{\partial g}{\Delta_2 x^2} &= 0, \\
\frac{\partial g_{11}}{\Delta_1 x^1} &= 0, \\
\frac{\partial g_{11}}{\Delta_2 x^2} &= 0, \\
\frac{\partial g_{12}}{\Delta_1 x^1} &= 0, \\
\frac{\partial g_{12}}{\Delta_2 x^2} &= 0, \\
\frac{\partial g_{21}}{\Delta_1 x^1} &= 0,
\end{aligned}$$

$$\frac{\partial g_{21}}{\varDelta_2 x^2} = 0,$$

$$\frac{\partial g_{22}}{\varDelta_1 x^1} = \sigma_1(x^1) + x^1,$$

$$\frac{\partial g_{22}}{\varDelta_2 x^2} = 0,$$

and thus

$$H_{21} = \det \begin{pmatrix} \frac{\partial g_{11}}{\varDelta_1 x^1} & \frac{\partial g_{12}}{\varDelta_1 x^1} \\ \frac{\partial g_{21}}{\varDelta_1 x^1} & \frac{\partial g_{22}}{\varDelta_1 x^1} \end{pmatrix}$$

$$= \det \begin{pmatrix} 0 & 0 \\ 0 & \sigma_1(x^1) + x^1 \end{pmatrix}$$

$$= 0,$$

$$H_{22} = \det \begin{pmatrix} \frac{\partial g_{11}}{\varDelta_2 x^2} & \frac{\partial g_{12}}{\varDelta_2 x^2} \\ \frac{\partial g_{21}}{\varDelta_2 x^2} & \frac{\partial g_{22}}{\varDelta_2 x^2} \end{pmatrix}$$

$$= \det \begin{pmatrix} 0 & 0 \\ 0 & 0 \end{pmatrix}$$

$$= 0.$$

Moreover, we have

$$x^1 \cos_2(x^2, x_0^2) = A_1$$
$$= \sqrt{g_{11}} A^1$$
$$= A^1,$$
$$-(x^1)^2 \sin_2(x^2, x_0^2) = A_2$$
$$= \sqrt{g_{22}} A^2$$
$$= x^1 A^2,$$

whereupon

$$A^1 = x^1 \cos_2(x^2, x_0^2),$$
$$A^2 = -x^1 \sin_2(x^2, x_0^2),$$

and thus

$$\frac{\partial A^1}{\varDelta_1 x^1} = \cos_2(x^2, x_0^2),$$

$$\frac{\partial A^2}{\varDelta_2 x^2} = -2x^1 \cos_2(x^2, x_0^2).$$

Consequently,

$$\operatorname{div} A^i = \frac{\partial A^1}{\varDelta_1 x^1} + \frac{\partial A^2}{\varDelta_2 x^2} + \left(\frac{1}{2g} \frac{\partial g}{\varDelta_1 x^1} - \frac{\mu_1}{g} H_{21} \right) A^1$$

$$+ \left(\frac{1}{2g} \frac{\partial g}{\varDelta_2 x^2} - \frac{\mu_2}{g} H_{22} \right) A^2$$

$$= \cos_2(x^2, x_0^2) - 2x^1 \cos_2(x^2, x_0^2) + \frac{\sigma_1(x^1) + x^1}{2(x^1)^2} x^1 \cos_2(x^2, x_0^2)$$

$$= \cos_2(x^2, x_0^2) - 2x^1 \cos_2(x^2, x_0^2) + \frac{\sigma_1(x^1) + x^1}{2x^1} \cos_2(x^2, x_0^2),$$

and

$$\operatorname{div} A_j = \operatorname{div} A^i - \mu_1 \frac{\partial g_{11}}{\varDelta_1 x^1} \frac{\partial A_1}{\varDelta_1 x^1} - \mu_2 \frac{\partial g_{22}}{\varDelta_2 x^2} \frac{\partial A_2}{\varDelta_2 x^2}$$

$$= \cos_2(x^2, x_0^2) - 2x^1 \cos_2(x^2, x_0^2) + \frac{\sigma_1(x^1) + x^1}{2x^1} \cos_2(x^2, x_0^2).$$

Example 4.22. We will express

$$\int_V \vec{\nabla} \cdot \vec{V} \varDelta V = \int_S \vec{A} \cdot \vec{n} \varDelta S$$

in a tensor form, where V is the volume bounded by a closed surface S, $\vec{A}$ is a vector function with continuous derivatives, and $\vec{n}$ is the positive normal to S.

Let A^k be a tensor field of rank 1, and let v denote the outward drawn unit normal to any point of the closed surface. Then

$$\vec{\nabla} \cdot \vec{V} = A^k_{,k},$$

and

$$\vec{A} \cdot \vec{n} = A^k v_k.$$

Thus we get

$$\int_V A^k_{,k} \varDelta V = \int_S A^k v_k \varDelta S.$$

Exercise 4.5. Let A^i and A_i be a contravariant vector and a covariant vector, respectively, and let α be a scalar. Prove that:
1. $\operatorname{div}(\alpha A^i) = \alpha \operatorname{div} A^i.$
2. $\operatorname{div}(\alpha A_i) = \alpha \operatorname{div} A_i.$

4.5 Gradient of an invariant

Definition 4.8. The partial derivatives of an invariant ϕ are a covariant vector, which is called the gradient of ϕ and denoted by

$$\operatorname{grad}\phi \quad \text{or} \quad \nabla\phi.$$

Thus

$$\operatorname{grad}\phi = \nabla\phi$$
$$= \phi_{,j},$$

which is a covariant tensor of the first order.

Theorem 4.5. *Let A be an arbitrary vector with components A^i, and let ϕ be a scalar function. Then*

$$\operatorname{div}(\phi A) = \nabla\phi A + \phi \operatorname{div} A + \frac{\partial\phi}{\Delta_i x^i}\frac{\partial A^i}{\Delta_i x^i}.$$

Proof. We have

$$\operatorname{div}(\phi A^i) = \frac{\partial(\phi A^i)}{\Delta_i x^i} + \left(\frac{1}{2g}\frac{\partial g}{\Delta_\alpha x^\alpha} - \frac{\mu_\alpha}{g}\sum_{l=2}^{N}H_{l\alpha}\right)\phi A^\alpha$$

$$= \frac{\partial\phi}{\Delta_i x^i}A^i + \phi^{\sigma_i}\frac{\partial A^i}{\Delta_i x^i} + \left(\frac{1}{2g}\frac{\partial g}{\Delta_\alpha x^\alpha} - \frac{\mu_\alpha}{g}\sum_{l=2}^{N}H_{l\alpha}\right)\phi A^\alpha$$

$$= \frac{\partial\phi}{\Delta_i x^i}A^i + \left(\phi + \frac{\partial\phi}{\Delta_i x^i}\right)\frac{\partial A^i}{\Delta_i x^i}$$

$$+ \left(\frac{1}{2g}\frac{\partial g}{\Delta_\alpha x^\alpha} - \frac{\mu_\alpha}{g}\sum_{l=2}^{N}H_{l\alpha}\right)\phi A^\alpha$$

$$= \frac{\partial\phi}{\Delta_i x^i}A^i + \phi\left(\frac{\partial A^i}{\Delta_i x^i} + \left(\frac{1}{2g}\frac{\partial g}{\Delta_\alpha x^\alpha} - \frac{\mu_\alpha}{g}\sum_{l=2}^{N}H_{l\alpha}\right)A^\alpha\right)$$

$$+ \frac{\partial\phi}{\Delta_i x^i}\frac{\partial A^i}{\Delta_i x^i}$$

$$= \nabla\phi A + \phi \operatorname{div} A + \frac{\partial\phi}{\Delta_i x^i}\frac{\partial A^i}{\Delta_i x^i}.$$

This completes the proof. $\qquad\square$

Exercise 4.6. Let ϕ and ψ be two invariants. Prove that

$$\nabla(\phi\psi) = \psi\nabla\phi + \phi\nabla\psi + \frac{\partial\phi}{\Delta_i x^i}\frac{\partial\psi}{\Delta_i x^i}.$$

4.6 The Laplacian operator

Let ϕ be a scalar function of coordinates x^i.

Definition 4.9. The divergence of grad ϕ is defined as the Laplacian of the invariant ϕ and is denoted by

$$\nabla^2 \phi.$$

Thus

$$
\begin{aligned}
\nabla^2 \phi &= \mathrm{div}(\nabla \phi) \\
&= \mathrm{div}(\mathrm{grad}\,\phi) \\
&= \mathrm{div}(\phi_{,i}) \\
&= \mathrm{div}(g^{ik}\phi_{,k}) - \mu_k \frac{\partial g_{jk}}{\Delta_k x^k}\frac{\partial(g^{ik}\phi_{,k})}{\Delta_k x^k} \\
&= \frac{\partial(g^{ik}\phi_{,k})}{\Delta_i x^i} + \left(\frac{1}{2g}\frac{\partial g}{\Delta_a x^a} - \frac{\mu_a}{g}\sum_{l=2}^{N} H_{la}\right)(g^{ak}\phi_{,k}) \\
&\quad - \mu_k \frac{\partial g_{jk}}{\Delta_k x^k}\frac{\partial(g^{ik}\phi_{,k})}{\Delta_k x^k}.
\end{aligned}
$$

Example 4.23. We will find an expression of $\nabla^2 \phi$ in cylindrical coordinates

$$\Delta s^2 = \left(\Delta_1 x^1\right)^2 + \left(x^1\right)^2\left(\Delta_2 x^2\right)^2 + \left(\Delta_3 x^3\right)^2.$$

Here

$$
\begin{aligned}
g_{11} &= 1, \\
g_{22} &= \left(x^1\right)^2, \\
g_{33} &= 1.
\end{aligned}
$$

We will use the computations in Example 4.18. We have

$$
\begin{aligned}
g^{11} &= \frac{1}{(x^1)^2}\det\begin{pmatrix}(x^1)^2 & 0 \\ 0 & 1\end{pmatrix} \\
&= \frac{1}{(x^1)^2}(x^1)^2 \\
&= 1, \\
g^{22} &= \frac{1}{(x^1)^2}\det\begin{pmatrix}1 & 0 \\ 0 & 1\end{pmatrix} \\
&= \frac{1}{(x^1)^2},
\end{aligned}
$$

$$g^{33} = \frac{1}{(x^1)^2} \det \begin{pmatrix} 1 & 0 \\ 1 & (x^1)^2 \end{pmatrix}$$

$$= \frac{1}{(x^1)^2}(x^1)^2$$

$$= 1.$$

Then

$$\nabla^2\phi = \frac{\partial(g^{11}\phi_{,1})}{\Delta_1 x^1} + \frac{\partial(g^{22}\phi_{,2})}{\Delta_2 x^2} + \frac{\partial(g^{33}\phi_{,3})}{\Delta_3 x^3}$$

$$+ \frac{1}{2g}\frac{\partial g}{\Delta_1 x^1}g^{11}\phi_{,1} + \frac{1}{2g}\frac{\partial g}{\Delta_2 x^2}g^{22}\phi_{,2} + \frac{1}{2g}\frac{\partial g}{\Delta_3 x^3}g^{33}\phi_{,3}$$

$$- \mu_1\frac{\partial g}{\Delta_1 x^1}\frac{\partial(g^{11}\phi_{,1})}{\Delta_1 x^1} - \mu_2\frac{\partial g}{\Delta_2 x^2}\frac{\partial(g^{22}\phi_{,2})}{\Delta_2 x^2} - \mu_3\frac{\partial g}{\Delta_3 x^3}\frac{\partial(g^{33}\phi_{,3})}{\Delta_3 x^3}$$

$$= \frac{\partial^2\phi}{(\Delta_1 x^1)^2} + \frac{1}{(x^1)^2}\frac{\partial^2\phi}{(\Delta_2 x^2)^2} + \frac{\partial^2\phi}{(\Delta_3 x^3)^2}$$

$$+ \frac{1}{2(x^1)^2}(\sigma_1(x^1) + x^1)\frac{1}{(x^1)^2}\frac{\partial\phi}{\Delta_1 x^1}$$

$$= \frac{\partial^2\phi}{(\Delta_1 x^1)^2} + \frac{1}{(x^1)^2}\frac{\partial^2\phi}{(\Delta_2 x^2)^2} + \frac{\partial^2\phi}{(\Delta_3 x^3)^2}$$

$$+ \frac{\sigma_1(x^1) + x^1}{2(x^1)^4}\frac{\partial\phi}{\Delta_1 x^1}.$$

Example 4.24. We will find $\nabla^2\phi$ in spherical coordinates

$$\Delta s^2 = (\Delta_1 x^1)^2 + (x^1)^2(\Delta_2 x^2)^2 + (x^1)^2(\sin_1(x^2, x_0^2))^2(\Delta_3 x^3)^2.$$

Here

$$g_{11} = 1,$$
$$g_{22} = (x^1)^2,$$
$$g_{33} = (x^1)^2(\sin_1(x^2, x_0^2))^2.$$

We will use the computations in Example 4.20. We have

$$g = (x^1)^4(\sin_1(x^2, x_0^2))^2$$

and

$$g^{11} = \frac{1}{(x^1)^4(\sin_1(x^2, x_0^2))^2}\det \begin{pmatrix} (x^1)^2 & 0 \\ 0 & (x^1)^2(\sin_1(x^2, x_0^2))^2 \end{pmatrix}$$

$$= \frac{1}{(x^1)^4(\sin_1(x^2, x_0^2))^2}(x^1)^4(\sin_1(x^2, x_0^2))^2$$

$$= 1,$$

$$g^{22} = \frac{1}{(x^1)^4(\sin_1(x^2,x_0^2))^2}\det\begin{pmatrix} 1 & 0 \\ 0 & (x^1)^2(\sin_1(x^2,x_0^2))^2 \end{pmatrix}$$

$$= \frac{1}{(x^1)^4(\sin_1(x^2,x_0^2))^2}(x^1)^2(\sin_1(x^2,x_0^2))^2$$

$$= \frac{1}{(x^1)^2},$$

$$g^{33} = \frac{1}{(x^1)^4(\sin_1(x^2,x_0^2))^2}\det\begin{pmatrix} 1 & 0 \\ 0 & (x^1)^2 \end{pmatrix}$$

$$= \frac{1}{(x^1)^4(\sin_1(x^2,x_0^2))^2}(x^1)^2$$

$$= \frac{1}{(x^1)^2(\sin_1(x^2,x_0^2))^2}.$$

Consequently,

$$\nabla^2\phi = \frac{\partial(g^{11}\phi_{,1})}{\Delta_1 x^1} + \frac{\partial(g^{22}\phi_{,2})}{\Delta_2 x^2} + \frac{\partial(g^{33}\phi_{,3})}{\Delta_3 x^3}$$

$$+ \frac{1}{2g}\frac{\partial g}{\Delta_1 x^1}g^{11}\phi_{,1} + \frac{1}{2g}\frac{\partial g}{\Delta_2 x^2}g^{22}\phi_{,2} + \frac{1}{2g}\frac{\partial g}{\Delta_3 x^3}g^{33}\phi_{,3} - \frac{\mu_1}{g}H_{31}g^{11}\phi_{,1}$$

$$- \mu_1\frac{\partial g}{\Delta_1 x^1}\frac{\partial(g^{11}\phi_{,1})}{\Delta_1 x^1} - \mu_2\frac{\partial g}{\Delta_2 x^2}\frac{\partial(g^{22}\phi_{,2})}{\Delta_2 x^2} - \mu_3\frac{\partial g}{\Delta_3 x^3}\frac{\partial(g^{33}\phi_{,3})}{\Delta_3 x^3}$$

$$= \frac{\partial^2\phi}{(\Delta_1 x^1)^2} + \frac{1}{(x^1)^2}\frac{\partial^2\phi}{(\Delta_2 x^2)^2} + \frac{1}{(x^1)^2(\sin_1(x^2,x_0^2))^2}\frac{\partial^2\phi}{(\Delta_3 x^3)^2}$$

$$+ \frac{1}{2(x^1)^4(\sin_1(x^2,x_0^2))^2}\left((\sigma_1(x^1))^3 + x^1(\sigma_1(x^1))^2 + (x^1)^2\sigma_1(x^1) + (x^1)^3\right)$$

$$\times (\sin_1(x^2,x_0^2))^2\frac{\partial\phi}{\Delta_1 x^1}$$

$$+ \frac{1}{2(x^1)^4(\sin_1(x^2,x_0^2))^2}(x^1)^4\cos_1(x^2,x_0^2)(\sin_1(x^2,x_0^2) + \sin_1(\sigma_2(x^2),x_0^2))\frac{1}{(x^1)^2}\frac{\partial\phi}{\Delta_2 x^2}$$

$$- \frac{\mu_1}{(x^1)^4(\sin_1(x^2,x_0^2))^2}(\sigma_1(x^1) + x^1)^2(\sin_1(x^2,x_0^2))^2\frac{\partial\phi}{\Delta_1 x^1}$$

$$- \frac{\partial^2\phi}{(\Delta_1 x^1)^2} + \frac{1}{(x^1)^2}\frac{\partial^2\phi}{(\Delta_2 x^2)^2} + \frac{1}{(x^1)^2(\sin_1(x^2,x_0^2))^2}\frac{\partial^2\phi}{(\Delta_3 x^3)^2}$$

$$+ \frac{(\sigma_1(x^1))^3 + x^1(\sigma_1(x^1))^2 + (x^1)^2\sigma_1(x^1) + (x^1)^3}{2(x^1)^4}\frac{\partial\phi}{\Delta_1 x^1}$$

$$+ \frac{\cos_1(x^2,x_0^2)(\sin_1(x^2,x_0^2 w) + \sin_1(\sigma_2(x^2),x_0^2))}{2(x^1)^2(\sin_1(x^2,x_0^2))^2}\frac{\partial\phi}{\Delta_2 x^2}$$

$$- \frac{\mu_1(x^1)(\sigma_1(x^1) + x^1)^2}{(x^1)^4}\frac{\partial\phi}{\Delta_1 x^1}$$

$$
\begin{aligned}
= \; & \frac{\partial^2 \phi}{(\Delta_1 x^1)^2} + \frac{1}{(x^1)^2} \frac{\partial^2 \phi}{(\Delta_2 x^2)^2} + \frac{1}{(x^1)^2 (\sin_1(x^2, x_0^2))^2} \frac{\partial^2 \phi}{(\Delta_3 x^3)^2} \\
& + \frac{(\sigma_1(x^1))^3 + x^1(\sigma_1(x^1))^2 + (x^1)^2 \sigma_1(x^1) + (x^1)^3 - 2\mu_1(\sigma_1(x^1) + x^1)^2}{2(x^1)^4} \frac{\partial \phi}{\Delta_1 x^1} \\
& + \frac{\cos_1(x^2, x_0^2)(\sin_1(x^2, x_0^2 w) + \sin_1(\sigma_2(x^2), x_0^2))}{2(x^1)^2(\sin_1(x^2, x_0^2))^2} \frac{\partial \phi}{\Delta_2 x^2}.
\end{aligned}
$$

Exercise 4.7. Let ϕ and ψ be scalar functions, and let α and β be constants. Prove that

$$
\nabla^2(\alpha \phi + \beta \psi) = \alpha \nabla^2 \phi + \beta \nabla^2 \psi.
$$

4.7 Curl

Consider a covariant vector A_i. Then $A_{i,j}$ is a tensor that is contravariant of the zero order and covariant of the second order. Hence

$$
A_{i,j} - A_{j,i} \tag{4.12}
$$

is a skew-symmetric tensor that is contravariant of the zero order and covariant of the second order.

Definition 4.10. Tensor (4.12) is called the curl, rotation, or rotor of the covariant vector A_i and is denoted

$$
\operatorname{curl} A_i.
$$

In fact, we have

$$
\begin{aligned}
\operatorname{curl} A_i = \; & A_{i,j} - A_{j,i} \\
= \; & \frac{\partial A_i}{\Delta_j x^j} - \Gamma_{ij}^\alpha A_\alpha - \left(\frac{\partial A_j}{\Delta_i x^i} - \Gamma_{ji}^\alpha A_\alpha \right) \\
= \; & \frac{\partial A_i}{\Delta_j x^j} - \Gamma_{ij}^\alpha A_\alpha - \frac{\partial A_j}{\Delta_i x^i} + \Gamma_{ji}^\alpha A_\alpha \\
= \; & \frac{\partial A_i}{\Delta_j x^j} - \frac{\partial A_j}{\Delta_i x^i} - \Gamma_{ij}^\alpha A_\alpha + \Gamma_{ji}^\alpha A_\alpha \\
= \; & \frac{\partial A_i}{\Delta_j x^j} - \frac{\partial A_j}{\Delta_i x^i} - \Gamma_{ij}^\alpha A_\alpha + \Gamma_{ij}^\alpha A_\alpha \\
= \; & \frac{\partial A_i}{\Delta_j x^j} - \frac{\partial A_j}{\Delta_i x^i}.
\end{aligned}
$$

Theorem 4.6. *A necessary and sufficient condition that the curl of a vector field vanishes is that the vector field is a gradient.*

Proof. Let A_i be a covariant vector.
1.   Assume that

$$\operatorname{curl} A_i = 0.$$

Then

$$\frac{\partial A_i}{\Delta_j x^j} = \frac{\partial A_j}{\Delta_i x^i},$$

whereupon

$$\frac{\partial A_i}{\Delta_j x^j} \Delta_j x^j = \frac{\partial A_j}{\Delta_i x^i} \Delta_j x^j,$$

and

$$\Delta A_i = \frac{\partial}{\Delta_j x^j} A_j \Delta_j x^j.$$

By integrating we find

$$A_i = \int \frac{\partial}{\Delta_j x^j} A_j \Delta_j x^j$$
$$= \frac{\partial}{\Delta_j x^j} \int A_j \Delta_j x^j.$$

Observe that

$$\int A_j \Delta_j x^j$$

is a scalar quantity. Set

$$\phi = \int A_j \Delta_j x^j.$$

Therefore

$$A_i = \frac{\partial \phi}{\Delta_i x^i},$$

i. e., A_i is the gradient of ϕ.
2.   Now assume that

$$A_i = \nabla \phi$$

for some scalar ϕ. Then we have

$$A_i = \frac{\partial \phi}{\Delta_i x^i},$$

$$A_j = \frac{\partial \phi}{\Delta_j x^j},$$

$$\frac{\partial A_i}{\Delta_j x^j} = \frac{\partial^2 \phi}{\Delta_j x^j \Delta_i x^i},$$

$$\frac{\partial A_j}{\Delta_i x^i} = \frac{\partial^2 \phi}{\Delta_i x^i \Delta_j x^j}.$$

Consequently,

$$\operatorname{curl} A_i = A_{i,j} - A_{j,i}$$

$$= \frac{\partial A_i}{\Delta_j x^j} - \frac{\partial A_j}{\Delta_i x^i}$$

$$= \frac{\partial^2 \phi}{\Delta_j x^j \Delta_i x^i} - \frac{\partial^2 \phi}{\Delta_i x^i \Delta_j x^j}$$

$$= \frac{\partial^2 \phi}{\Delta_i x^i \Delta_j x^j} - \frac{\partial^2 \phi}{\Delta_i x^i \Delta_j x^j}$$

$$= 0.$$

This completes the proof. $\qquad\square$

Example 4.25. Suppose that A_{ij} is the curl of a covariant vector B_i. We will show that

$$A_{ij,k} + A_{jk,i} + A_{ki,j} = 0.$$

By the definition of A_{ij} we find

$$A_{ij} = \operatorname{curl} B_i$$

$$= B_{i,j} - B_{j,i}$$

$$= \frac{\partial B_i}{\Delta_j x^j} - \frac{\partial B_j}{\Delta_i x^i},$$

$$A_{jk} = \operatorname{curl} B_j$$

$$= B_{j,k} - B_{k,j}$$

$$= \frac{\partial B_j}{\Delta_k x^k} - \frac{\partial B_k}{\Delta_j x^j},$$

$$A_{ki} = \operatorname{curl} B_k$$

$$= B_{k,i} - B_{i,k}$$

$$= \frac{\partial B_k}{\Delta_i x^i} - \frac{\partial B_i}{\Delta_k x^k}.$$

Hence

$$A_{ij,k} = \frac{\partial A_{ij}}{\Delta_k x^k} - \Gamma_{ik}^{\alpha} A_{\alpha j} - \Gamma_{kj}^{\alpha} A_{i\alpha}$$

$$= \frac{\partial}{\Delta_k x^k} \left(\frac{\partial B_i}{\Delta_j x^j} - \frac{\partial B_j}{\Delta_i x^i} \right) - \Gamma_{ik}^{\alpha} A_{\alpha j} - \Gamma_{kj}^{\alpha} A_{i\alpha},$$

$$= \frac{\partial^2 B_i}{\Delta_k x^k \Delta_j x^j} - \frac{\partial^2 B_j}{\Delta_k x^k \Delta_i x^i} - \Gamma_{ik}^{\alpha} A_{\alpha j} - \Gamma_{kj}^{\alpha} A_{i\alpha}$$

and

$$A_{jk,i} = \frac{\partial A_{jk}}{\Delta_i x^i} - \Gamma_{ji}^{\alpha} A_{\alpha k} - \Gamma_{ki}^{\alpha} A_{j\alpha}$$

$$= \frac{\partial}{\Delta_i x^i} \left(\frac{\partial B_j}{\Delta_k x^k} - \frac{\partial B_k}{\Delta_j x^j} \right) - \Gamma_{ji}^{\alpha} A_{\alpha k} - \Gamma_{ki}^{\alpha} A_{j\alpha}$$

$$= \frac{\partial^2 B_j}{\Delta_i x^i \Delta_k x^k} - \frac{\partial^2 B_k}{\Delta_i x^i \Delta_j x^j} - \Gamma_{ji}^{\alpha} A_{\alpha k} - \Gamma_{ki}^{\alpha} A_{j\alpha}$$

$$= \frac{\partial^2 B_j}{\Delta_i x^i \Delta_k x^k} - \frac{\partial^2 B_k}{\Delta_i x^i \Delta_j x^j} + \Gamma_{ji}^{\alpha} A_{\alpha k} - \Gamma_{ki}^{\alpha} A_{\alpha j},$$

where we have used that

$$A_{j\alpha} = -A_{\alpha j},$$

and

$$A_{ki,j} = \frac{\partial A_{ki}}{\Delta_j x^j} - \Gamma_{kj}^{\alpha} A_{\alpha i} - \Gamma_{ji}^{\alpha} A_{k\alpha}$$

$$= \frac{\partial}{\Delta_j x^j} \left(\frac{\partial B_k}{\Delta_i x^i} - \frac{\partial B_i}{\Delta_k x^k} \right) + \Gamma_{kj}^{\alpha} A_{\alpha i} + \Gamma_{ji}^{\alpha} A_{k\alpha}$$

$$= \frac{\partial^2 B_k}{\Delta_j x^j \Delta_i x^i} - \frac{\partial^2 B_i}{\Delta_j x^j \Delta_k x^k} + \Gamma_{kj}^{\alpha} A_{\alpha i} + \Gamma_{ji}^{\alpha} A_{k\alpha},$$

where we have used that

$$A_{\alpha i} = -A_{i\alpha},$$
$$A_{k\alpha} = -A_{\alpha k}.$$

Consequently,

$$A_{ij,k} + A_{jk,i} + A_{ki,j} = \frac{\partial^2 B_i}{\Delta_k x^k \Delta_j x^j} - \frac{\partial^2 B_j}{\Delta_k x^k \Delta_i x^i} - \Gamma_{ik}^{\alpha} A_{\alpha j} - \Gamma_{kj}^{\alpha} A_{i\alpha}$$

$$+ \frac{\partial^2 B_j}{\Delta_i x^i \Delta_k x^k} - \frac{\partial^2 B_k}{\Delta_i x^i \Delta_j x^j} + \Gamma_{ji}^{\alpha} A_{\alpha k} - \Gamma_{ki}^{\alpha} A_{\alpha j}$$

$$+ \frac{\partial^2 B_k}{\Delta_j x^j \Delta_i x^i} - \frac{\partial^2 B_i}{\Delta_j x^j \Delta_k x^k} + \Gamma_{kj}^{\alpha} A_{\alpha i} + \Gamma_{ji}^{\alpha} A_{k\alpha},$$

where we have used that

$$\frac{\partial^2 B_i}{\Delta_k x^k \Delta_j x^j} = \frac{\partial^2 B_i}{\Delta_j x^j \Delta_k x^k},$$

$$\frac{\partial^2 B_j}{\Delta_k x^k \Delta_i x^i} = \frac{\partial^2 B_j}{\Delta_i x^i \Delta_k x^k},$$

$$\frac{\partial^2 B_k}{\Delta_i x^i \Delta_j x^j} = \frac{\partial^2 B_k}{\Delta_j x^j \Delta_i x^i}.$$

4.8 Curl of covariant vectors in V_3

Before to define curl of covariant vectors in V_3, we will introduce the ε-systems.

Definition 4.11. The covariant ε-tensor of the second order is defined by

$$\varepsilon_{11} = 0,$$
$$\varepsilon_{12} = \sqrt{g},$$
$$\varepsilon_{21} = -\sqrt{g},$$
$$\varepsilon_{22} = 0.$$

By the definition it follows that

$$\varepsilon_{ij} = \sqrt{g}\, e_{ij}.$$

This tensor is skew-symmetric, and

$$\bar{\varepsilon}_{ij} = \frac{\Delta_p x^p}{\Delta_{(i)} \bar{x}^i} \frac{\Delta_q x^q}{\Delta_{(j)} \bar{x}^j} \varepsilon_{pq}$$

$$= \frac{\Delta_1 x^1}{\Delta_{(i)} \bar{x}^i} \frac{\Delta_1 x^1}{\Delta_{(j)} \bar{x}^j} \varepsilon_{11} + \frac{\Delta_1 x^1}{\Delta_{(i)} \bar{x}^i} \frac{\Delta_2 x^2}{\Delta_{(j)} \bar{x}^j} \varepsilon_{12}$$

$$+ \frac{\Delta_2 x^2}{\Delta_{(i)} \bar{x}^i} \frac{\Delta_1 x^1}{\Delta_{(j)} \bar{x}^j} \varepsilon_{21} + \frac{\Delta_2 x^2}{\Delta_{(i)} \bar{x}^i} \frac{\Delta_2 x^2}{\Delta_{(j)} \bar{x}^j} \varepsilon_{22}$$

$$= \sqrt{g}\,\frac{\Delta_1 x^1}{\Delta_{(i)}\overline{x}^i}\,\frac{\Delta_2 x^2}{\Delta_{(j)}\overline{x}^j} - \sqrt{g}\,\frac{\Delta_1 x^1}{\Delta_{(j)}\overline{x}^j}\,\frac{\Delta_2 x^2}{\Delta_{(i)}\overline{x}^i}$$

$$= \sqrt{g}\left(\frac{\Delta_1 x^1}{\Delta_{(i)}\overline{x}^i}\,\frac{\Delta_2 x^2}{\Delta_{(j)}\overline{x}^j} - \frac{\Delta_1 x^1}{\Delta_{(j)}\overline{x}^j}\,\frac{\Delta_2 x^2}{\Delta_{(i)}\overline{x}^i}\right)$$

$$= \sqrt{g}\,\det\begin{pmatrix} \dfrac{\Delta_1 x^1}{\Delta_{(i)}\overline{x}^i} & \dfrac{\Delta_2 x^2}{\Delta_{(i)}\overline{x}^i} \\[2mm] \dfrac{\Delta_1 x^1}{\Delta_{(j)}\overline{x}^j} & \dfrac{\Delta_2 x^2}{\Delta_{(j)}\overline{x}^j} \end{pmatrix}$$

$$= \sqrt{g}\,\frac{\Delta(x^1, x^2)}{\Delta(\overline{x}^1, \overline{x}^2)}.$$

In particular, we have

$$\overline{\varepsilon}_{11} = \sqrt{g}\left(\frac{\Delta_1 x^1}{\Delta_{(1)}\overline{x}^1}\,\frac{\Delta_2 x^2}{\Delta_{(1)}\overline{x}^1} - \frac{\Delta_1 x^1}{\Delta_{(1)}\overline{x}^1}\,\frac{\Delta_2 x^2}{\Delta_{(1)}\overline{x}^1}\right)$$

$$= 0,$$

$$\overline{\varepsilon}_{12} = \sqrt{g}\left(\frac{\Delta_1 x^1}{\Delta_{(1)}\overline{x}^1}\,\frac{\Delta_2 x^2}{\Delta_{(2)}\overline{x}^2} - \frac{\Delta_1 x^1}{\Delta_{(2)}\overline{x}^2}\,\frac{\Delta_2 x^2}{\Delta_{(1)}\overline{x}^1}\right)$$

$$= \sqrt{g},$$

$$\overline{\varepsilon}_{21} = \sqrt{g}\left(\frac{\Delta_1 x^1}{\Delta_{(2)}\overline{x}^2}\,\frac{\Delta_2 x^2}{\Delta_{(1)}\overline{x}^1} - \frac{\Delta_1 x^1}{\Delta_{(1)}\overline{x}^1}\,\frac{\Delta_2 x^2}{\Delta_{(2)}\overline{x}^2}\right)$$

$$= -\sqrt{g},$$

$$\overline{\varepsilon}_{22} = \sqrt{g}\left(\frac{\Delta_1 x^1}{\Delta_{(2)}\overline{x}^2}\,\frac{\Delta_2 x^2}{\Delta_{(2)}\overline{x}^2} - \frac{\Delta_1 x^1}{\Delta_{(2)}\overline{x}^2}\,\frac{\Delta_2 x^2}{\Delta_{(2)}\overline{x}^2}\right)$$

$$= 0.$$

Definition 4.12. Define

$$\varepsilon^{ij} = \varepsilon_{rs}g^{ir}g^{js}.$$

We have

$$\varepsilon^{ij} = \varepsilon_{11}g^{i1}g^{j1} + \varepsilon_{12}g^{i1}g^{j2} + \varepsilon_{21}g^{i2}g^{j1} + \varepsilon_{22}g^{i2}g^{j2}$$
$$= \varepsilon_{12}g^{i1}g^{j2} - \varepsilon_{12}g^{i2}g^{j1}$$
$$= \sqrt{g}\,(g^{i1}g^{j2} - g^{i2}g^{j1}).$$

In particular, we have

$$\varepsilon^{11} = \sqrt{g}\,(g^{11}g^{12} - g^{12}g^{11})$$
$$= 0,$$
$$\varepsilon^{12} = \sqrt{g}\,(g^{11}g^{22} - g^{12}g^{21})$$

$$= \sqrt{g}\left(\frac{1}{g}\right)$$

$$= \frac{1}{\sqrt{g}},$$

$$\varepsilon_{21} = \sqrt{g}\left(g^{21}g^{12} - g^{22}g^{11}\right)$$

$$= -\varepsilon^{12}$$

$$= -\frac{1}{\sqrt{g}},$$

$$\varepsilon^{22} = \sqrt{g}\left(g^{21}g^{22} - g^{22}g^{21}\right)$$

$$= 0.$$

Therefore

$$\varepsilon^{ij} = \frac{1}{\sqrt{g}}e^{ij}.$$

Definition 4.13. Define

$$\varepsilon_{ijk} = \sqrt{g}\,e_{ijk},$$
$$\varepsilon^{ijk} = \frac{1}{\sqrt{g}}e^{ijk}.$$

More generally, we define

$$\varepsilon_{i_1 i_2 \cdots i_p} = \sqrt{g}\,e_{i_1 i_2 \cdots i_p},$$
$$\varepsilon^{j_1 j_2 \cdots j_q} = \frac{1}{\sqrt{g}}e^{j_1 j_2 \cdots j_q}.$$

Example 4.26. Let a_i be a vector. We will show that

$$\varepsilon_{ijk}a_j a_k = 0.$$

Indeed, using the definition of ε_{ijk}, we obtain

$$
\begin{aligned}
\varepsilon_{ijk}a_j a_k &= \sqrt{g}\,e_{ijk}a_j a_k \\
&= \sqrt{g}\,(e_{111}a_1 a_1 + e_{112}a_1 a_2 + e_{113}a_1 a_3 \\
&\quad + e_{121}a_2 a_1 + e_{122}a_2 a_2 + e_{123}a_2 a_3 \\
&\quad + e_{131}a_3 a_1 + e_{132}a_3 a_2 + e_{133}a_3 a_3 \\
&\quad + e_{211}a_1 a_1 + e_{212}a_1 a_2 + e_{213}a_1 a_3 \\
&\quad + e_{221}a_2 a_1 + e_{222}a_2 a_2 + e_{223}a_2 a_3 \\
&\quad + e_{231}a_3 a_1 + e_{232}a_3 a_2 + e_{233}a_3 a_3 \\
&\quad + e_{311}a_1 a_1 + e_{312}a_1 a_2 + e_{313}a_1 a_3
\end{aligned}
$$

$$+ e_{321}a_2a_1 + e_{322}a_2a_2 + e_{323}a_2a_3$$
$$+ e_{331}a_3a_1 + e_{332}a_3a_2 + e_{333}a_3a_3)$$
$$= \sqrt{g}(a_2a_3 - a_3a_2 - a_1a_3 + a_3a_1 + a_1a_2 - a_1a_2)$$
$$= 0.$$

Exercise 4.8. Prove that

$$\varepsilon_{iks}\varepsilon^{mps} = \delta_{im}\delta_{kp} - \delta_{ip}\delta_{km}.$$

Definition 4.14. Let A_i be a covariant vector. Then the inner product

$$\varepsilon^{ikl}A_{l,k} = B^i$$

is called the curl of A_i.

Thus, in V_3, we have

$$\operatorname{curl} A_i = \varepsilon^{ikl}A_{l,k}$$
$$= \frac{1}{\sqrt{g}}e^{ikl}A_{l,k}.$$

In particular, we get

$$B^1 = \frac{1}{\sqrt{g}}e^{2kl}A_{l,k}$$
$$= \frac{1}{\sqrt{g}}(e^{111}A_{1,1} + e^{112}A_{1,2} + e^{113}A_{1,3}$$
$$+ e^{121}A_{2,1} + e^{122}A_{2,2} + e^{123}A_{2,3}$$
$$+ e^{131}A_{3,1} + e^{132}A_{3,2} + e^{133}A_{3,3})$$
$$= \frac{1}{\sqrt{g}}(A_{3,2} - A_{2,3}),$$
$$B^2 = \frac{1}{\sqrt{g}}e^{2kl}A_{l,k}$$
$$= \frac{1}{\sqrt{g}}(e^{211}A_{1,1} + e^{212}A_{1,2} + e^{213}A_{1,3}$$
$$+ e^{221}A_{2,1} + e^{222}A_{2,2} + e^{223}A_{2,3}$$
$$+ e^{231}A_{3,1} + e^{232}A_{3,2} + e^{233}A_{3,3})$$
$$= \frac{1}{\sqrt{g}}(A_{1,3} - A_{3,1}),$$
$$B^3 = \frac{1}{\sqrt{g}}e^{3kl}A_{l,k}$$
$$= \frac{1}{\sqrt{g}}(e^{311}A_{1,1} + e^{312}A_{1,2} + e^{313}A_{1,3}$$

$$+ e^{321} A_{2,1} + e^{322} A_{2,2} + e^{323} A_{2,3}$$
$$+ e^{331} A_{3,1} + e^{332} A_{3,2} + e^{333} A_{3,3})$$
$$= \frac{1}{\sqrt{g}} (A_{2,1} - A_{1,2}).$$

Thus

$$\operatorname{curl} A_i = \frac{1}{\sqrt{g}} (A_{3,2} - A_{2,3}, A_{1,3} - A_{3,1}, A_{2,1} - A_{1,2})$$
$$= \frac{1}{\sqrt{g}} \left(\frac{\partial A_3}{\Delta_2 x^2} - \frac{\partial A_2}{\Delta_3 x^3}, \frac{\partial A_1}{\Delta_3 x^3} - \frac{\partial A_3}{\Delta_1 x^1}, \frac{\partial A_2}{\Delta_1 x^1} - \frac{\partial A_1}{\Delta_2 x^2} \right).$$

Definition 4.15. The vector product of two covariant vectors A_i and B_j in V_3 is defined by

$$\varepsilon^{ijk} A_j B_k = C^i.$$

In fact, we have

$$C^i = \frac{1}{\sqrt{g}} e^{ijk} A_j B_k$$

and

$$C^1 = \frac{1}{\sqrt{g}} e^{1jk} A_j B_k$$
$$= \frac{1}{\sqrt{g}} (e^{111} A_1 B_1 + e^{112} A_1 B_2 + e^{113} A_1 B_3$$
$$+ e^{121} A_2 B_1 + e^{122} A_2 B_2 + e^{123} A_2 B_3$$
$$+ e^{131} A_3 B_1 + e^{132} A_3 B_2 + e^{133} A_3 B_3)$$
$$= \frac{1}{\sqrt{g}} \frac{1}{\sqrt{g}} (A_2 B_3 - A_3 B_2),$$
$$C^2 = \frac{1}{\sqrt{g}} e^{2jk} A_j B_k$$
$$= \frac{1}{\sqrt{g}} (e^{211} A_1 B_1 + e^{212} A_1 B_2 + e^{213} A_1 B_3$$
$$+ e^{221} A_2 B_1 + e^{222} A_2 B_2 + e^{223} A_2 B_3$$
$$+ e^{231} A_3 B_1 + e^{232} A_3 B_2 + e^{233} A_3 B_3)$$
$$= \frac{1}{\sqrt{g}} \frac{1}{\sqrt{g}} (A_3 B_1 - A_1 B_3),$$
$$C^3 = \frac{1}{\sqrt{g}} e^{3jk} A_j B_k$$
$$= \frac{1}{\sqrt{g}} (e^{311} A_1 B_1 + e^{312} A_1 B_2 + e^{313} A_1 B_3$$

$$+ e^{321} A_2 B_1 + e^{322} A_2 B_2 + e^{323} A_2 B_3$$
$$+ e^{331} A_3 B_1 + e^{332} A_3 B_2 + e^{333} A_3 B_3)$$
$$= \frac{1}{\sqrt{g}} \frac{1}{\sqrt{g}} (A_2 B_1 - A_1 B_2).$$

Thus the components of the vector product of covariant vectors A_j and B_k are

$$\frac{1}{\sqrt{g}} \frac{1}{\sqrt{g}} (A_2 B_3 - A_3 B_2), \quad \frac{1}{\sqrt{g}} \frac{1}{\sqrt{g}} (A_3 B_1 - A_1 B_3), \quad \frac{1}{\sqrt{g}} \frac{1}{\sqrt{g}} (A_2 B_1 - A_1 B_2).$$

Exercise 4.9. Let S be an open two-sided surface bounded by a closed nonintersecting closed curve C, let $\frac{\Delta_p x^p}{\Delta s}$ denote the tangent vector to C, and let v^p denote the positive unit normal to S. Suppose that $\vec{A}$ has continuous derivatives. Prove that:

1.

$$\int_C \vec{A} \cdot d\vec{r} = \int_C A_p \frac{\Delta_p x^p}{\Delta s} \Delta s.$$

2.

$$\int_S (\nabla \times \vec{A}) \cdot d\vec{s} = - \int_S \varepsilon^{ijk} A_{j,k} v^i \Delta s.$$

4.9 Intrinsic derivatives

Let C be a certain space curve described by the parametric equations in V_N, a Riemann space, as

$$C : x^i = x^i(t), \quad i \in \{1, 2, \ldots, N\},$$

where x^i are N functions of a single parameter t. In general, it is sufficient that the derivatives exist up to any order and are continuous. Let

$$\frac{\Delta_k x^k}{\Delta t}$$

denote the tangent vector of the curve C.

Definition 4.16. The absolute or intrinsic derivative of a tensor $A_{j_1 j_2 j_q}^{i_1 i_2 \cdots i_p}$ is defined as

$$\frac{\delta(A_{j_1 j_2 \ldots l_q}^{i_1 i_2 \cdots i_p})}{\delta t} = (A_{j_1 j_2 \cdots j_q, k}^{i_1 i_2 \cdots i_p})^{\sigma_0 \sigma_1 \cdots \sigma_{k-1}} \frac{\Delta_k x^k}{\Delta t}.$$

In fact, we have

$$\frac{\delta A^{i_1 i_2 \cdots i_p}_{j_1 j_2 \cdots j_q}}{\delta t} = \left(\frac{\partial A^{i_1 i_2 \cdots i_p}_{j_1 j_2 \cdots j_q}}{\Delta_k x^k} + A^{a i_2 \cdots i_p}_{j_1 j_2 \cdots j_q} \Gamma^{i_1}_{ak} + A^{i_1 a i_3 \cdots i_p}_{j_1 j_2 j_3 \cdots j_q, k} \Gamma^{i_2}_{ak} + \cdots + A^{i_1 i_2 \cdots i_{p-1} a}_{j_1 j_2 \cdots j_{q-1} j_q} \Gamma^{i_p}_{ak} \right.$$
$$\left. - A^{i_1 i_2 \cdots i_p}_{\beta j_2 \cdots j_q} \Gamma^{\beta}_{j_1 k} - A^{i_1 i_2 \cdots i_p}_{j_1 \beta j_3 \cdots j_q} \Gamma^{\beta}_{j_2 k} - \cdots - A^{i_1 i_2 \cdots i_p}_{j_1 j_2 \cdots j_{q-1} \beta} \Gamma^{\beta}_{j_q k} \right)^{\sigma_0 \sigma_1 \cdots \sigma_{k-1}} \frac{\Delta_k x^k}{\Delta t}$$

$$= \left(\frac{\partial A^{i_1 i_2 \cdots i_p}_{j_1 j_2 \cdots j_q}}{\Delta_k x^k} \right)^{\sigma_0 \sigma_1 \cdots \sigma_{k-1}} \frac{\Delta_k x^k}{\Delta t} + \left(A^{a i_2 \cdots i_p}_{j_1 j_2 \cdots j_q} \Gamma^{i_1}_{ak} + A^{i_1 a i_3 \cdots i_p}_{j_1 j_2 j_3 \cdots j_q, k} \Gamma^{i_2}_{ak} + \cdots + A^{i_1 i_2 \cdots i_{p-1} a}_{j_1 j_2 \cdots j_{q-1} j_q} \Gamma^{i_p}_{ak} \right.$$
$$\left. - A^{i_1 i_2 \cdots i_p}_{\beta j_2 \cdots j_q} \Gamma^{\beta}_{j_1 k} - A^{i_1 i_2 \cdots i_p}_{j_1 \beta j_3 \cdots j_q} \Gamma^{\beta}_{j_2 k} - \cdots - A^{i_1 i_2 \cdots i_p}_{j_1 j_2 \cdots j_{q-1} \beta} \Gamma^{\beta}_{j_q k} \right)^{\sigma_0 \sigma_1 \cdots \sigma_{k-1}} \frac{\Delta_k x^k}{\Delta t}$$

$$= \frac{\partial A^{i_1 i_2 \cdots i_p}_{j_1 j_2 \cdots j_q}}{\Delta t} + \left(A^{a i_2 \cdots i_p}_{j_1 j_2 \cdots j_q} \Gamma^{i_1}_{ak} + A^{i_1 a i_3 \cdots i_p}_{j_1 j_2 j_3 \cdots j_q, k} \Gamma^{i_2}_{ak} + \cdots + A^{i_1 i_2 \cdots i_{p-1} a}_{j_1 j_2 \cdots j_{q-1} j_q} \Gamma^{i_p}_{ak} \right.$$
$$\left. - A^{i_1 i_2 \cdots i_p}_{\beta j_2 \cdots j_q} \Gamma^{\beta}_{j_1 k} - A^{i_1 i_2 \cdots i_p}_{j_1 \beta j_3 \cdots j_q} \Gamma^{\beta}_{j_2 k} - \cdots - A^{i_1 i_2 \cdots i_p}_{j_1 j_2 \cdots j_{q-1} \beta} \Gamma^{\beta}_{j_q k} \right)^{\sigma_0 \sigma_1 \cdots \sigma_{k-1}} \frac{\Delta_k x^k}{\Delta t}.$$

Example 4.27. Let $I(x^i)$ be a scalar field such that $I = I(t)$ along C. Thus I is an invariant, and

$$I_{,k} = 0.$$

Hence, for its intrinsic derivative, we find

$$\frac{\delta I}{\delta t} = I^{\sigma_0 \sigma_1 \cdots \sigma_{k-1}}_{,k} \frac{\Delta_k x^k}{\Delta t}$$
$$= 0.$$

Example 4.28. We know that

$$g_{ij,k} = 0.$$

Then

$$\frac{\delta g^{ij}}{\delta t} = g^{\sigma_0 \sigma_1 \cdots \sigma_{k-1}}_{ij,k} \frac{\Delta_k x^k}{\Delta t}$$
$$= 0.$$

Example 4.29. We know that

$$\delta^i_{j,k} = 0.$$

Then

$$\frac{\delta \delta^i_j}{\delta t} = \delta^{i \sigma_0 \sigma_1 \cdots \sigma_{k-1}}_{j,k} \frac{\Delta_k x^k}{\Delta t}$$
$$= 0.$$

Example 4.30. We know that

$$g_{ij}g^{jk} = \delta_i^k.$$

Then

$$
\begin{aligned}
0 &= \frac{\partial(g_{ij}g^{jk})}{\Delta_m x^m} \\
&= \frac{\partial g_{ij}}{\Delta_m x^m}g^{jk\sigma_m} + g_{ij}\frac{\partial g^{jk}}{\Delta_m x^m} \\
&= \left(g^{jk} + \mu_m\frac{\partial g^{jk}}{\Delta_m x^m}\right)(\Gamma_{im,j} + \Gamma_{jm,i}) + g_{ij}\frac{\partial g^{jk}}{\Delta_m x^m} \\
&= g^{jk}(\Gamma_{im,j} + \Gamma_{jm,i}) + g_{ij}\frac{\partial g^{jk}}{\Delta_m x^m} \\
&\quad + \mu_m\frac{\partial g^{jk}}{\Delta_m x^m}(\Gamma_{im,j} + \Gamma_{jm,i}) \\
&= g^{jk}\Gamma_{im,j} + g^{jk}\Gamma_{jm,i} + g_{ij}\frac{\partial g^{jk}}{\Delta_m x^m} \\
&\quad + \mu_m\frac{\partial g^{jk}}{\Delta_m x^m}\frac{\partial g_{ij}}{\Delta_m x^m} \\
&= \Gamma_{im}^k + g^{jk}\Gamma_{jm,i} + g_{ij}\frac{\partial g^{jk}}{\Delta_m x^m} \\
&\quad + \mu_m\frac{\partial g^{jk}}{\Delta_m x^m}\frac{\partial g_{ij}}{\Delta_m x^m},
\end{aligned}
$$

i. e.,

$$
\begin{aligned}
0 = \Gamma_{im}^k + g^{jk}\Gamma_{jm,i} + g_{ij}\frac{\partial g^{jk}}{\Delta_m x^m} \\
+ \mu_m\frac{\partial g^{jk}}{\Delta_m x^m}\frac{\partial g_{ij}}{\Delta_m x^m}.
\end{aligned}
$$

Multiply the last equation by g^{li}, we get

$$
\begin{aligned}
0 &= g^{li}\Gamma_{im}^k + g^{li}g^{jk}\Gamma_{jm,i} + g^{li}g_{ij}\frac{\partial g^{jk}}{\Delta_m x^m} \\
&\quad + \mu_m\frac{\partial g^{jk}}{\Delta_m x^m}\frac{\partial g_{ij}}{\Delta_m x^m} \\
&= \delta_j^l\frac{\partial g^{jk}}{\Delta_m x^m} + g^{li}\Gamma_{im}^k + g^{jk}\Gamma_{jm}^l \\
&\quad + \mu_m\frac{\partial g^{jk}}{\Delta_m x^m}\frac{\partial g_{ij}}{\Delta_m x^m}
\end{aligned}
$$

$$= \frac{\partial g^{lk}}{\Delta_m x^m} + g^{li}\Gamma - im^k + g^{jk}\Gamma^l_{jm}$$

$$+ \mu_m \frac{\partial g^{jk}}{\Delta_m x^m} \frac{\partial g_{ij}}{\Delta_m x^m}$$

$$= g^{lk}_{,m} + \mu_m \frac{\partial g^{jk}}{\Delta_m x^m} \frac{\partial g_{ij}}{\Delta_m x^m},$$

whereupon

$$g^{lk}_{,m} = -\mu_m \frac{\partial g^{jk}}{\Delta_m x^m} \frac{\partial g_{ij}}{\Delta_m x^m},$$

and

$$\frac{\delta g^{lk}}{\delta t} = g^{lk\sigma_0\sigma_1\cdots\sigma_{m-1}}_{,m} \frac{\Delta_m x^m}{\Delta t}$$

$$= -\mu_m^{\sigma_0\sigma_1\cdots\sigma_{m-1}} \left(\frac{\partial g^{jk}}{\Delta_m x^m} \right)^{\sigma_0\sigma_1\cdots\sigma_{m-1}} \left(\frac{\partial g_{ij}}{\Delta_m x^m} \right)^{\sigma_0\sigma_1\cdots\sigma_{m-1}}.$$

Example 4.31. Consider

$$\frac{\Delta_i x^i}{\Delta t}.$$

We have

$$\frac{\delta}{\delta t}\left(\frac{\Delta_i x^i}{\Delta t} \right) = \left(\frac{\Delta_i x^i}{\Delta t} \right)^{\sigma_0\sigma_1\cdots\sigma_{k-1}}_{,k} \frac{\Delta_k x^k}{\Delta t}$$

$$= \left(\frac{\partial}{\Delta_k x^k}\left(\frac{\Delta_i x^i}{\Delta t} \right) + \Gamma^i_{ak}\frac{\Delta_a x a}{\Delta t} \right)^{\sigma_0\sigma_1\cdots\sigma_{k-1}} \frac{\Delta_k x^k}{\Delta t}$$

$$= \left(\frac{\partial}{\Delta_k x^k}\left(\frac{\Delta_i x^i}{\Delta t} \right) \right)^{\sigma_0\sigma_1\cdots\sigma_{k-1}} \frac{\Delta_k x^k}{\Delta t} + \left(\Gamma^i_{ak}\frac{\Delta_a x^\alpha}{\Delta t} \right)^{\sigma_0\sigma_1\cdots\sigma_{k-1}} \frac{\Delta_k x^k}{\Delta t}$$

$$= \frac{\Delta_i^2 x^i}{\Delta t^2} + \left(\Gamma^i_{ak}\frac{\Delta_a x^\alpha}{\Delta t} \right)^{\sigma_0\sigma_1\cdots\sigma_{k-1}} \frac{\Delta_k x^k}{\Delta t}.$$

Exercise 4.10. Let A^i and B^i be two contravariant vectors. Check if

$$\frac{\delta(A^i B^i)}{\delta t} = A^i \frac{\delta B^i}{\delta t} + B^i \frac{\delta A^i}{\delta t}.$$

Answer 4.7. In general, it is not true.

4.10 Advanced practical problems

Problem 4.1. Compute the nonzero Christoffel symbols corresponding to the metric

1.

$$\Delta s^2 = -a(\Delta_1 x^1)^2 - b(\Delta_2 x^2)^2 - c(\Delta_3 x^3)^2 - d(\Delta_4 x^4)^2,$$

where a, b, c, and d are constants.

2.

$$\Delta s^2 = ((x^1)^2 + (x^2)^2)((\Delta_1 x^1)^2 + (\Delta_2 x^2)^2) + (\Delta_3 x^3)^2.$$

3.

$$\Delta s^2 = f(x^3)(\Delta_1 x^1)^2 + (\Delta_2 x^2)^2 + (\Delta_3 x^3)^2,$$

where f is continuously differentiable with respect to x^3.

4.

$$\Delta s^2 = (\Delta_1 x^1)^2 + (f(x^1))^2 (\Delta_2 x^2)^2 + (\Delta_3 x^3)^2,$$

where f is continuously differentiable with respect to x^1.

5.

$$\Delta s^2 = (\Delta_1 x^1)^2 + (\Delta_2 x^2)^2 + (f(x^1))^4 (\Delta_3 x^3)^2,$$

where f is continuously differentiable with respect to x^1.

6.

$$\Delta s^2 = f(x^1, x^2)(\Delta_1 x^1)^2 + (\Delta_2 x^2)^2 + (\Delta_3 x^3)^2 + (g(x^1, x^2, x^3))^4 (\Delta_4 x^4)^2,$$

where f is continuously differentiable with respect to x^1 and x^2, and g is continuously differentiable with respect to x^1, x^2, and x^3.

Problem 4.2. Let A^{pqr} be a skew-symmetric tensor. Prove that:

1.

$$A^{pqr} \Gamma^l_{qr} = 0.$$

2.

$$A^{pqr} \Gamma^l_{pr} = 0.$$

1. *Solution.* Since A^{pqr} is a skew-symmetric tensor, we have

$$A^{pqr} = -A^{prq}.$$

Now exchanging the dummy indices q and r in

$$A^{pqr}\Gamma^{l}_{qr},$$

we find

$$A^{pqr}\Gamma^{l}_{qr} = A^{prq}\Gamma^{l}_{rq}$$
$$= -A^{pqr}\Gamma^{l}_{rq}$$
$$= -A^{pqr}\Gamma^{l}_{qr}.$$

Hence

$$2A^{pqr}\Gamma^{l}_{qr} = 0,$$

or

$$A^{pqr}\Gamma^{l}_{qr} = 0.$$

2. *Hint.* Use the solution of 1.

Problem 4.3. Let

$$B^{m}_{np} = \Gamma^{m}_{np} + 2\delta^{m}_{n}A_{k},$$

where A_{k} is a covariant vector. Prove that

$$g_{tn}B^{t}_{mp} + g_{tm}B^{t}_{np} - 4g_{mn}A_{k} = \frac{\partial g_{mn}}{\Delta_{p}x^{p}}.$$

Solution. We have

$$\frac{\partial g_{mn}}{\Delta_{p}x^{p}} = \Gamma_{mp,n} + \Gamma_{np,m}.$$

Then

$$\begin{aligned}
g_{tn}B^{t}_{mp} &+ g_{tm}B^{t}_{np} - 4g_{mn}A_{k} \\
&= g_{tn}\left(\Gamma^{t}_{mp} + 2\delta^{t}_{m}A_{k}\right) + g_{tm}\left(\Gamma^{t}_{np} + 2\delta^{t}_{n}A_{k}\right) - 4g_{mn}A_{k} \\
&= g_{tn}\Gamma^{t}_{mp} + 2g_{tn}\delta^{t}_{m}A_{k} + g_{tm}\Gamma^{t}_{np} + 2g_{tm}\delta^{t}_{n}A_{k} - 4g_{mn}A_{k} \\
&= \Gamma_{mp,n} + 2g_{mn}A_{k} + \Gamma_{np,m} + 2g_{mn}A_{k} - 4g_{mn}A_{k} \\
&= \Gamma_{mp,n} + \Gamma_{np,m} + 4g_{mn}A_{k} - 4g_{mn}A_{k} \\
&= \Gamma_{mp,n} + \Gamma_{np,m} \\
&= \frac{\partial g_{mn}}{\Delta_{p}x^{p}}.
\end{aligned}$$

Problem 4.4. Let A^{pqr} be a skew-symmetric tensor. Prove that

$$A^{pqr}\Gamma^{l}_{pq} = 0,$$
$$A^{pqr}\Gamma^{l}_{pr} = 0,$$
$$A^{pqr}\Gamma^{l}_{qr} = 0.$$

Solution. Since A^{pqr} is skew-symmetric, we have

$$A^{pqr} = -A^{qpr},$$
$$A^{pqr} = -A^{rqp},$$
$$A^{pqr} = -A^{prq}.$$

Because Γ^{k}_{ij} is symmetric with respect to the lower indices, we find

$$\Gamma^{l}_{pq} = \Gamma^{l}_{qp},$$
$$\Gamma^{l}_{rp} = \Gamma^{l}_{pr},$$
$$\Gamma^{l}_{rq} = \Gamma^{l}_{qr}.$$

Then

$$A^{pqr}\Gamma^{l}_{pq} = A^{qpr}\Gamma^{l}_{qp}$$
$$= -A^{pqr}\Gamma^{l}_{qp}$$
$$= -A^{pqr}\Gamma^{l}_{pq},$$

whereupon

$$A^{pqr}\Gamma^{l}_{pq} = 0.$$

Next, we have

$$A^{pqr}\Gamma^{l}_{pr} = A^{rqp}\Gamma^{l}_{rp}$$
$$= -A^{pqr}\Gamma^{l}_{rp}$$
$$= -A^{pqr}\Gamma^{l}_{pr},$$

from which we get

$$A^{pqr}\Gamma^{l}_{pr} = 0.$$

Moreover, we find

$$A^{pqr}\Gamma^l_{qr} = A^{prw}\Gamma^l_{rq}$$
$$= -A^{pqr}\Gamma^l_{rq}$$
$$= -A^{pqr}\Gamma^l_{qr},$$

whereupon

$$A^{pqr}\Gamma^l_{qr} = 0.$$

Problem 4.5. Let

$$B^m_{np} = \Gamma^m_{np} + 2\delta^m_n A_k,$$

where A_k is a covariant vector. Prove that

$$g_{tn}B^t_{mp} + g_{tm}B^t_{np} - 4g_{mn}A_k = \frac{\partial g_{mn}}{\Delta_k x^k}.$$

Solution. We have

$$g_{tn}B^t_{mp} = g_{tn}(\Gamma^t_{mp} + 2\delta^t_m A_k)$$
$$= g_{tn}\Gamma^t_{mp} + 2g_{tn}\delta^t_m A_k$$
$$= \Gamma_{mp,n} + 2g_{mn}A_k$$

and

$$g_{tm}B^t_{np} = g_{tm}(\Gamma^t_{np} + 2\delta^t_n A_k)$$
$$= g_{tm}\Gamma^t_{np} + 2g_{tm}\delta^t_n A_k$$
$$= \Gamma_{np,m} + 2g_{mn}A_k.$$

Hence

$$g_{tn}B^t_{mp} + g_{tm}B^t_{np} - 4g_{mn}A_k = \Gamma_{mp,n} + 2g_{mn}A_k + \Gamma_{np,m} + 2g_{mn}A_k$$
$$- 4g_{mn}A_k$$
$$= \frac{1}{2}\left(\frac{\partial g_{mn}}{\Delta_p x^p} + \frac{\partial g_{pn}}{\Delta_m x^m} - \frac{\partial g_{mp}}{\Delta_n x^n}\right)$$
$$+ \frac{1}{2}\left(\frac{\partial g_{nm}}{\Delta_p x^p} + \frac{\partial g_{pm}}{\Delta_n x^n} - \frac{\partial g_{np}}{\Delta_m x^m}\right)$$
$$= \frac{1}{2}\left(\frac{\partial g_{mn}}{\Delta_p x^p} + \frac{\partial g_{pn}}{\Delta_m x^m} - \frac{\partial g_{mp}}{\Delta_n x^n}\right)$$
$$+ \frac{1}{2}\left(\frac{\partial g_{mn}}{\Delta_p x^p} + \frac{\partial g_{mp}}{\Delta_n x^n} - \frac{\partial g_{pn}}{\Delta_m x^m}\right)$$
$$= \frac{\partial g_{mn}}{\Delta_p x^p}.$$

Problem 4.6. Prove that

$$A^r_{ijk,l} = \frac{\partial A^r_{ijk}}{\Delta_l x^l} - \Gamma^a_{il} A^r_{ajk} + \Gamma^a_{jl}\Gamma^r_{iak} - \Gamma^a_{kl}A^r_{ija}$$
$$+ \Gamma^r_{al}A^a_{ijk}.$$

Problem 4.7. Find the covariant derivative of

$$C_j = g_{jk}A^k$$

with respect to x^q, where A^k is a contravariant vector.

Solution. Note that C_j is a covariant vector. Then using the definition of covariant derivatives, we get

$$C_{j,q} = \left(g_{jk}A^k\right)_q$$
$$= \frac{\partial(g_{jk}A^k)}{\Delta_q x^q} - \Gamma^a_{jq}C_a$$
$$= \frac{\partial g_{jk}}{\Delta_q x^q}A^k + g^{\sigma_q}_{jk}\frac{\partial A^k}{\Delta_q x^q} - \Gamma^a_{jq}g_{al}A^l$$
$$= \frac{\partial g_{jk}}{\Delta_q x^q}A^k + g^{\sigma_q}_{jk}\frac{\partial A^k}{\Delta_q x^q} - \Gamma^a_{jq}g_{ak}A^k$$
$$= \left(\frac{\partial g_{jk}}{\Delta_q x^q} - \Gamma^a_{jq}g_{ak}\right) + g^{\sigma_q}_{jk}\frac{\partial A^k}{\Delta_q x^q}$$
$$= g_{jk,q} + g^{\sigma_q}_{jk}\frac{\partial A^k}{\Delta_q x^q}.$$

Problem 4.8. Find the covariant derivative of

$$B_k = \delta^j_k A_j,$$

with respect to x^p, where A_j is a covariant vector.

Solution. Note that B_k is a covariant vector. Then using the definition of covariant derivatives, we find

$$B_{k,p} = \frac{\partial B_k}{\Delta_p x^p} - \Gamma^a_{kp}B_a$$
$$= \frac{\partial(\delta^j_k A_j)}{\Delta_p x^p} - \Gamma^a_{kp}\delta^a_k A_a$$
$$= \frac{\partial \delta^j_k}{\Delta_p x^p}A_j + \delta^{j\sigma_p}_k\frac{\partial A_j}{\Delta_p x^p} - \Gamma^k_{kq}A_k$$

$$= \delta_k^j \frac{\partial A_j}{\Delta_p x^p} - \Gamma_{kp}^k A_k$$

$$= \frac{\partial A_k}{\Delta_p x^p} - \Gamma_{kp}^k A_k$$

$$= \frac{\partial A_k}{\Delta_p x^p} - \left(\frac{1}{2} \frac{\partial g}{\Delta_p x^p} - \frac{\mu_p}{g} \sum_{l=2}^{N} H_{lp} \right) A_k.$$

Problem 4.9. Find the covariant derivative of

$$C_k^j = A^j B_k$$

with respect to x_q, where A^j is a contravariant vector, and B_k is a covariant vector.

Solution. Note that C_k^j is a mixed tensor, contravariant and covariant of the first order. Now applying the definition of a covariant derivative of a mixed tensor, we arrive at

$$C_{k,q}^j = \frac{\partial C_k^j}{\Delta_q x^q} + \Gamma_{aq}^j C_k^a - \Gamma_{jq}^a C_a^j$$

$$= \frac{\partial (A^j B_k)}{\Delta_q x^q} + \Gamma_{aq}^j A^a B_k - \Gamma_{jq}^a A^j B_a$$

$$= \frac{\partial A^j}{\Delta_q x^q} B_k + A^{j\sigma_q} \frac{\partial B_k}{\Delta_q x^q} + \Gamma_{aq}^j A^a B_k - \Gamma_{jq}^a A^j B_a$$

$$= \left(\frac{\partial A^j}{\Delta_q x^q} + \Gamma_{aq}^j A^a \right) B - k + \left(A^j + \mu_q \frac{\partial A^j}{\Delta_q x^q} \right) \frac{\partial B_k}{\Delta_q x^q} - \Gamma_{jq}^a A^j B_a$$

$$= A^j_{,q} B_k + A^j \left(\frac{\partial B_k}{\Delta_q x^q} - \Gamma_{jq}^a B_a \right) + \mu_q \frac{\partial A^j}{\Delta_q x^q} \frac{\partial B_k}{\Delta_q x^q}$$

$$= A^j_{,k} B_k + A^j B_{j,q} + \mu_q \frac{\partial A^j}{\Delta_q x^q} \frac{\partial B_k}{\Delta_q x^q}.$$

Problem 4.10. Let a_{ij} be a skew-symmetric tensor. Prove that $a_{ij,k}$ is a skew-symmetric tensor with respect to the indices i and j.

Solution. We have

$$a_{ij,k} = \frac{\partial a_{ij}}{\Delta_k x^k} - \Gamma_{ik}^a a_{aj} - \Gamma_{kj}^a a_{ia}$$

$$= -\frac{\partial a_{ji}}{\Delta_k x^k} + \Gamma_{ki}^a a_{ja} + \Gamma_{jk}^a a_{ai}$$

$$= -\left(\frac{\partial a_{ji}}{\Delta_k x^k} - \Gamma_{ki}^a a_{ja} - \Gamma_{jk}^a a_{ai} \right)$$

$$= -a_{ji,k}.$$

Problem 4.11. Prove that:

1.

$$\varepsilon_{iks}\varepsilon_{mks} = 2\delta_{im}.$$

2.

$$\varepsilon_{ijk}\varepsilon^{ist} = \det\begin{pmatrix} \delta_j^s & \delta_j^t \\ \delta_k^s & \delta_k^t \end{pmatrix}.$$

3.

$$\varepsilon_{ijk}\varepsilon^{ijt} = 2\delta_k^t.$$

4.

$$\varepsilon_{ijk}\varepsilon^{ijk} = 3!.$$

Problem 4.12. Let A^{ij} and B^{ij} be two contravariant vectors. Check if

$$\frac{\delta(A^{ij}B^{ij})}{\delta t} = A^{ij}\frac{\delta B^{ij}}{\delta t} + B^{ij}\frac{\delta A^{ij}}{\delta t}.$$

Answer 4.8. In general, it is not true.

5 Riemannian geometry

In this chapter, we define the Riemann–Christoffel tensor, Ricci tensor, covariant curvature tensor, Riemann curvature, and Einstein tensor and deduct some of their properties.

5.1 Riemann–Christoffel tensor

Let B_i be an arbitrary covariant vector. Then its covariant derivative with respect to x^j is given by

$$B_{i,j} = \frac{\partial B_i}{\Delta_j x^j} - \Gamma_{ij}^\alpha B_\alpha,$$

which is a covariant second-order tensor. Further, covariant differentiation gives

$$
\begin{aligned}
(B_{i,j})_{,k} &= B_{i,jk} \\
&= \frac{\partial B_{i,j}}{\Delta_k x^k} - \Gamma_{ik}^\alpha B_{\alpha,j} - \Gamma_{kj}^\alpha B_{i,\alpha} \\
&= \frac{\partial}{\Delta_k x^k}\left(\frac{\partial B_i}{\Delta_j x^j} - \Gamma_{ij}^\alpha B_\alpha\right) - \Gamma_{ik}^\alpha\left(\frac{\partial B_\alpha}{\Delta_j x^j} - \Gamma_{\alpha j}^\beta B_\beta\right) \\
&\quad - \Gamma_{kj}^\alpha\left(\frac{\partial B_i}{\Delta_\alpha x^\alpha} - \Gamma_{i\alpha}^\beta B_\beta\right) \\
&= \frac{\partial^2 B_i}{\Delta_k x^k \Delta_j x^j} - \Gamma_{ij}^{\alpha\sigma_k}\frac{\partial B_\alpha}{\Delta_k x^k} - \frac{\partial \Gamma_{ij}^\alpha}{\Delta_k x^k}B_\alpha - \Gamma_{ik}^\alpha\frac{\partial B_\alpha}{\Delta_j x^j} \\
&\quad + \Gamma_{ik}^\alpha \Gamma_{\alpha j}^\beta B_\beta - \Gamma_{kj}^\alpha\frac{\partial B_i}{\Delta_\alpha x^\alpha} + \Gamma_{kj}^\alpha \Gamma_{i\alpha}^\beta B_\beta \\
&= \frac{\partial^2 B_i}{\Delta_k x^k \Delta_j x^j} - \Gamma_{ij}^\alpha\frac{\partial B_\alpha}{\Delta_k x^k} - \Gamma_{ik}^\alpha\frac{\partial B_\alpha}{\Delta_j x^j} - \Gamma_{kj}^\alpha\frac{\partial B_i}{\Delta_\alpha x^\alpha} \\
&\quad - \frac{\partial \Gamma_{ij}^\alpha}{\Delta_k x^k}B_\alpha + \Gamma_{ik}^\alpha \Gamma_{\alpha j}^\beta B_\beta + \Gamma_{kj}^\alpha \Gamma_{i\alpha}^\beta B_\beta - \mu_k\frac{\partial \Gamma_{ij}^\alpha}{\Delta_k x^k}\frac{\partial B_\alpha}{\Delta_k x^k}.
\end{aligned}
$$

Interchanging the dummy indices α and β in the terms

$$\Gamma_{ik}^\alpha \Gamma_{\alpha j}^\beta B_\beta \quad \text{and} \quad \Gamma_{kj}^\alpha \Gamma_{i\alpha}^\beta B_\beta,$$

we get

$$
\begin{aligned}
(B_{i,j})_{,k} &= B_{i,jk} \\
&= \frac{\partial^2 B_i}{\Delta_k x^k \Delta_j x^j} - \Gamma_{ij}^\alpha\frac{\partial B_\alpha}{\Delta_k x^k} - \Gamma_{ik}^\alpha\frac{\partial B_\alpha}{\Delta_j x^j} - \Gamma_{kj}^\alpha\frac{\partial B_i}{\Delta_\alpha x^\alpha} \\
&\quad + \Gamma_{ik}^\beta \Gamma_{\beta j}^\alpha B_\alpha + \Gamma_{kj}^\beta \Gamma_{i\beta}^\alpha B_\alpha - \frac{\partial \Gamma_{ij}^\alpha}{\Delta_k x^k}B_\alpha - \mu_k\frac{\partial \Gamma_{ij}^\alpha}{\Delta_k x^k}\frac{\partial B_\alpha}{\Delta_k x^k}.
\end{aligned}
$$

https://doi.org/10.1515/9783112228494-005

Interchanging j and k in the last equation, we arrive at

$$
\begin{aligned}
B_{i,kj} = {} & \frac{\partial^2 B_i}{\Delta_j x^j \Delta_k x^k} - \Gamma_{ik}^\alpha \frac{\partial B_\alpha}{\Delta_j x^j} - \Gamma_{ij}^\alpha \frac{\partial B_\alpha}{\Delta_k x^k} - \Gamma_{kj}^\alpha \frac{\partial B_i}{\Delta_\alpha x^\alpha} \\
& + \Gamma_{ij}^\beta \Gamma_{\beta k}^\alpha B_\alpha + \Gamma_{kj}^\beta \Gamma_{i\beta}^\alpha B_\alpha - \frac{\partial \Gamma_{ik}^\alpha}{\Delta_j x^j} B_\alpha - \mu_j \frac{\partial \Gamma_{ik}^\alpha}{\Delta_j x^j} \frac{\partial B_\alpha}{\Delta_j x^j}.
\end{aligned}
$$

Hence

$$
\begin{aligned}
B_{i,jk} - B_{i,kj} = {} & \frac{\partial^2 B_i}{\Delta_k x^k \Delta_j x^j} - \Gamma_{ij}^\alpha \frac{\partial B_\alpha}{\Delta_k x^k} - \Gamma_{ik}^\alpha \frac{\partial B_\alpha}{\Delta_j x^j} - \Gamma_{kj}^\alpha \frac{\partial B_i}{\Delta_\alpha x^\alpha} \\
& + \Gamma_{ik}^\beta \Gamma_{\beta j}^\alpha B_\alpha + \Gamma_{kj}^\beta \Gamma_{i\beta}^\alpha B_\alpha - \frac{\partial \Gamma_{ij}^\alpha}{\Delta_k x^k} B_\alpha - \mu_k \frac{\partial \Gamma_{ij}^\alpha}{\Delta_k x^k} \frac{\partial B_\alpha}{\Delta_k x^k} \\
& - \left(\frac{\partial^2 B_i}{\Delta_j x^j \Delta_k x^k} - \Gamma_{ik}^\alpha \frac{\partial B_\alpha}{\Delta_j x^j} - \Gamma_{ij}^\alpha \frac{\partial B_\alpha}{\Delta_k x^k} - \Gamma_{kj}^\alpha \frac{\partial B_i}{\Delta_\alpha x^\alpha} \right. \\
& \left. + \Gamma_{ij}^\beta \Gamma_{\beta k}^\alpha B_\alpha + \Gamma_{kj}^\beta \Gamma_{i\beta}^\alpha B_\alpha - \frac{\partial \Gamma_{ik}^\alpha}{\Delta_j x^j} B_\alpha - \mu_j \frac{\partial \Gamma_{ik}^\alpha}{\Delta_j x^j} \frac{\partial B_\alpha}{\Delta_j x^j} \right) \\
= {} & \left(-\frac{\partial \Gamma_{ij}^\alpha}{\Delta_k x^k} + \frac{\partial \Gamma_{ik}^\alpha}{\Delta_j x^j} + \Gamma_{\beta j}^\alpha \Gamma_{ik}^\beta - \Gamma_{\beta k}^\alpha \Gamma_{ij}^\beta \right) B_\alpha \\
& + \mu_j \frac{\partial \Gamma_{ik}^\alpha}{\Delta_j x^j} \frac{\partial B_\alpha}{\Delta_j x^j} - \mu_k \frac{\partial \Gamma_{ij}^\alpha}{\Delta_k x^k} \frac{\partial B_\alpha}{\Delta_k x^k}.
\end{aligned}
$$

Let

$$
R_{ijk}^\alpha = -\frac{\partial \Gamma_{ij}^\alpha}{\Delta_k x^k} + \frac{\partial \Gamma_{ik}^\alpha}{\Delta_j x^j} + \Gamma_{\beta j}^\alpha \Gamma_{ik}^\beta - \Gamma_{\beta k}^\alpha \Gamma_{ij}^\beta
$$

and

$$
\widetilde{R}_{ijk} = \mu_j \frac{\partial \Gamma_{ik}^\alpha}{\Delta_j x^j} \frac{\partial B_\alpha}{\Delta_j x^j} - \mu_k \frac{\partial \Gamma_{ij}^\alpha}{\Delta_k x^k} \frac{\partial B_\alpha}{\Delta_k x^k}.
$$

Then

$$
R_{ijk}^\alpha = \det \begin{pmatrix} \frac{\partial}{\Delta_j x^j} & \frac{\partial}{\Delta_k x^k} \\ \Gamma_{ij}^\alpha \Gamma_{ik}^\alpha & \Gamma_{ik}^\beta \end{pmatrix} + \det \begin{pmatrix} \Gamma_{\beta j}^\alpha & \Gamma_{\beta k}^\alpha \\ \Gamma_{ij}^\beta & \Gamma_{ik}^\beta \end{pmatrix},
$$

and

$$
\widetilde{R}_{ijk} = \det \begin{pmatrix} \frac{\partial \Gamma_{ik}^\alpha}{\Delta_j x^j} & \frac{\partial \Gamma_{ij}^\alpha}{\Delta_k x^k} \\ \mu_k \frac{\partial B_\alpha}{\Delta_k x^k} & \mu_j \frac{\partial B_\alpha}{\Delta_j x^j} \end{pmatrix}.
$$

Thus

$$B_{i,jk} - B_{i,kj} = R^{\alpha}_{ijk} B_{\alpha} + \widetilde{R}_{ijk}.$$

Therefore the covariant differentiation is not commutative. Since

$$B_{i,jk} - B_{i,kj}$$

is a third-order covariant tensor, and the sum of two third-order covariant tensors is a third-order covariant tensor. Then we conclude that $\widetilde{R}_{ijk}$ is a third-order covariant tensor and

$$R^{\alpha}_{ijk} B_{\alpha}$$

is a third-order covariant tensor. From here, we get that F^{α}_{ijk} is a mixed tensor that is contravariant of the first order and covariant of the third order.

Definition 5.1. The tensor R^{α}_{ijk} is called the Riemann–Christoffel tensor of the second kind or the curvature tensor of the Riemann space. The symbol R^{α}_{ijk} is referred to as the Riemann symbol of the second kind.

Definition 5.2. The tensor $\widetilde{R}_{ijk}$ is called the rest Riemann–Christoffel tensor of the second kind or the rest curvature tensor of the Riemann space. The symbol R^{α}_{ijk} is referred to as the rest Riemann symbol of the second kind.

Theorem 5.1. *The Riemann–Christoffel tensors R^{α}_{ijk} and $\widetilde{R}_{ijk}$ are skew-symmetric with respect to the indices j and k.*

Proof. We have

$$R^{\alpha}_{ijk} = -\frac{\partial \Gamma^{\alpha}_{ij}}{\Delta_k x^k} + \frac{\partial \Gamma^{\alpha}_{ik}}{\Delta_j x^j} + \Gamma^{\alpha}_{\beta j} \Gamma^{\beta}_{ik} - \Gamma^{\alpha}_{\beta k} \Gamma^{\beta}_{ij}$$

and

$$R^{\alpha}_{ikj} = \frac{\partial \Gamma^{\alpha}_{ij}}{\Delta_k x^k} - \frac{\partial \Gamma^{\alpha}_{ik}}{\Delta_j x^j} + \Gamma^{\alpha}_{\beta k} \Gamma^{\beta}_{ij} - \Gamma^{\beta}_{ik} \Gamma^{\alpha}_{\beta j}$$

$$= -\left(-\frac{\partial \Gamma^{\alpha}_{ij}}{\Delta_k x^k} + \frac{\partial \Gamma^{\alpha}_{ik}}{\Delta_j x^j} + \Gamma^{\alpha}_{\beta j} \Gamma^{\beta}_{ik} - \Gamma^{\alpha}_{\beta k} \Gamma^{\beta}_{ij} \right)$$

$$= -R^{\alpha}_{ijk}.$$

Next, we have

$$\widetilde{R}_{ijk} = \mu_j \frac{\partial \Gamma^{\alpha}_{ik}}{\Delta_j x^j} - \mu_k \frac{\partial \Gamma^{\alpha}_{ij}}{\Delta_k x^k} \frac{\partial B_{\alpha}}{\Delta_k x^k}$$

and

$$\widetilde{R}_{ikj} = \mu_k \frac{\partial \Gamma_{ij}^{\alpha}}{\Delta_k x^k} \frac{\partial B_{\alpha}}{\Delta_k x^k} - \mu_j \frac{\partial \Gamma_{ik}^{\alpha}}{\Delta_j x^j} \frac{\partial B_{\alpha}}{\Delta_j x^j}$$

$$= -\left(\mu_j \frac{\partial \Gamma_{ik}^{\alpha}}{\Delta_j x^j} - \mu_k \frac{\partial \Gamma_{ij}^{\alpha}}{\Delta_k x^k} \frac{\partial B_{\alpha}}{\Delta_k x^k} \right)$$

$$= -\widetilde{R}_{ikj}.$$

This completes the proof. $\square$

Theorem 5.2. *The curvature tensors R_{ijk}^{α} and $\widetilde{R}_{ijk}$ satisfy the cyclic property:*

$$R_{ijk}^{\alpha} + R_{jki}^{\alpha} + R_{kij}^{\alpha} = 0,$$

and

$$\widetilde{R}_{ijk} + \widetilde{R}_{jki} + \widetilde{R}_{kij} = 0.$$

Proof. We have

$$R_{ijk}^{\alpha} = \frac{\partial \Gamma_{ik}^{\alpha}}{\Delta_j x^j} - \frac{\partial \Gamma_{ij}^{\alpha}}{\Delta_k x^k} + \Gamma_{\beta j}^{\alpha} \Gamma_{ik}^{\beta} - \Gamma_{ij}^{\beta} \Gamma_{\beta k}^{\alpha},$$

$$R_{jki}^{\alpha} = \frac{\partial \Gamma_{ji}^{\alpha}}{\Delta_k x^k} - \frac{\partial \Gamma_{jk}^{\alpha}}{\Delta_i x^i} + \Gamma_{\beta k}^{\alpha} \Gamma_{ji}^{\beta} - \Gamma_{jk}^{\beta} \Gamma_{\beta i}^{\alpha}$$

$$= \frac{\partial \Gamma_{ij}^{\alpha}}{\Delta_k x^k} - \frac{\partial \Gamma_{jk}^{\alpha}}{\Delta_i x^i} + \Gamma_{\beta k}^{\alpha} \Gamma_{ij}^{\beta} - \Gamma_{jk}^{\beta} \Gamma_{\beta i}^{\alpha},$$

$$R_{kij}^{\alpha} = \frac{\partial \Gamma_{kj}^{\alpha}}{\Delta_i x^i} - \frac{\partial \Gamma_{ki}^{\alpha}}{\Delta_j x^j} + \Gamma_{\beta i}^{\alpha} \Gamma_{kj}^{\beta} - \Gamma_{ki}^{\beta} \Gamma_{\beta j}^{\alpha}$$

$$= \frac{\partial \Gamma_{jk}^{\alpha}}{\Delta_i x^i} - \frac{\partial \Gamma_{ik}^{\alpha}}{\Delta_j x^j} + \Gamma_{\beta i}^{\alpha} \Gamma_{jk}^{\beta} - \Gamma_{ik}^{\beta} \Gamma_{\beta j}^{\alpha}.$$

Hence

$$R_{ijk}^{\alpha} + R_{jki}^{\alpha} + R_{kij}^{\alpha} = \frac{\partial \Gamma_{ik}^{\alpha}}{\Delta_j x^j} - \frac{\partial \Gamma_{ij}^{\alpha}}{\Delta_k x^k} + \Gamma_{\beta j}^{\alpha} \Gamma_{ik}^{\beta} - \Gamma_{ij}^{\beta} \Gamma_{\beta k}^{\alpha}$$

$$+ \frac{\partial \Gamma_{ij}^{\alpha}}{\Delta_k x^k} - \frac{\partial \Gamma_{jk}^{\alpha}}{\Delta_i x^i} + \Gamma_{\beta k}^{\alpha} \Gamma_{ij}^{\beta} - \Gamma_{jk}^{\beta} \Gamma_{\beta i}^{\alpha}$$

$$+ \frac{\partial \Gamma_{jk}^{\alpha}}{\Delta_i x^i} - \frac{\partial \Gamma_{ik}^{\alpha}}{\Delta_j x^j} + \Gamma_{\beta i}^{\alpha} \Gamma_{jk}^{\beta} - \Gamma_{ik}^{\beta} \Gamma_{\beta j}^{\alpha}$$

$$= 0.$$

Next, we find

$$\tilde{R}_{ijk} = \mu_j \frac{\partial \Gamma^\alpha_{ik}}{\Delta_j x^j} \frac{\partial B_\alpha}{\Delta_j x^j} - \mu_k \frac{\partial \Gamma^\alpha_{ij}}{\Delta_k x^k} \frac{\partial B_\alpha}{\Delta_k x^k},$$

$$\tilde{R}_{jki} = \mu_k \frac{\partial \Gamma^\alpha_{ji}}{\Delta_k x^k} \frac{\partial B_\alpha}{\Delta_k x^k} - \mu_i \frac{\partial \Gamma^\alpha_{jk}}{\Delta_i x^i} \frac{\partial B_\alpha}{\Delta_i x^i}$$

$$= \mu_k \frac{\partial \Gamma^\alpha_{ij}}{\Delta_k x^k} \frac{\partial B_\alpha}{\Delta_k x^k} - \mu_i \frac{\partial \Gamma^\alpha_{jk}}{\Delta_i x^i} \frac{\partial B_\alpha}{\Delta_i x^i},$$

$$\tilde{R}_{kij} = \mu_i \frac{\partial \Gamma^\alpha_{kj}}{\Delta_i x^i} \frac{\partial B_\alpha}{\Delta_i x^i} - \mu_j \frac{\partial \Gamma^\alpha_{ki}}{\Delta_j x^j} \frac{\partial B_\alpha}{\Delta_j x^j}$$

$$= \mu_i \frac{\partial \Gamma^\alpha_{jk}}{\Delta_i x^i} \frac{\partial B_\alpha}{\Delta_i x^i} - \mu_j \frac{\partial \Gamma^\alpha_{ik}}{\Delta_j x^j} \frac{\partial B_\alpha}{\Delta_j x^j}.$$

Therefore

$$\tilde{R}_{ijk} + \tilde{R}_{jki} + \tilde{R}_{kij} = \mu_j \frac{\partial \Gamma^\alpha_{ik}}{\Delta_j x^j} \frac{\partial B_\alpha}{\Delta_j x^j} - \mu_k \frac{\partial \Gamma^\alpha_{ij}}{\Delta_k x^k} \frac{\partial B_\alpha}{\Delta_k x^k}$$

$$+ \mu_k \frac{\partial \Gamma^\alpha_{ij}}{\Delta_k x^k} \frac{\partial B_\alpha}{\Delta_k x^k} - \mu_i \frac{\partial \Gamma^\alpha_{jk}}{\Delta_i x^i} \frac{\partial B_\alpha}{\Delta_i x^i}$$

$$+ \mu_i \frac{\partial \Gamma^\alpha_{jk}}{\Delta_i x^i} \frac{\partial B_\alpha}{\Delta_i x^i} - \mu_j \frac{\partial \Gamma^\alpha_{ik}}{\Delta_j x^j} \frac{\partial B_\alpha}{\Delta_j x^j}$$

$$= 0.$$

This completes the proof. $\qquad\square$

5.2 The Ricci tensor

Definition 5.3. The tensor

$$R_{ij} = R^\alpha_{ij\alpha}$$

is called the Ricci tensor of the first kind.

We have

$$R_{ij} = \frac{\partial \Gamma^\alpha_{i\alpha}}{\Delta_j x^j} - \frac{\Gamma^\alpha_{ij}}{\Delta_\alpha x^\alpha} + \Gamma^\alpha_{\beta j}\Gamma^\beta_{i\alpha} - \Gamma^\alpha_{\beta\alpha}\Gamma^\beta_{ij}$$

$$= \frac{\partial}{\Delta_j x^j}\left(\frac{1}{2g}\frac{\partial g}{\Delta_i x^i} - \frac{\mu_i}{g}\sum_{l=2}^{N} H_{li} \right) - \frac{\partial \Gamma^\alpha_{ij}}{\Delta_\alpha x^\alpha}$$

$$+ \Gamma^\alpha_{\beta j}\Gamma^\beta_{i\alpha} - \Gamma^\beta_{ij}\left(\frac{1}{2g}\frac{\partial g}{\Delta_\beta x^\beta} - \frac{\mu_\beta}{g}\sum_{l=2}^{N} H_{l\beta} \right).$$

Moreover,

$$R_{ji} = \frac{\partial \Gamma^{\alpha}_{ja}}{\Delta_i x^i} - \frac{\partial \Gamma^{\alpha}_{ji}}{\Delta_{\alpha} x^{\alpha}} + \Gamma^{\alpha}_{\beta i}\Gamma^{\beta}_{ja} - \Gamma^{\alpha}_{\beta a}\Gamma^{\beta}_{ij}$$

$$= \frac{\partial}{\Delta_i x^i}\left(\frac{1}{2g}\frac{\partial g}{\Delta_j x^j} - \frac{\mu_j}{g}\sum_{l=2}^{N}H_{lj}\right) - \frac{\partial \Gamma^{\alpha}_{ij}}{\Delta_{\alpha} x^{\alpha}}$$

$$+ \Gamma^{\alpha}_{\beta i}\Gamma^{\beta}_{ja} - \Gamma^{\beta}_{ij}\left(\frac{1}{2g}\frac{\partial g}{\Delta_{\beta} x^{\beta}} - \frac{\mu_j}{g}\sum_{l=2}^{N}H_{l\beta}\right).$$

Since in the general case,

$$\frac{\partial}{\Delta_j x^j}\left(\frac{1}{2g}\frac{\partial g}{\Delta_i x^i} - \frac{\mu_i}{g}\sum_{l=2}^{N}H_{li}\right) \neq \frac{\partial}{\Delta_i x^i}\left(\frac{1}{2g}\frac{\partial g}{\Delta_j x^j} - \frac{\mu_j}{g}\sum_{l=2}^{N}H_{lj}\right),$$

we conclude that in the general case the Ricci tensor is not symmetric, i. e.,

$$R_{ij} \neq R_{ji}.$$

Definition 5.4. The tensor

$$R^i_j = g^{i\alpha}R_{\alpha j}$$

is called the Ricci tensor of the second kind.

Definition 5.5. The scalar curvature or curvature invariant R is defined by

$$R = R^i_i.$$

Example 5.1. We will find the scalar curvature for the metric

$$\Delta s^2 = a^2(\Delta_1 x^1)^2 + a^2(\sin_1(x^1, x^1_0))^2(\Delta_2 x^2)^2.$$

Here

$$g_{11} = a^2,$$
$$g_{12} = 0,$$
$$g_{21} = 0,$$
$$g_{22} = a^2(\sin_1(x^1, x^1_0))^2.$$

Then

$$g = \det \begin{pmatrix} a^2 & 0 \\ 0 & a^2(\sin_1(x^1, x_0^1))^2 \end{pmatrix}$$
$$= a^4(\sin_1(x^1, x_0^2))^2,$$

and

$$g^{11} = \frac{g_{22}}{g}$$
$$= \frac{a^2(\sin_1(x^1, x_0^1))^2}{a^4(\sin_1(x^1, x_0^1))^2}$$
$$= \frac{1}{a^2},$$
$$g^{12} = -\frac{g_{21}}{g}$$
$$= 0,$$
$$g^{21} = -\frac{g_{12}}{g}$$
$$= 0,$$
$$g^{22} = \frac{g_{11}}{g}$$
$$= \frac{a^2}{a^4(\sin_1(x^1, x_0^1))^2}$$
$$= \frac{1}{a^2(\sin_1(x^1, x_0^1))^2},$$

and

$$\frac{\partial g_{11}}{\varDelta_1 x^1} = 0,$$
$$\frac{\partial g_{11}}{\varDelta_2 x^2} = 0,$$
$$\frac{\partial g_{22}}{\varDelta_1 x^1} = \cos_1(x^1, x_0^1)(\sin_1(\sigma_1(x^1), x_0^1) + \sin_1(x^1, x_0^1)),$$
$$\frac{\partial g_{22}}{\varDelta_2 x^2} = 0.$$

Hence

$$\Gamma_{11,1} = \frac{1}{2}\left(\frac{\partial g_{11}}{\varDelta_1 x^1} + \frac{\partial g_{11}}{\varDelta_1 x^1} - \frac{\partial g_{11}}{\varDelta_1 x^1}\right)$$
$$= \frac{1}{2}\frac{\partial g_{11}}{\varDelta_1 x^1}$$
$$= 0,$$

$$\Gamma_{11,2} = \frac{1}{2}\left(\frac{\partial g_{12}}{\Delta_1 x^1} + \frac{\partial g_{12}}{\Delta_1 x^1} - \frac{\partial g_{11}}{\Delta_2 x^2} \right)$$
$$= 0,$$

$$\Gamma_{12,1} = \frac{1}{2}\left(\frac{\partial g_{11}}{\Delta_2 x^2} + \frac{\partial g_{21}}{\Delta_1 x^1} - \frac{\partial g_{12}}{\Delta_1 x^1} \right)$$
$$= 0,$$

$$\Gamma_{12,2} = \frac{1}{2}\left(\frac{\partial g_{12}}{\Delta_2 x^2} + \frac{\partial g_{22}}{\Delta_1 x^1} - \frac{\partial g_{12}}{\Delta_2 x^2} \right)$$
$$= \frac{a^2}{2} \cos_1(x^1, x_0^1)(\sin_1(x^1, x_0^1) + \sin_1(\sigma_1(x^1), x_0^1)),$$

$$\Gamma_{21,1} = \frac{1}{2}\left(\frac{\partial g_{21}}{\Delta_1 x^1} + \frac{\partial g_{11}}{\Delta_2 x^2} - \frac{\partial g_{21}}{\Delta_1 x^1} \right)$$
$$= 0,$$

$$\Gamma_{21,2} = \frac{1}{2}\left(\frac{\partial g_{22}}{\Delta_1 x^1} + \frac{\partial g_{12}}{\Delta_2 x^2} - \frac{\partial g_{21}}{\Delta_2 x^2} \right)$$
$$= \frac{a^2}{2} \cos_1(x^1, x_0^1)(\sin_1(x^1, x_0^1) + \sin_1(\sigma_1(x^1), x_0^1)),$$

$$\Gamma_{22,1} = \frac{1}{2}\left(\frac{\partial g_{21}}{\Delta_2 x^2} + \frac{\partial g_{21}}{\Delta_2 x^2} + \frac{\partial g_{22}}{\Delta_1 x^1} \right)$$
$$= \frac{a^2}{2} \cos_1(x^1, x_0^1)(\sin_1(x^1, x_0^1) + \sin_1(\sigma_1(x^1), x_0^1)),$$

$$\Gamma_{22,2} = \frac{1}{2}\left(\frac{\partial g_{22}}{\Delta_2 x^2} + \frac{\partial g_{22}}{\delta_2 x^2} - \frac{\partial g_{22}}{\Delta_2 x^2} \right)$$
$$= 0,$$

and

$$\Gamma^1_{11} = g^{\alpha 1}\Gamma_{11,\alpha}$$
$$= g^{11}\Gamma_{11,1} + g^{12}\Gamma_{11,2}$$
$$= 0,$$

$$\Gamma^1_{12} = g^{\alpha 1}\Gamma_{12,\alpha}$$
$$= g^{11}\Gamma_{12,1} + g^{12}\Gamma_{12,2}$$
$$= 0,$$

$$\Gamma^1_{21} = g^{\alpha 1}\Gamma_{21,\alpha}$$
$$= g^{11}\Gamma_{21,1} + g^{21}\Gamma_{21,2}$$
$$= 0,$$

$$\Gamma^1_{22} = g^{\alpha 1}\Gamma_{22,\alpha}$$
$$= g^{11}\Gamma_{22,1} + g^{21}\Gamma_{22,2}$$
$$= \frac{1}{2} \cos_1(x^1, x_0^1)(\sin_1(x^1, x_0^1) + \sin_1(\sigma_1(x^1), x_0^1)),$$

$$\Gamma_{11}^2 = g^{a1}\Gamma_{11,a}$$
$$= g^{12}\Gamma_{11,1} + g^{22}\Gamma_{11,2}$$
$$= 0,$$
$$\Gamma_{12}^2 = g^{a2}\Gamma_{12,a}$$
$$= g^{12}\Gamma_{12,1} + g^{22}\Gamma_{12,2}$$
$$= \frac{\cos_1(x^1, x_0^1)(\sin_1(x^1, x_0^1) + \sin_1(\sigma_1(x^1), x_0^1))}{2(\sin_1(x^1, x_0^1))^2},$$
$$\Gamma_{21}^2 = g^{a2}\Gamma_{21,a}$$
$$= g^{12}\Gamma_{21,1} + g^{22}\Gamma_{21,2}$$
$$= \frac{\cos_1(x^1, x_0^1)(\sin_1(x^1, x_0^1) + \sin_1(\sigma_1(x^1), x_0^1))}{2(\sin_1(x^1, x_0^1))^2},$$
$$\Gamma_{22}^2 = g^{a2}\Gamma_{22,a}$$
$$= g^{12}\Gamma_{22,1} + g^{22}\Gamma_{22,2}$$
$$= 0.$$

Moreover, we get

$$H_{21} = \det\begin{pmatrix} \frac{\partial g_{11}}{\Delta_1 x^1} & \frac{\partial g_{12}}{\Delta_1 x^1} \\ \frac{\partial g_{21}}{\Delta_1 x^1} & \frac{\partial g_{22}}{\Delta_1 x^1} \end{pmatrix}$$
$$= \det\begin{pmatrix} 0 & 0 \\ 0 & \cos_1(x^1, x_0^1)(\sin_1(x^1, x_0^1) + \sin_1(\sigma_1(x^1), x_0^1)) \end{pmatrix}$$
$$= 0,$$
$$H_{22} = \det\begin{pmatrix} \frac{\partial g_{11}}{\Delta_2 x^2} & \frac{\partial g_{12}}{\Delta_2 x^2} \\ \frac{\partial g_{21}}{\Delta_2 x^2} & \frac{\partial g_{22}}{\Delta_2 x^2} \end{pmatrix}$$
$$= \det\begin{pmatrix} 0 & 0 \\ 0 & 0 \end{pmatrix}$$
$$= 0.$$

Therefore

$$R_{11} = \frac{\partial}{\Delta_1 x^1}\left(\frac{1}{2a^4(\sin_1(x^1, x_0^1))^2} \frac{\partial}{\Delta_1 x^1}(a^4(\sin_1(x^1, x_0^1))^2) \right)$$
$$- \frac{\partial \Gamma_{11}^1}{\Delta_1 x^1} - \frac{\partial \Gamma_{11}^2}{\Delta_2 x^2} + \Gamma_{11}^1\Gamma_{11}^1 + \Gamma_{11}^2\Gamma_{12}^1 + \Gamma_{21}^1\Gamma_{11}^1 + \Gamma_{21}^2\Gamma - 12^2$$
$$- \Gamma_{11}^1\left(\frac{1}{2a^4(\sin_1(x^1, x_0^1))^2} \frac{\partial}{\Delta_1 x^1}(a^4(\sin_1(x^1, x_0^1))^2) \right)$$

$$-\Gamma^2_{11}\left(\frac{1}{2a^4(\sin_1(x^1,x_0^1))^2}\frac{\partial}{\Delta_2 x^2}(a^4(\sin_1(x^1,x_0^1))^2)\right)$$

$$=\frac{1}{2(\sin_1(x^1,x_0^1))^2(\sin_1(\sigma_1(x^1),x_0^1))^2}$$

$$\times\left((\sin_1(x^1,x_0^1))^2\left(-\sin_1(x^1,x_0^1)(\sin_1(\sigma_1(x^1),x_0^1)+\sin_1(x^1,x_0^1))\right.\right.$$

$$\left.+\cos_1(\sigma_1(x^1),x_0^1)\left(\frac{\partial\sin_1(\sigma_1(x^1),x_0^1)}{\Delta_1 x^1}+\cos_1(x^1,x_0^1)\right)\right)$$

$$-(\cos_1(x^1,x_0^1))^2(\sin_1(\sigma_1(x^1),x_0^1)+\sin_1(x^1,x_0^1))^2\bigg)$$

$$+\frac{(\cos_1(x^1,x_0^1))^2(\sin_1(\sigma_1(x^1),x_0^1)+\sin_1(x^1,x_0^1))}{4(\sin_1(x^1,x_0^1))^4},$$

$$R_{12}=\frac{\partial}{\Delta_2 x^1}\left(\frac{1}{2g}\frac{\partial g}{\Delta_1 x^1}\right)-\frac{\partial\Gamma^\alpha_{12}}{\Delta_\alpha x^\alpha}+\Gamma^\alpha_{\beta 2}\Gamma^\beta_{1\alpha}$$

$$-\Gamma^\beta_{12}\left(\frac{1}{2g}\frac{\partial g}{\Delta_\beta x^\beta}\right)$$

$$=-\frac{\partial\Gamma^1_{12}}{\Delta_1 x^1}-\frac{\partial\Gamma^2_{12}}{\Delta_2 x^2}+\Gamma^1_{12}\Gamma^1_{11}+\Gamma^2_{12}\Gamma^1_{12}$$

$$+\Gamma^1_{22}\Gamma^2_{11}+\Gamma^2_{22}\Gamma^2_{12}-\Gamma^1_{12}\left(\frac{1}{2g}\frac{\partial g}{\Delta_1 x^1}\right)-\Gamma^2_{12}\left(\frac{1}{2g}\frac{\partial g}{\Delta_2 x^2}\right)$$

$$=0,$$

$$R_{21}=\frac{\partial}{\Delta_1 x^1}\left(\frac{1}{2g}\frac{\partial g}{\Delta_2 x^2}\right)-\frac{\partial\Gamma-12^\alpha}{\Delta_\alpha x^\alpha}+\Gamma^\alpha_{\beta 2}\Gamma^\beta_{1\alpha}$$

$$-\Gamma^\beta_{12}\left(\frac{1}{2g}\frac{\partial g}{\Delta_\beta x^\beta}\right)$$

$$=-\frac{\partial\Gamma^1_{12}}{\Delta_1 x^1}-\frac{\partial\Gamma^2_{12}}{\Delta_2 x^2}+\Gamma^1_{12}\Gamma^1_{11}+\Gamma^2_{12}\Gamma^1_{12}$$

$$+\Gamma^1_{22}\Gamma^2_{11}+\Gamma^2_{22}\Gamma^2_{12}-\Gamma^1_{12}\left(\frac{1}{2g}\frac{\partial g}{\Delta_1 x^1}\right)-\Gamma^2_{12}\left(\frac{1}{2g}\frac{\partial g}{\Delta_2 x^2}\right)$$

$$=0,$$

$$R_{22}=\frac{\partial}{\Delta_2 x^2}\left(\frac{1}{2g}\frac{\partial g}{\Delta_2 x^2}\right)-\frac{\partial\Gamma^\alpha_{22}}{\Delta_\alpha x^\alpha}+\Gamma^\alpha_{\beta 2}\Gamma^\beta_{2\alpha}$$

$$-\Gamma^\beta_{22}\left(\frac{1}{2g}\frac{\partial g}{\Delta_\beta x^\beta}\right)$$

$$=-\frac{\partial\Gamma^1_{22}}{\Delta_1 x^1}-\frac{\partial\Gamma^2_{22}}{\Delta_2 x^2}+\Gamma^1_{12}\Gamma^1_{21}+\Gamma^2_{12}\Gamma^1_{22}$$

$$+\Gamma^1_{22}\Gamma^2_{21}+\Gamma^2_{22}\Gamma^2_{22}-\Gamma^1_{22}\left(\frac{1}{2g}\frac{\partial g}{\Delta_1 x^1}\right)-\Gamma^2_{22}\left(\frac{1}{2g}\frac{\partial g}{\Delta_2 x^2}\right)$$

$$= -\frac{\partial}{\Delta_1 x^1}\left(\frac{1}{2}\cos_1(x^1,x_0^1)(\sin_1(x^1,x_0^1) + \sin_1(\sigma_1(x^1),x_0^1))\right)$$

$$+ 2\left(\frac{1}{2}\cos_1(x^1,x_0^1)(\sin_1(x^1,x_0^1) + \sin_1(\sigma_1(x^1),x_0^1))\right)$$

$$\times \left(\frac{\cos_1(x^1,x_0^1)(\sin_1(x^1,x_0^1) + \sin_1(\sigma_1(x^1),x_0^1))}{2(\sin_1(x^1,x_0^1))^2}\right)$$

$$- \frac{1}{2}\cos_1(x^1,x_0^1)(\sin_1(x^1,x_0^1) + \sin_1(\sigma_1(x^1),x_0^1))$$

$$\times \left(\frac{1}{2a^4(\sin_1(x^1,x_0^1))^2}a^4\cos_1(x^1,x_0^1)(\sin_1(x^1,x_0^1) + \sin_1(\sigma_1(x^1),x_0^1))\right)$$

$$= \frac{1}{2}\sin_1(x^1,x_0^1)(\sin_1(x^1,x_0^1) + \sin_1(\sigma_1(x^1),x_0^1))$$

$$- \frac{1}{2}\cos_1(x^1,x_0^1)\left(\cos_1(x^1,x_0^1) + \frac{\partial}{\Delta_1 x^1}\sin_1(\sigma_1(x^1),x_0^1)\right)$$

$$+ \frac{(\cos_1(x^1,x_0^1))^2(\sin_1(x^1,x_0^1) + \sin_1(\sigma_1(x^1),x_0^1))^2}{(\sin_1(x^1,x_0^1))^2}$$

$$- \frac{(\cos_1(x^1,x_0^1))^2(\sin_1(x^1,x_0^1) + \sin_1(\sigma_1(x^1),x_0^1))^2}{4(\sin_1(x^1,x_0^1))^2}$$

$$= \frac{1}{2}\sin_1(x^1,x_0^1)(\sin_1(x^1,x_0^1) + \sin_1(\sigma_1(x^1),x_0^1))$$

$$- \frac{1}{2}\cos_1(x^1,x_0^1)\left(\cos_1(x^1,x_0^1) + \frac{\partial}{\Delta_1 x^1}\sin_1(\sigma_1(x^1),x_0^1)\right)$$

$$+ \frac{3(\cos_1(x^1,x_0^1))^2(\sin_1(x^1,x_0^1) + \sin_1(\sigma_1(x^1),x_0^1))^2}{4(\sin_1(x^1,x_0^1))^2}.$$

Consequently,

$$R = g^{11}R_{11} + g^{22}R_{22}$$

$$= \frac{1}{2a^2(\sin_1(x^1,x_0^1))^2(\sin_1(\sigma_1(x^1),x_0^1))^2}$$

$$\times \left((\sin_1(x^1,x_0^1))^2\left(-\sin_1(x^1,x_0^1)(\sin_1(\sigma_1(x^1),x_0^1) + \sin_1(x^1,x_0^1))\right.\right.$$

$$+ \cos_1(\sigma_1(x^1),x_0^1)\left(\frac{\partial\sin_1(\sigma_1(x^1),x_0^1)}{\Delta_1 x^1} + \cos_1(x^1,x_0^1)\right)\right)$$

$$\left.- (\cos_1(x^1,x_0^1))^2(\sin_1(\sigma_1(x^1),x_0^1) + \sin_1(x^1,x_0^1))^2\right)$$

$$+ \frac{(\cos_1(x^1,x_0^1))^2(\sin_1(\sigma_1(x^1),x_0^1) + \sin_1(x^1,x_0^1))}{4a^2(\sin_1(x^1,x_0^1))^4}$$

$$+ \frac{1}{a^2(\sin_1(x^1,x_0^1))^2}\left(\frac{1}{2}\sin_1(x^1,x_0^1)(\sin_1(x^1,x_0^1) + \sin_1(\sigma_1(x^1),x_0^1))\right.$$

$$-\frac{1}{2}\cos_1(x^1,x_0^1)\left(\cos_1(x^1,x_0^1)+\frac{\partial}{\Delta_1 x^1}\sin_1(\sigma_1(x^1),x_0^1)\right)$$

$$+\frac{3(\cos_1(x^1,x_0^1))^2(\sin_1(x^1,x_0^1)+\sin_1(\sigma_1(x^1),x_0^1))^2}{4(\sin_1(x^1,x_0^1))^2}\Bigg).$$

5.3 The covariant curvature tensor

Definition 5.6. The completely covariant curvature tensor R_{hijk} is defined as

$$R_{hijk}=g_{ha}R^a_{ijk}.$$

Note that

$$g_{ah}\frac{\partial\Gamma^a_{ij}}{\Delta_k x^k}=\frac{\partial}{\Delta_k x^k}(g_{ah}\Gamma^a_{ij})-\Gamma^{a\sigma_k}_{ij}\frac{\partial g_{ah}}{\Delta_k x^k}$$

$$=\frac{\partial\Gamma_{ij,h}}{\Delta_k x^k}-\Gamma^{a\sigma_k}_{ij}\frac{\partial g_{ah}}{\Delta_k x^k}$$

and

$$g_{ah}\frac{\Gamma^a_{ik}}{\Delta_j x^j}=\frac{\partial\Gamma_{ik,h}}{\Delta_j x^j}-\Gamma^{a\sigma_j}_{ik}\frac{\partial g_{ah}}{\Delta_j x^j}.$$

Therefore

$$R_{hijk}=g_{ha}R^a_{ijk}$$

$$=g_{ha}\left(\frac{\partial\Gamma^a_{ik}}{\Delta_j x^j}-\frac{\partial\Gamma^a_{ij}}{\Delta_k x^k}+\Gamma^a_{\beta j}\Gamma^\beta_{ik}-\Gamma^a_{\beta k}\Gamma^\beta_{ij}\right)$$

$$=g_{ha}\frac{\partial\Gamma^a_{ik}}{\Delta_j x^j}-g_{ha}\frac{\partial\Gamma^a_{ij}}{\Delta_k x^k}+g_{ha}\Gamma^a_{\beta j}\Gamma^\beta_{ik}-g_{ha}\Gamma^a_{\beta k}\Gamma^\beta_{ij}$$

$$=\frac{\partial\Gamma_{ik,h}}{\Delta_j x^j}-\Gamma^{a\sigma_j}_{ik}\frac{\partial g_{ah}}{\Delta_j x^j}-\frac{\partial\Gamma_{ij,h}}{\Delta_k x^k}+\Gamma^{a\sigma_k}_{ij}\frac{\partial g_{ah}}{\Delta_k x^k}$$

$$+\Gamma_{\beta j,h}\Gamma^\beta_{ik}-\Gamma_{\beta k,h}\Gamma^\beta_{ij}$$

$$=\frac{1}{2}\frac{\partial}{\Delta_j x^j}\left(\frac{\partial g_{ih}}{\Delta_k x^k}+\frac{\partial g_{kh}}{\Delta_i x^i}-\frac{\partial g_{ik}}{\Delta_h x^h}\right)$$

$$-\frac{1}{2}\frac{\partial}{\Delta_k x^k}\left(\frac{\partial g_{ih}}{\Delta_j x^j}+\frac{\partial g_{jh}}{\Delta_i x^i}-\frac{\partial g_{ij}}{\Delta_h x^h}\right)$$

$$-\left(\Gamma^a_{ik}+\mu_j\frac{\partial\Gamma^a_{ik}}{\Delta_j x^j}\right)\frac{\partial g_{ah}}{\Delta_j x^j}+\left(\Gamma^a_{ij}+\mu_k\frac{\partial\Gamma^a_{ij}}{\Delta_k x^k}\right)\frac{\partial g_{ah}}{\Delta_k x^k}$$

$$+\Gamma_{\beta j,h}\Gamma^\beta_{ik}-\Gamma_{\beta k,h}\Gamma^\beta_{ij}$$

$$
= \frac{1}{2}\left(\frac{\partial^2 g_{ih}}{\Delta_j x^j \Delta_k x^k} + \frac{\partial^2 g_{kh}}{\Delta_j x^j \Delta_i x^i} - \frac{\partial^2 g_{ik}}{\Delta_j x^j \Delta_h x^h} \right.
$$
$$
\left. - \frac{\partial^2 g_{ih}}{\Delta_k x^k \Delta_j x^j} - \frac{\partial^2 g_{jh}}{\Delta_k x^k \Delta_i x^i} + \frac{\partial^2 g_{ij}}{\Delta_k x^k \Delta_h x^h} \right)
$$
$$
- \Gamma_{ik}^{\alpha} \frac{\partial g_{\alpha h}}{\Delta_j x^j} + \Gamma_{ij}^{\alpha} \frac{\partial g_{\alpha h}}{\Delta_k x^k} + \Gamma_{\beta j,h}\Gamma_{ik}^{\beta} - \Gamma_{\beta k,h}\Gamma_{ij}^{\beta}
$$
$$
- \mu_j \frac{\partial \Gamma_{ik}^{\alpha}}{\Delta_j x^j}\frac{\partial g_{\alpha h}}{\Delta_j x^j} + \mu_k \frac{\partial \Gamma_{ij}^{\alpha}}{\Delta_k x^k}\frac{\partial g_{\alpha h}}{\Delta_k x^k}
$$

$$
= \frac{1}{2}\left(\frac{\partial^2 g_{kh}}{\Delta_j x^j \Delta_i x^i} + \frac{\partial^2 g_{ij}}{\Delta_k x^k \Delta_h x^h} - \frac{\partial^2 g_{ik}}{\Delta_j x^j \Delta_h x^h} - \frac{\partial^2 g_{jh}}{\Delta_k x^k \Delta_i x^i} \right)
$$
$$
- \Gamma_{ik}^{\alpha}(\Gamma_{\alpha j,h} + \Gamma_{hj,\alpha}) + \Gamma_{ij}^{\alpha}(\Gamma_{\alpha k,h} + \Gamma_{hk,\alpha}) + \Gamma_{\beta j,h}\Gamma_{ik}^{\beta} - \Gamma_{\beta k,h}\Gamma_{ij}^{\beta}
$$
$$
- \mu_j \frac{\partial \Gamma_{ik}^{\alpha}}{\Delta_j x^j}\frac{\partial g_{\alpha h}}{\Delta_j x^j} + \mu_k \frac{\partial \Gamma_{ij}^{\alpha}}{\Delta_k x^k}\frac{\partial g_{\alpha h}}{\Delta_k x^k}
$$

$$
= \frac{1}{2}\left(\frac{\partial^2 g_{kh}}{\Delta_j x^j \Delta_i x^i} + \frac{\partial^2 g_{ij}}{\Delta_k x^k \Delta_h x^h} - \frac{\partial^2 g_{ik}}{\Delta_j x^j \Delta_h x^h} - \frac{\partial^2 g_{jh}}{\Delta_k x^k \Delta_i x^i} \right)
$$
$$
- \Gamma_{ik}^{\alpha}\Gamma_{\alpha j,h} - \Gamma_{ik}^{\alpha}\Gamma_{hj,\alpha} + \Gamma_{ij}^{\alpha}\Gamma_{\alpha k,h} + \Gamma_{ij}^{\alpha}\Gamma_{hk,\alpha} + \Gamma_{\beta j,h}\Gamma_{ik}^{\beta} - \Gamma_{\beta k,h}\Gamma_{ij}^{\beta}
$$
$$
- \mu_j \frac{\partial \Gamma_{ik}^{\alpha}}{\Delta_j x^j}\frac{\partial g_{\alpha h}}{\Delta_j x^j} + \mu_k \frac{\partial \Gamma_{ij}^{\alpha}}{\Delta_k x^k}\frac{\partial g_{\alpha h}}{\Delta_k x^k}
$$

$$
= \frac{1}{2}\left(\frac{\partial^2 g_{kh}}{\Delta_j x^j \Delta_i x^i} + \frac{\partial^2 g_{ij}}{\Delta_k x^k \Delta_h x^h} - \frac{\partial^2 g_{ik}}{\Delta_j x^j \Delta_h x^h} - \frac{\partial^2 g_{jh}}{\Delta_k x^k \Delta_i x^i} \right)
$$
$$
+ \Gamma_{ij}^{\alpha}\Gamma_{hk,\alpha} - \Gamma_{ik}^{\alpha}\Gamma_{hj,\alpha}
$$
$$
- \mu_j \frac{\partial \Gamma_{ik}^{\alpha}}{\Delta_j x^j}\frac{\partial g_{\alpha h}}{\Delta_j x^j} + \mu_k \frac{\partial \Gamma_{ij}^{\alpha}}{\Delta_k x^k}\frac{\partial g_{\alpha h}}{\Delta_k x^k}
$$

$$
= -\frac{\partial \Gamma_{ij}^{h}}{\Delta_k x^k} + \frac{\partial \Gamma_{ik}^{h}}{\Delta_j x^j} + \Gamma_{ij}^{\alpha}\Gamma_{hk,\alpha} - \Gamma_{ik}^{\alpha}\Gamma_{hj,\alpha}
$$
$$
- \mu_j \frac{\partial \Gamma_{ik}^{\alpha}}{\Delta_j x^j}\frac{\partial g_{\alpha h}}{\Delta_j x^j} + \mu_k \frac{\partial \Gamma_{ij}^{\alpha}}{\Delta_k x^k}\frac{\partial g_{\alpha h}}{\Delta_k x^k}
$$

$$
= \det\begin{pmatrix} \dfrac{\partial}{\Delta_j x^j} & \dfrac{\partial}{\Delta_k x^k} \\[2mm] \Gamma_{ij}^{h} & \Gamma_{ik}^{h} \end{pmatrix} + \det\begin{pmatrix} \Gamma_{ij}^{\alpha} & \Gamma_{ik}^{\alpha} \\[2mm] \Gamma_{hj,\alpha} & \Gamma_{hk,\alpha} \end{pmatrix}
$$
$$
+ \det\begin{pmatrix} \dfrac{\partial \Gamma_{ij}^{\alpha}}{\Delta_k x^k} & \dfrac{\partial \Gamma_{ik}^{\alpha}}{\Delta_j x^j} \\[2mm] \mu_j \dfrac{\partial g_{\alpha h}}{\Delta_j x^j} & \mu_k \dfrac{\partial g_{\alpha h}}{\Delta_k x^k} \end{pmatrix}.
$$

Theorem 5.3. *The covariant curvature tensor R_{hijk} is skew-symmetric with respect to the indices j and k, i. e.,*

$$R_{hijk} = -R_{hikj}.$$

Proof. We have

$$R_{hijk} = -\frac{\partial \Gamma_{ij}^{h}}{\Delta_k x^k} + \frac{\partial \Gamma_{ik}^{h}}{\Delta_j x^j} + \Gamma_{ij}^{a}\Gamma_{hk,a} - \Gamma_{ik}^{a}\Gamma_{hj,a}$$
$$- \mu_j \frac{\partial \Gamma_{ik}^{a}}{\Delta_j x^j}\frac{\partial g_{ah}}{\Delta_j x^j} + \mu_k \frac{\partial \Gamma_{ij}^{a}}{\Delta_k x^k}\frac{\partial g_{ah}}{\Delta_k x^k}$$

and

$$R_{hikj} = -\frac{\partial \Gamma_{ik}^{h}}{\Delta_j x^j} + \frac{\partial \Gamma_{ij}^{h}}{\Delta_k x^k} + \Gamma_{ik}^{a}\Gamma_{hj,a} - \Gamma_{ij}^{a}\Gamma_{hk,a}$$
$$- \mu_k \frac{\partial \Gamma_{ij}^{a}}{\Delta_k x^k}\frac{\partial g_{ah}}{\Delta_k x^k} + \mu_j \frac{\partial \Gamma_{ik}^{a}}{\Delta_j x^j}\frac{\partial g_{ah}}{\Delta_j x^j}$$
$$= -\left(-\frac{\partial \Gamma_{ij}^{h}}{\Delta_k x^k} + \frac{\partial \Gamma_{ik}^{h}}{\Delta_j x^j} + \Gamma_{ij}^{a}\Gamma_{hk,a} - \Gamma_{ik}^{a}\Gamma_{hj,a} \right.$$
$$\left. - \mu_j \frac{\partial \Gamma_{ik}^{a}}{\Delta_j x^j}\frac{\partial g_{ah}}{\Delta_j x^j} + \mu_k \frac{\partial \Gamma_{ij}^{a}}{\Delta_k x^k}\frac{\partial g_{ah}}{\Delta_k x^k} \right)$$
$$= -R_{hikj}.$$

This completes the proof. $\square$

Theorem 5.4. *We have*

$$R_{hijk} + R_{hjki} + R_{hkij} = 0.$$

Proof. We have

$$R_{hijk} = -\frac{\partial \Gamma_{ij}^{h}}{\Delta_k x^k} + \frac{\partial \Gamma_{ik}^{h}}{\Delta_j x^j} + \Gamma_{ij}^{a}\Gamma_{hk,a} - \Gamma_{ik}^{a}\Gamma_{hj,a}$$
$$- \mu_j \frac{\partial \Gamma_{ik}^{a}}{\Delta_j x^j}\frac{\partial g_{ah}}{\Delta_j x^j} + \mu_k \frac{\partial \Gamma_{ij}^{a}}{\Delta_k x^k}\frac{\partial g_{ah}}{\Delta_k x^k},$$
$$R_{hjki} = -\frac{\partial \Gamma_{jk}^{h}}{\Delta_i x^i} + \frac{\partial \Gamma_{ji}^{h}}{\Delta_k x^k} + \Gamma_{jk}^{a}\Gamma_{hi,a} - \Gamma_{ji}^{a}\Gamma_{hk,a}$$
$$- \mu_k \frac{\partial \Gamma_{ji}^{a}}{\Delta_k x^k}\frac{\partial g_{ah}}{\Delta_k x^k} + \mu_i \frac{\partial \Gamma_{jk}^{h}}{\Delta_i x^i}\frac{\partial g_{ah}}{\Delta_i x^i},$$

and

$$R_{hkij} = -\frac{\partial \Gamma_{ki}^h}{\Delta_j x^j} + \frac{\partial \Gamma_{kj}^h}{\Delta_i x^i} + \Gamma_{ki}^a \Gamma_{hj,a} - \Gamma_{kj}^a \Gamma_{hi,a}$$

$$- \mu_i \frac{\partial \Gamma_{kj}^a}{\Delta_i x^i} \frac{\partial g_{ah}}{\Delta_i x^i} + \mu_j \frac{\partial \Gamma_{ki}^a}{\Delta_j x^j} \frac{\partial g_{ah}}{\Delta_j x^j}.$$

Hence

$$R_{hijk} + R_{hjki} + R_{hkij} = -\frac{\partial \Gamma_{ij}^h}{\Delta_k x^k} + \frac{\partial \Gamma_{ik}^h}{\Delta_j x^j} + \Gamma_{ij}^a \Gamma_{hk,a} - \Gamma_{ik}^a \Gamma_{hj,a}$$

$$- \mu_j \frac{\partial \Gamma_{ik}^a}{\Delta_j x^j} \frac{\partial g_{ah}}{\Delta_j x^j} + \mu_k \frac{\partial \Gamma_{ij}^a}{\Delta_k x^k} \frac{\partial g_{ah}}{\Delta_k x^k}$$

$$+ -\frac{\partial \Gamma_{jk}^h}{\Delta_i x^i} + \frac{\partial \Gamma_{ji}^h}{\Delta_k x^k} + \Gamma_{jk}^a \Gamma_{hi,a} - \Gamma_{ji}^a \Gamma_{hk,a}$$

$$- \mu_k \frac{\partial \Gamma_{ji}^a}{\Delta_k x^k} \frac{\partial g_{ah}}{\Delta_k x^k} + \mu_i \frac{\partial \Gamma_{jk}^h}{\Delta_i x^i} \frac{\partial g_{ah}}{\Delta_i x^i}$$

$$+ -\frac{\partial \Gamma_{ki}^h}{\Delta_j x^j} + \frac{\partial \Gamma_{kj}^h}{\Delta_i x^i} + \Gamma_{ki}^a \Gamma_{hj,a} - \Gamma_{kj}^a \Gamma_{hi,a}$$

$$- \mu_i \frac{\partial \Gamma_{kj}^a}{\Delta_i x^i} \frac{\partial g_{ah}}{\Delta_i x^i} + \mu_j \frac{\partial \Gamma_{ki}^a}{\Delta_j x^j} \frac{\partial g_{ah}}{\Delta_j x^j}$$

$$= 0.$$

This completes the proof. $\square$

Definition 5.7. In a V_N, it is always possible to choose a coordinate system such that all the Christoffel symbols vanish at a particular point P_0. Such a coordinate system is called a geodesic coordinate system with a pole P_0.

Theorem 5.5. *The curvature tensor R_{hijk} satisfies the differential properties*

$$R_{ijk,m}^a + R_{ikm,j}^a + R_{imj,k}^a = 0$$

and

$$R_{hijk,l} + R_{hikl,j} + R_{hilj,k} = 0.$$

Proof. Let us choose a geodesic coordinate system with a pole P_0. Then at P_0 the Christoffel symbols vanish, and the covariant derivatives at P_0 coincide with the ordinary derivatives at P_0. Thus, at P_0, we have

$$R_{ijk}^a = \frac{\partial \Gamma_{ik}^a}{\Delta_j x^j} - \frac{\partial \Gamma_{ij}^a}{\Delta_k x^k},$$

and hence

$$R^a_{ijk,m} = \frac{\partial^2 \Gamma^a_{ik}}{\Delta_m x^m \Delta_j x^j} - \frac{\partial^2 \Gamma^a_{ij}}{\Delta_m x^m \Delta_k x^k},$$

$$R^a_{ikm,j} = \frac{\partial^2 \Gamma^a_{im}}{\Delta_j x^j \Delta_k x^k} - \frac{\partial^2 \Gamma^a_{ik}}{\Delta_j x^j \Delta_m x^m},$$

and

$$R^a_{imj,k} = \frac{\partial^2 \Gamma^a_{ij}}{\Delta_k x^k \Delta_m x^m} - \frac{\partial^2 \Gamma^a_{im}}{\Delta_k x^k \Delta_j x^j}.$$

Therefore

$$R^a_{ijk,m} + R^a_{ikm,j} + R^a_{imj,k} = \frac{\partial^2 \Gamma^a_{ik}}{\Delta_m x^m \Delta_j x^j} - \frac{\partial^2 \Gamma^a_{ij}}{\Delta_m x^m \Delta_k x^k}$$

$$+ \frac{\partial^2 \Gamma^a_{im}}{\Delta_j x^j \Delta_k x^k} - \frac{\partial^2 \Gamma^a_{ik}}{\Delta_j x^j \Delta_m x^m}$$

$$+ \frac{\partial^2 \Gamma^a_{ij}}{\Delta_k x^k \Delta_m x^m} - \frac{\partial^2 \Gamma^a_{im}}{\Delta_k x^k \Delta_j x^j}$$

$$= 0.$$

Now multiplying the last equation by g_{ha}, we find

$$0 = g_{ha} R^a_{ijk,m} + g_{ha} R^a_{ikm,j} + g_{ha} R^a_{imj,k}$$

$$= R_{hijk,m} + R_{hikm,j} + R_{himj,k}.$$

This completes the proof. $\qquad\square$

Example 5.2. We will find the components R_{hik} of the Riemann tensor for the metric

$$\Delta s^2 = \left(\Delta_1 x^1\right)^2 - \frac{1}{(x^2)^2}\left(\Delta_2 x^2\right)^2.$$

Here

$$g_{11} = 1,$$
$$g_{12} = 0,$$
$$g_{21} = 0,$$
$$g_{22} = -\frac{1}{(x^2)^2}.$$

Then

$$g = \det \begin{pmatrix} g_{11} & g_{12} \\ g_{21} & g_{22} \end{pmatrix}$$

$$= \det \begin{pmatrix} 1 & 0 \\ 0 & -\frac{1}{(x^2)^2} \end{pmatrix}$$

$$= -\frac{1}{(x^2)^2},$$

$$g^{11} = \frac{g^{22}}{g}$$

$$= \frac{-\frac{1}{(x^2)^2}}{-\frac{1}{(x^2)^2}}$$

$$= 1,$$

$$g^{12} = -\frac{g_{21}}{g}$$

$$= 0,$$

$$g^{21} = -\frac{g_{12}}{g}$$

$$= 0,$$

$$g^{22} = \frac{g_{11}}{g}$$

$$= \frac{1}{-\frac{1}{(x^2)^2}}$$

$$= -(x^2)^2.$$

Next, for the Christoffel symbols, we get

$$\Gamma_{11,1} = \frac{1}{2}\left(\frac{\partial g_{11}}{\Delta_1 x^1} + \frac{\partial g_{11}}{\Delta_1 x^1} - \frac{\partial g_{11}}{\Delta_1 x^1} \right)$$

$$= 0,$$

$$\Gamma_{11,2} = \frac{1}{2}\left(\frac{\partial g_{12}}{\Delta_1 x^1} + \frac{\partial g_{12}}{\Delta_1 x^1} - \frac{\partial g_{11}}{\Delta_2 x^2} \right)$$

$$= 0,$$

$$\Gamma_{12,1} = \frac{1}{2}\left(\frac{\partial g_{11}}{\Delta_2 x^2} + \frac{\partial g_{21}}{\Delta_1 x^1} - \frac{\partial g_{12}}{\Delta_1 x^1} \right)$$

$$= 0,$$

$$\Gamma_{12,2} = \frac{1}{2}\left(\frac{\partial g_{12}}{\Delta_2 x^2} + \frac{\partial g_{22}}{\Delta_1 x^1} - \frac{\partial g_{12}}{\Delta_2 x^2} \right)$$

$$= 0,$$

$$\Gamma_{21,1} = \frac{1}{2}\left(\frac{\partial g_{21}}{\Delta_1 x^1} + \frac{\partial g_{11}}{\delta_2 x^2} - \frac{\partial g_{21}}{\Delta_1 x^1} \right)$$

$$= 0,$$

$$\Gamma_{21,2} = \frac{1}{2}\left(\frac{\partial g_{22}}{\Delta_1 x^1} + \frac{\partial g_{12}}{\Delta_2 x^2} - \frac{\partial g_{21}}{\Delta_2 x^2} \right)$$
$$= 0,$$

$$\Gamma_{22,1} = \frac{1}{2}\left(\frac{\partial g_{21}}{\Delta_2 x^2} + \frac{\partial g_{21}}{\Delta_2 x^2} - \frac{\partial g_{22}}{\Delta_1 x^1} \right)$$
$$= 0,$$

$$\Gamma_{22,2} = \frac{1}{2}\left(\frac{\partial g_{22}}{\Delta_2 x^2} + \frac{\partial g_{22}}{\Delta_2 x^2} - \frac{\partial g_{22}}{\Delta_2 x^2} \right)$$
$$= \frac{x^2 + \sigma_2(x^2)}{2(x^2)^2(\sigma_2(x^2))^2}$$

and

$$\Gamma_{11}^1 = g^{1a}\Gamma_{11,a}$$
$$= g^{11}\Gamma_{11,1} + g^{12}\Gamma_{11,2}$$
$$= 0,$$

$$\Gamma_{12}^1 = g^{1a}\Gamma_{12,a}$$
$$= g^{11}\Gamma_{12,1} + g^{12}\Gamma_{12,2}$$
$$= 0,$$

$$\Gamma_{21}^1 = g^{1a}\Gamma_{21,a}$$
$$= g^{11}\Gamma_{21,1} + g^{12}\Gamma_{21,2}$$
$$= 0,$$

$$\Gamma_{22}^1 = g^{1a}\Gamma_{22,a}$$
$$= g^{11}\Gamma_{22,1} + g^{12}\Gamma_{22,2}$$
$$= 0,$$

$$\Gamma_{11}^2 = g^{2a}\Gamma_{11,a}$$
$$= g^{21}\Gamma_{1,1} + g^{22}\Gamma_{11,2}$$
$$= 0,$$

$$\Gamma_{12}^2 = g^{2a}\Gamma_{12,a}$$
$$= g^{P21}\Gamma_{12,1} + g^{22}\Gamma_{12,2}$$
$$= 0,$$

$$\Gamma_{21}^2 = g^{2a}\Gamma_{21,a}$$
$$= g^{21}\Gamma_{21,1} + g^{22}\Gamma_{21,2}$$
$$= 0,$$

$$\Gamma_{22}^2 = g^{2a}\Gamma_{22,a}$$
$$= g^{21}\Gamma_{22,1} + g^{22}\Gamma_{22,2}$$

$$= -(x^2)^2 \left(\frac{x^2 + \sigma_2(x^2)}{(x^2)^2 (\sigma_2(x^2))^2} \right)$$

$$= -\frac{x^2 + \sigma_2(x^2)}{2(\sigma_2(x^2))^2}.$$

Therefore

$$R^1_{111} = \frac{\partial \Gamma^1_{11}}{\Delta_1 x^1} - \frac{\partial \Gamma^1_{11}}{\Delta_1 x^1} + \Gamma^1_{\beta 1}\Gamma^\beta_{11} - \Gamma^1_{1\beta}\Gamma^\beta_{11}$$

$$= \Gamma^1_{11}\Gamma^1_{11} + \Gamma^1_{21}\Gamma^2_{11} - \Gamma^1_{11}\Gamma^1_{11} - \Gamma^1_{12}\Gamma^2_{11}$$

$$= 0,$$

$$R^1_{112} = \frac{\partial \Gamma^1_{12}}{\Delta_1 x^1} - \frac{\partial \Gamma^1_{11}}{\Delta_2 x^2} + \Gamma^1_{\beta 1}\Gamma^\beta_{12} - \Gamma^1_{\beta 2}\Gamma^\beta_{11}$$

$$= \Gamma^1_{11}\Gamma^1_{12} + \Gamma^1_{21}\Gamma^2_{12} - \Gamma^1_{12}\Gamma^1_{11} - \Gamma^1_{22}\Gamma^2_{11}$$

$$= 0,$$

$$R^1_{121} = \frac{\partial \Gamma^1_{11}}{\Delta_2 x^2} - \frac{\partial \Gamma^1_{12}}{\Delta_1 x^1} - \Gamma^1_{\beta 2}\Gamma^\beta_{11} - \Gamma^1_{\beta 1}\Gamma^\beta_{12}$$

$$= -\Gamma^1_{12}\Gamma^1_{11} - \Gamma^1_{22}\Gamma^2_{11} - \Gamma^1_{11}\Gamma^1_{12} - \Gamma^1_{21}\Gamma^2_{12}$$

$$= 0,$$

$$R^1_{122} = \frac{\partial \Gamma^1_{12}}{\Delta_2 x^2} - \frac{\partial \Gamma^1_{12}}{\Delta_2 x^2} = \Gamma^1_{\beta 2}\Gamma^\beta_{12} - \Gamma^1_{\beta 2}\Gamma^\beta_{12}$$

$$= \Gamma^1_{12}\Gamma^1_{12} + \Gamma^1_{22}\Gamma^2_{12} - \Gamma^1_{12}\Gamma^1_{12} - \Gamma^1_{22}\Gamma^2_{12}$$

$$= 0,$$

$$R^1_{211} = \frac{\partial \Gamma^1_{21}}{\Delta_1 x^1} - \frac{\partial \Gamma^1_{21}}{\Delta_2 x^2} + \Gamma^1_{\beta 1}\Gamma^\beta_{21} - \Gamma^1_{\beta 1}\Gamma^\beta_{21}$$

$$= \Gamma^1_{11}\Gamma^1_{21} + \Gamma^1_{21}\Gamma^2_{21} - \Gamma^1_{11}\Gamma^1_{21} - \Gamma^1_{21}\Gamma^2_{21}$$

$$= 0,$$

$$R^1_{212} = \frac{\partial \Gamma^1_{22}}{\Delta_1 x^1} - \frac{\partial \Gamma^1_{21}}{\Delta_2 x^2} + \Gamma^1_{\beta 1}\Gamma^\beta_{22} - \Gamma^1_{\beta 1}\Gamma^\beta_{21}$$

$$= \Gamma^1_{11}\Gamma^1_{22} + \Gamma^1_{21}\Gamma^2_{22} - \Gamma^1_{11}\Gamma^1_{21} - \Gamma^1_{21}\Gamma^2_{21}$$

$$= 0,$$

$$R^1_{221} = \frac{\partial \Gamma^1_{21}}{\Delta_2 x^2} - \frac{\partial \Gamma^1_{22}}{\Delta_1 x^1} + \Gamma^1_{\beta 2}\Gamma^\beta_{21} - \Gamma^1_{\beta 1}\Gamma^\beta_{22}$$

$$= \Gamma^1_{12}\Gamma^1_{21} + \Gamma^1_{22}\Gamma^2_{21} - \Gamma^1_{11}\Gamma^1_{22} - \Gamma^1_{21}\Gamma^2_{22}$$

$$= 0,$$

$$R^1_{222} = \frac{\partial \Gamma^1_{22}}{\Delta_2 x^2} - \frac{\partial \Gamma^1_{22}}{\Delta_2 x^2} + \Gamma^1_{\beta 2}\Gamma^\beta_{22} - \Gamma^1_{\beta 2}\Gamma^\beta_{22}$$

$$= \Gamma^1_{12}\Gamma^1_{22} + \Gamma^1_{22}\Gamma^2_{22} - \Gamma^1_{12}\Gamma^1_{22} - \Gamma^1_{22}\Gamma^2_{22}$$

$$= 0,$$

$$R^2_{111} = \frac{\partial \Gamma^2_{11}}{\Delta_1 x^1} - \frac{\partial \Gamma^2_{11}}{\Delta_1 x^1} + \Gamma^2_{\beta 1}\Gamma^\beta_{11} - \Gamma^2_{\beta 1}\Gamma^\beta_{11}$$
$$= \Gamma^2_{11}\Gamma^1_{11} + \Gamma^2_{21}\Gamma^2_{11} - \Gamma^2_{11}\Gamma^1_{11} - \Gamma^2_{21}\Gamma^2_{11}$$
$$= 0,$$

$$R^2_{112} = \frac{\partial \Gamma^2_{12}}{\Delta_1 x^1} - \frac{\partial \Gamma^2_{11}}{\Delta_2 x^2} + \Gamma^2_{\beta 1}\Gamma^\beta_{12} - \Gamma^2_{\beta 2}\Gamma^\beta_{11}$$
$$= \Gamma^2_{11}\Gamma^1_{12} + \Gamma^2_{21}\Gamma^2_{12} - \Gamma^2_{12}\Gamma^1_{11} - \Gamma^2_{22}\Gamma^2_{11}$$
$$= 0,$$

$$R^2_{121} = \frac{\partial \Gamma^2_{11}}{\Delta_2 x^2} - \frac{\partial \Gamma^2_{12}}{\Delta_1 x^1} + \Gamma^2_{\beta 2}\Gamma^\beta_{11} - \Gamma^2_{\beta 1}\Gamma^\beta_{12}$$
$$= \Gamma^2_{12}\Gamma^1_{11} + \Gamma^2_{22}\Gamma^2_{11} - \Gamma^2_{11}\Gamma^1_{12} - \Gamma^2_{21}\Gamma^2_{12}$$
$$= 0,$$

$$R^2_{122} = \frac{\partial \Gamma^2_{12}}{\Delta_2 x^2} - \frac{\partial \Gamma^2_{12}}{\Delta_2 x^2} + \Gamma^2_{\beta 2}\Gamma^\beta_{12} - \Gamma^2_{\beta 2}\Gamma^\beta_{12}$$
$$= \Gamma^2_{12}\Gamma^1_{12} + \Gamma^2_{22}\Gamma^2_{22} - \Gamma^2_{12}\Gamma^1_{12} - \Gamma^2_{22}\Gamma^2_{12}$$
$$= 0,$$

$$R^2_{211} = \frac{\partial \Gamma^2_{21}}{\Delta_1 x^1} - \frac{\partial \Gamma^2_{21}}{\Delta_1 x^1} + \Gamma^2_{\beta 1}\Gamma^\beta_{21} - \Gamma^2_{\beta 1}\Gamma^\beta_{21}$$
$$= \Gamma^2_{11}\Gamma^1_{21} + \Gamma^2_{21}\Gamma^2_{21} - \Gamma^2_{11}\Gamma^1_{21} - \Gamma^2_{21}\Gamma^2_{21}$$
$$= 0,$$

$$R^2_{212} = \frac{\partial \Gamma^2_{22}}{\Delta_1 x^1} - \frac{\partial \Gamma^2_{21}}{\Delta_1 x^1} + \Gamma^2_{\beta 1}\Gamma^\beta_{22} - \Gamma^2_{\beta 2}\Gamma^\beta_{21}$$
$$= \Gamma^2_{11}\Gamma^1_{22} + \Gamma^2_{21}\Gamma^2_{22} - \Gamma^2_{12}\Gamma^1_{21} - \Gamma^2_{22}\Gamma^2_{21}$$
$$= 0,$$

$$R^2_{221} = \frac{\partial \Gamma^2_{21}}{\Delta_2 x^2} - \frac{\partial \Gamma^2_{22}}{\Delta_1 x^1} + \Gamma^2_{\beta 2}\Gamma^\beta_{21} - \Gamma^2_{\beta 1}\Gamma^\beta_{22}$$
$$= \Gamma^2_{12}\Gamma^1_{21} + \Gamma^2_{22}\Gamma^2_{21} - \Gamma^2_{11}\Gamma^1_{22} - \Gamma^2_{21}\Gamma^2_{22}$$
$$= 0,$$

$$R^2_{222} = \frac{\partial \Gamma^2_{22}}{\Delta_2 x^2} - \frac{\partial \Gamma^2_{22}}{\Delta_2 x^2} + \Gamma^2_{\beta 2}\Gamma^\beta_{22} - \Gamma^2_{\beta 2}\Gamma^\beta_{22}$$
$$= \Gamma^2_{12}\Gamma^1_{22} + \Gamma^2_{22}\Gamma^2_{22} - \Gamma^2_{12}\Gamma^2_{22} - \Gamma^2_{22}\Gamma^2_{22}$$
$$= 0.$$

From here we conclude that

$$R_{hijk} = g_{ah}R^a_{ijk}$$
$$= 0$$

for all $h, i, j, k \in \{1, 2\}$.

Example 5.3. We will find R_{1212} for the metric

$$ds^2 = (\Delta_1 x^1)^2 + (x^1)^2 (\Delta_2 x^2)^2 + (x^2)^2 (\Delta_3 x^3)^2.$$

Here

$$g_{11} = 1,$$
$$g_{12} = 0,$$
$$g_{13} = 0,$$
$$g_{21} = 0,$$
$$g_{22} = (x^1)^2,$$
$$g_{23} = 0,$$
$$g_{31} = 0,$$
$$g_{32} = 0,$$
$$g_{33} = (x^2)^2,$$

and

$$g = \det \begin{pmatrix} g_{11} & g_{12} & g_{13} \\ g_{21} & g_{22} & g_{23} \\ g_{31} & g_{32} & g_{33} \end{pmatrix}$$

$$= \det \begin{pmatrix} 1 & 0 & 0 \\ 0 & (x^1)^2 & 0 \\ 0 & 0 & (x^2)^2 \end{pmatrix}$$

$$= (x^1)^2 (x^2)^2,$$

$$g^{11} = \frac{1}{(x^1)^2(x^2)^2} \det \begin{pmatrix} (x^1)^2 & 0 \\ 0 & (x^2)^2 \end{pmatrix}$$

$$= \frac{1}{(x^1)^2(x^2)^2} (x^1)^2 (x^2)^2$$

$$= 1,$$

$$g^{12} = -\frac{1}{(x^1)^2(x^2)^2} \det \begin{pmatrix} 0 & 0 \\ 0 & (x^2)^2 \end{pmatrix}$$

$$= 0,$$

$$g^{13} = \frac{1}{(x^1)^2(x^2)^2} \det \begin{pmatrix} 0 & (x^1)^2 \\ 0 & 0 \end{pmatrix}$$

$$= 0,$$

$$g^{21} = -\frac{1}{(x^1)^2(x^2)^2} \det \begin{pmatrix} 0 & 0 \\ 0 & (x^2)^2 \end{pmatrix}$$

$$= 0,$$

$$g^{22} = \frac{1}{(x^1)^2(x^2)^2} \det \begin{pmatrix} 1 & 0 \\ 0 & (x^2)^2 \end{pmatrix}$$

$$= \frac{1}{(x^1)^2(x^2)^2}(x^2)^2$$

$$= \frac{1}{(x^1)^2},$$

$$g^{23} = -\frac{1}{(x^1)^2(x^2)^2} \det \begin{pmatrix} 1 & 0 \\ 0 & 0 \end{pmatrix}$$

$$= 0,$$

$$g^{31} = g^{13}$$

$$= 0,$$

$$g^{32} = g^{23}$$

$$= 0,$$

$$g^{33} = \frac{1}{(x^1)^2(x^2)^2} \det \begin{pmatrix} 1 & 0 \\ 0 & (x^1)^2 \end{pmatrix}$$

$$= \frac{1}{(x^1)^2(x^2)^2}(x^1)^2$$

$$= \frac{1}{(x^2)^2}.$$

Then

$$\frac{\partial g_{11}}{\Delta_1 x^1} = 0,$$

$$\frac{\partial g_{11}}{\Delta_2 x^2} = 0,$$

$$\frac{\partial g_{11}}{\delta_3 x^3} = 0,$$

$$\frac{\partial g_{12}}{\Delta_1 x^1} = 0,$$

$$\frac{\partial g_{12}}{\Delta_2 x^2} = 0,$$

$$\frac{\partial g_{12}}{\delta_3 x^3} = 0,$$

$$\frac{\partial g_{13}}{\Delta_1 x^1} = 0,$$

$$\frac{\partial g_{13}}{\Delta_2 x^2} = 0,$$

$$\frac{\partial g_{13}}{\delta_3 x^3} = 0,$$

$$\frac{\partial g_{21}}{\Delta_1 x^1} = 0,$$

$$\frac{\partial g_{21}}{\Delta_2 x^2} = 0,$$

$$\frac{\partial g_{21}}{\delta_3 x^3} = 0,$$

$$\frac{\partial g_{22}}{\Delta_1 x^1} = \sigma_1(x^1) + x^1,$$

$$\frac{\partial g_{22}}{\Delta_2 x^2} = 0,$$

$$\frac{\partial g_{22}}{\delta_3 x^3} = 0,$$

$$\frac{\partial g_{23}}{\Delta_1 x^1} = 0,$$

$$\frac{\partial g_{23}}{\Delta_2 x^2} = 0,$$

$$\frac{\partial g_{23}}{\delta_3 x^3} = 0,$$

$$\frac{\partial g_{31}}{\Delta_1 x^1} = 0,$$

$$\frac{\partial g_{31}}{\Delta_2 x^2} = 0,$$

$$\frac{\partial g_{31}}{\delta_3 x^3} = 0,$$

$$\frac{\partial g_{32}}{\Delta_1 x^1} = 0,$$

$$\frac{\partial g_{32}}{\Delta_2 x^2} = 0,$$

$$\frac{\partial g_{32}}{\delta_3 x^3} = 0,$$

$$\frac{\partial g_{33}}{\Delta_1 x^1} = 0,$$

$$\frac{\partial g_{33}}{\Delta_2 x^2} = \sigma_2(x^2) + x^2,$$

$$\frac{\partial g_{33}}{\delta_3 x^3} = 0.$$

Hence

$$\Gamma_{11,1} = \frac{1}{2}\left(\frac{\partial g_{11}}{\Delta_1 x^1} + \frac{\partial g_{11}}{\delta_1 x^1} + \frac{\partial g_{11}}{\Delta_1 x^1}\right)$$
$$= 0,$$
$$\Gamma_{11,2} = \frac{1}{2}\left(\frac{\partial g_{12}}{\Delta_1 x^1} + \frac{\partial g_{12}}{\Delta_1 x^1} - \frac{\partial g_{11}}{\Delta_2 x^2}\right)$$
$$= 0,$$

$$\Gamma_{21,1} = \frac{1}{2}\left(\frac{\partial g_{21}}{\Delta_1 x^1} + \frac{\partial g_{11}}{\Delta_2 x^2} - \frac{\partial g_{21}}{\Delta_1 x^1} \right)$$
$$= 0,$$

$$\Gamma_{21,2} = \frac{1}{2}\left(\frac{\partial g_{22}}{\Delta_1 x^1} + \frac{\partial g_{12}}{\Delta_2 x^2} - \frac{\partial g_{21}}{\Delta_2 x^2} \right)$$
$$= \frac{1}{2}\left(x^1 + \sigma_1(x^1) \right),$$

$$\Gamma_{22,1} = \frac{1}{2}\left(\frac{\partial g_{21}}{\Delta_2 x^2} + \frac{\partial g_{21}}{\Delta_2 x^2} - \frac{\partial g_{22}}{\Delta_1 x^1} \right)$$
$$= 0,$$

$$\Gamma_{22,2} = \frac{1}{2}\left(\frac{\partial g_{22}}{\Delta_2 x^2} + \frac{\partial g_{22}}{\Delta_2 x^2} - \frac{\partial g_{22}}{\Delta_2 x^2} \right)$$
$$= 0$$

and

$$\Gamma^1_{11} = g^{11}\Gamma_{11,1} + g^{12}\Gamma_{11,2}$$
$$= 0,$$

$$\Gamma^1_{21} = g^{11}\Gamma_{21,1} + g^{12}\Gamma_{21,2}$$
$$= 0,$$

$$\Gamma^2_{21} = g^{21}\Gamma_{21,1} + g^{22}\Gamma_{21,2}$$
$$= 0,$$

$$\Gamma^1_{22} = g^{11}\Gamma_{22,1} + g^{12}\Gamma_{22,2}$$
$$= 0,$$

$$\Gamma^2_{22} = g^{21}\Gamma_{22,1} + g^{22}\Gamma_{22,2}$$
$$= 0.$$

Therefore

$$R^1_{212} = \frac{\partial \Gamma^1_{22}}{\Delta_1 x^1} - \frac{\partial \Gamma^1_{21}}{\Delta_2 x^2} + \Gamma^1_{11}\Gamma^1_{22} + \Gamma^1_{21}\Gamma^2_{22} - \Gamma^1_{11}\Gamma^1_{21} - \Gamma^1_{21}\Gamma^2_{21}$$
$$= 0,$$

and

$$R_{1212} = g_{1\alpha}R^\alpha_{212}$$
$$= g_{11}R^1_{212} + g_{12}R^2_{212} + g_{13}R^3_{212}$$
$$= g_{11}R^1_{212}$$
$$= 0.$$

Exercise 5.1. Prove that

$$\operatorname{div}(R^{\alpha}_{ijk}) = R_{ij,k} - R_{ik,j}.$$

Solution. By the identity

$$R^{\alpha}_{ijk,m} + R^{\alpha}_{ikm,j} + R^{\alpha}_{imj,k} = 0,$$

contracting α and m, we find

$$R^{\alpha}_{ijk,\alpha} + R^{\alpha}_{ika,j} + R^{\alpha}_{iaj,k} = 0,$$

or

$$\begin{aligned}
R^{\alpha}_{ijk,\alpha} &= -R^{\alpha}_{ika,j} - R^{\alpha}_{iaj,k} \\
&= -R_{ik,j} + R^{\alpha}_{ija,k} \\
&= -R_{ik,j} + R_{ij,k},
\end{aligned}$$

where we have used that R^{α}_{ijk} is skew-symmetric with respect to the indices j and k, i. e.,

$$R^{\alpha}_{ijk} = -R^{\alpha}_{ikj}.$$

5.4 The Riemannian curvature

Let p^j and q^j be two unit vectors defined at a point $P_0 \in V_N$. Consider the vector

$$t^j = \alpha p^j + \beta q^j,$$

where α and β are parameters. This vector determines pencils of the direction of p^j. Similarly, one and only one geodesic will pass through in the direction of q^j. These geodesics through P_0 determine a two-dimensional surface through P_0 determined by the orientation of the unit vectors p^j and q^j. Denote this surface by S.

Introduce the Riemannian coordinates y^j with the origin at P_0. The equation of the surface S in terms of y^j is

$$y^j = (\alpha p^j + \beta q^j)s, \tag{5.1}$$

where s denotes the arc length measured from P_0 to any point along the geodesic equation (5.1) through P_0 in the direction of t^j. Set

$$\alpha s = u^1,$$
$$\beta s = u^2.$$

Then

$$y^j = p^j u^1 + q^j u^2$$

with fixed p^j and q^j. Let

$$\Delta s^2 = a_{ij}\Delta_{(i)}y^i\Delta_{(j)}y^j$$

be the metric for the surface S. Let also, Γ^k_{ijg} and $\Gamma^\gamma_{\alpha\beta a}$ be the Christoffel symbols of the second kind corresponding to the coordinates y^i and u^α, respectively. Let $R_{\alpha\beta\gamma\delta}$ and R_{hijk} be the curvature tensors corresponding to the metrics

$$a_{\alpha\beta}\Delta_\alpha u^\alpha \Delta_\beta u^\beta \quad \text{and} \quad g_{ij}\Delta_{(i)}y^i\Delta_{(j)}y^j,$$

respectively, where $\alpha,\beta,\gamma,\delta \in \{1,2\}$, $h,i,j,k \in \{1,\dots,N\}$. Since a_{ij} is a covariant tensor, we have

$$\bar{a}_{\alpha\beta} = \frac{\Delta_\gamma u^\gamma}{\Delta_{(\alpha)}\bar{u}^\alpha}\frac{\Delta_\delta u^\delta}{\Delta_{(\beta)}\bar{u}^\beta}a_{\gamma\delta},$$

and hence

$$|\bar{a}_{\alpha\beta}| = \left|\frac{\Delta_\gamma u^\gamma}{\Delta_{(\alpha)}\bar{u}^\alpha}\right|\left|\frac{\Delta_\delta u^\delta}{\Delta_{(\beta)}\bar{u}^\beta}\right||a_{\gamma\delta}|,$$

or

$$\bar{a} = J^2 a,$$

where

$$J = \det\begin{pmatrix} \dfrac{\Delta_1 u^1}{\Delta_{(1)}\bar{u}^1} & \dfrac{\Delta_1 u^1}{\Delta_{(2)}\bar{u}^2} \\[2mm] \dfrac{\Delta_2 u^2}{\Delta_{(1)}\bar{u}^1} & \dfrac{\Delta_2 u^2}{\Delta_{(2)}\bar{u}^2} \end{pmatrix}.$$

Definition 5.8. The quantity

$$\kappa = \frac{R_{1212}}{\bar{a}}$$

is called the Riemann curvature of the surface S at P_0.

Remark 5.1. In general time scale the Riemann curvature is not invariant.

Example 5.4. Let $\mathbb{T}_1 = 2^{\mathbb{N}_0}$ and $\mathbb{T}_2 = 3^{\mathbb{N}_0}$. We will compute the Riemann curvature κ for the Riemann metric

$$\Delta s^2 = (x^1)^{-2}(\Delta_1 x^1)^2 - (x^1)^{-2}(\Delta_2 x^2)^2.$$

We have

$$\sigma_1(x^1) = 2x^1, \quad x^1 \in \mathbb{T}_1,$$
$$\sigma_2(x^2) = 3x^2, \quad x^2 \in \mathbb{T}_2,$$

and

$$g_{11} = (x^1)^{-2},$$
$$g_{12} = 0,$$
$$g_{21} = 0,$$
$$g_{22} = -(x^1)^{-2}.$$

Then

$$
\begin{aligned}
g &= \det \begin{pmatrix} g_{11} & g_{12} \\ g_{21} & g_{22} \end{pmatrix} \\
&= \det \begin{pmatrix} (x^1)^{-2} & 0 \\ 0 & -(x^1)^{-2} \end{pmatrix} \\
&= -(x^1)^{-4},
\end{aligned}
$$

and

$$g^{11} = -(x^1)^{-2},$$
$$g^{12} = 0,$$
$$g^{21} = 0,$$
$$g^{22} = (x^1)^{-2},$$
$$
\begin{aligned}
\frac{\partial g_{11}}{\Delta_1 x^1} &= -\frac{\sigma_1(x^1) + x^1}{(x^1)^2(\sigma_1(x^1))^2} \\
&= -\frac{2x^1 + x^1}{(x^1)^2(2x^1)^2} \\
&= -\frac{3}{4(x^1)^3},
\end{aligned}
$$
$$\frac{\partial g_{11}}{\Delta_2 x^2} = 0,$$
$$
\begin{aligned}
\frac{\partial g_{22}}{\Delta_1 x^1} &= \frac{\sigma_1(x^1) + x^1}{(x^1)^2(\sigma_1(x^1))^2} \\
&= \frac{3}{4(x^1)^3},
\end{aligned}
$$

$$\frac{\partial g_{22}}{\Delta_2 x^2} = 0,$$

$$\frac{\partial g_{ij}}{\Delta_k x^k} = 0, \quad i,j,k \in \{1,2\}, \ i \neq j.$$

Hence

$$\Gamma_{11,1} = \frac{1}{2}\left(\frac{\partial g_{11}}{\Delta_1 x^1} + \frac{\partial g_{11}}{\Delta_1 x^1} - \frac{\partial g_{11}}{\Delta_1 x^1}\right)$$

$$= \frac{1}{2}\frac{\partial g_{11}}{\Delta_1 x^1}$$

$$= \frac{1}{2}\left(-\frac{3}{4(x^1)^3}\right)$$

$$= -\frac{3}{8(x^1)^3},$$

$$\Gamma_{11,2} = \frac{1}{2}\left(\frac{\partial g_{12}}{\Delta_1 x^1} + \frac{\partial g_{12}}{\Delta_1 x^1} - \frac{\partial g_{11}}{\Delta_2 x^2}\right)$$

$$= 0,$$

$$\Gamma_{12,1} = \frac{1}{2}\left(\frac{\partial g_{11}}{\Delta_2 x^2} + \frac{\partial g_{21}}{\Delta_1 x^1} - \frac{\partial g_{12}}{\Delta_1 x^1}\right)$$

$$= 0,$$

$$\Gamma_{12,2} = \frac{1}{2}\left(\frac{\partial g_{12}}{\Delta_2 x^2} + \frac{\partial g_{22}}{\Delta_1 x^1} - \frac{\partial g_{12}}{\Delta_2 x^2}\right)$$

$$= \frac{1}{2}\frac{\partial g_{22}}{\Delta_1 x^1}$$

$$= \frac{1}{2}\left(\frac{3}{4(x^1)^3}\right)$$

$$= \frac{3}{8(x^1)^3},$$

$$\Gamma_{21,1} = \frac{1}{2}\left(\frac{\partial g_{21}}{\Delta_1 x^1} + \frac{\partial g_{11}}{\Delta_2 x^2} - \frac{\partial g_{21}}{\Delta_1 x^1}\right)$$

$$= 0,$$

$$\Gamma_{21,2} = \frac{1}{2}\left(\frac{\partial g_{22}}{\Delta_1 x^1} + \frac{\partial g_{12}}{\Delta_2 x^2} - \frac{\partial g_{21}}{\Delta_2 x^2}\right)$$

$$= \frac{1}{2}\frac{\partial g_{22}}{\Delta_1 x^1}$$

$$= \frac{1}{2}\left(\frac{3}{4(x^1)^3}\right)$$

$$= \frac{3}{8(x^1)^3},$$

$$\Gamma_{22,1} = \frac{1}{2}\left(\frac{\partial g_{21}}{\Delta_2 x^2} + \frac{\partial g_{21}}{\Delta_2 x^2} - \frac{\partial g_{22}}{\Delta_1 x^1}\right)$$

$$= -\frac{1}{2}\frac{\partial g_{22}}{\Delta_1 x^1}$$

$$= -\frac{1}{2}\left(\frac{3}{4(x^1)^3}\right)$$

$$= -\frac{3}{8(x^1)^3},$$

$$\Gamma_{22,2} = \frac{1}{2}\left(\frac{\partial g_{22}}{\Delta_2 x^2} + \frac{\partial g_{22}}{\Delta_2 x^2} - \frac{\partial g_{22}}{\Delta_2 x^2}\right)$$

$$= 0$$

and

$$\Gamma_{11}^1 = g^{1a}\Gamma_{11,a}$$
$$= g^{11}\Gamma_{11,1} + g^{12}\Gamma_{11,2}$$
$$= -(x^1)^{-2}\left(-\frac{3}{8(x^1)^3}\right)$$
$$= \frac{3}{8(x^1)^5},$$

$$\Gamma_{11}^2 = g^{2a}\Gamma_{11,a}$$
$$= g^{21}\Gamma_{11,1} + g^{22}\Gamma_{11,2}$$
$$= 0,$$

$$\Gamma_{12}^1 = g^{1a}\Gamma_{12,a}$$
$$= g^{11}\Gamma_{12,1} + g^{12}\Gamma_{12,2}$$
$$= 0,$$

$$\Gamma_{12}^2 = g^{2a}\Gamma_{12,a}$$
$$= g^{21}\Gamma_{12,1} + g^{22}\Gamma_{12,2}$$
$$= (x^1)^{-2}\left(\frac{3}{8(x^1)^3}\right)$$
$$= \frac{3}{8(x^1)^5},$$

$$\Gamma_{21}^1 = g^{1a}\Gamma - 21, a$$
$$= g^{11}\Gamma_{21,1} + g^{12}\Gamma_{21,2}$$
$$= 0,$$

$$\Gamma_{21}^2 = g^{2a}\Gamma_{21,a}$$
$$= g^{21}\Gamma_{21,1} + g^{22}\Gamma_{21,2}$$
$$= (x^1)^{-2}\left(\frac{3}{8(x^1)^3}\right)$$
$$= \frac{3}{8(x^1)^5},$$

$$\Gamma^1_{22} = g^{1a}\Gamma_{22,a}$$
$$= g^{11}\Gamma_{22,1} + g^{12}\Gamma_{22,2}$$
$$= (x^1)^{-2}\left(-\frac{3}{8(x^1)^3}\right)$$
$$= -\frac{3}{8(x^1)^5},$$
$$\Gamma^2_{22} = g^{2a}\Gamma_{22,a}$$
$$= g^{21}\Gamma_{22,1} + g^{22}\Gamma_{22,2}$$
$$= 0.$$

Therefore

$$
\begin{aligned}
R_{1212} &= \frac{1}{2}\left(\frac{\partial^2 g_{12}}{\Delta_1 x^1 \Delta_2 x^2} + \frac{\partial^2 g_{12}}{\Delta_1 x^1 \Delta_2 x^2} - \frac{\partial^2 g_{22}}{(\Delta_1 x^1)^2} - \frac{\partial^2 g_{11}}{(\Delta_2 x^2)^2}\right) \\
&\quad + \Gamma^a_{21}\Gamma - 12, a - \Gamma^a_{22}\Gamma_{11,a} \\
&\quad - \mu_1 \frac{\partial \Gamma^a_{22}}{\Delta_1 x^1}\frac{\partial g_{21}}{\Delta_1 x^1} + \mu_2 \frac{\partial \Gamma^a_{21}}{\Delta_2 x^2}\frac{\partial g_{a1}}{\Delta_2 x^2} \\
&= \frac{1}{2}\left(\frac{\partial^2 g_{12}}{\Delta_1 x^1 \Delta_2 x^2} + \frac{\partial^2 g_{12}}{\Delta_1 x^1 \Delta_2 x^2} - \frac{\partial^2 g_{22}}{(\Delta_1 x^1)^2} - \frac{\partial^2 g_{11}}{(\Delta_2 x^2)^2}\right) \\
&\quad + \Gamma^1_{21}\Gamma_{12,1} + \Gamma^2_{21}\Gamma_{12,2} - \Gamma^1_{22}\Gamma_{11,1} - \Gamma^2_{22}\Gamma_{11,2} \\
&\quad - \mu_1 \frac{\partial \Gamma^1_{22}}{\Delta_1 x^1}\frac{\partial g_{11}}{\Delta_1 x^1} - \mu_1 \frac{\partial \Gamma^2_{22}}{\Delta_1 x^1}\frac{\partial g_{21}}{\Delta_1 x^1} \\
&\quad + \mu_2 \frac{\partial \Gamma^1_{21}}{\Delta_2 x^2}\frac{\partial g_{11}}{\Delta_2 x^2} + \mu_2 \frac{\partial \Gamma - 21^2}{\Delta_2 x^2}\frac{\partial g_{21}}{\Delta_2 x^2} \\
&= -\frac{1}{2}\frac{\partial}{\Delta_1 x^1}\left(\frac{3}{4(x^1)^3}\right) + \left(\frac{3}{8(x^1)^5}\right)\left(\frac{3}{8(x^1)^5}\right) - \left(-\frac{3}{8(x^1)^5}\right)\left(\frac{3}{8(x^1)^5}\right) \\
&\quad - x^1 \frac{\partial}{\Delta_1 x^1}\left(-\frac{3}{8(x^1)^5}\right)\left(-\frac{3}{4(x^1)^5}\right) \\
&= \frac{3}{8}\frac{(\sigma_1(x^1))^2 + x^1\sigma_1(x^1) + (x^1)^2}{(x^1)^3(\sigma_1(x^1))^3} + \frac{9}{32(x^1)^{10}} \\
&\quad - \frac{9}{32}x^1\left(\frac{(\sigma_1(x^1))^4 + x^1(\sigma_1(x^1))^3 + (x^1)^2(\sigma_1(x^1))^2 + (x^1)^3\sigma_1(x^1) + (x^1)^4}{(x^1)^5(\sigma_1(x^1))^5}\right)\left(-\frac{3}{4(x^1)^5}\right) \\
&= \frac{3}{8}\frac{(2x^1)^2 + x^1(2x^1) + (x^1)^2}{(x^1)^3(2x^1)^3} + \frac{9}{32(x^1)^{10}} \\
&\quad + \frac{27}{128}x^1\frac{(2x^1)^4 + x^1(2x^1)^3 + (x^1)^2(2x^1)^2 + (x^1)^3(2x^1) + (x^1)^4}{(x^1)^{10}(2x^1)^5} \\
&= \frac{3}{8}\frac{7(x^1)^2}{(x^1)^3(2x^1)^3} + \frac{9}{32(x^1)^{10}} + \frac{27}{128}\frac{47(x^1)^3}{32(x^1)^5} \\
&= \frac{21}{64(x^1)^4} + \frac{9}{32(x^1)^{10}} + \frac{1269}{4096(x^1)^{12}}.
\end{aligned}
$$

Consequently,

$$\kappa = \frac{1}{g}R_{1212}$$

$$= -(x^1)^4\left(\frac{21}{64(x^1)^4} + \frac{9}{32(x^1)^{10}} + \frac{1269}{4096(x^1)^{12}}\right)$$

$$= -\frac{21}{64} - \frac{9}{32(x^1)^6} - \frac{1269}{4096(x^1)^8}.$$

Example 5.5. We will compute the Riemann curvature κ for the Riemann metric

$$\Delta s^2 = (\Delta_1 x^1)^2 + 2x^1(\Delta_2 x^2)^2 + 2x^2(\Delta_3 x^3)^2.$$

Here

$$g_{11} = 1,$$
$$g_{22} = 2x^1,$$
$$g_{33} = 2x^2,$$
$$g_{ij} = 0, \quad i,j \in \{1,2,3\}, \ i \neq j.$$

Then

$$g = \det\begin{pmatrix} g_{11} & g_{12} & g_{13} \\ g_{21} & g_{22} & g_{23} \\ g_{31} & g_{32} & g_{33} \end{pmatrix}$$

$$= \begin{pmatrix} 1 & 0 & 0 \\ 0 & 2x^1 & 0 \\ 0 & 0 & 2x^2 \end{pmatrix}$$

$$= 4x^1 x^2,$$

and

$$g^{11} = \det\begin{pmatrix} 2x^1 & 0 \\ 0 & 2x^2 \end{pmatrix}$$

$$= 4x^1 x^2,$$

$$g^{22} = \det\begin{pmatrix} 1 & 0 \\ 0 & 2x^2 \end{pmatrix}$$

$$= 2x^2,$$

$$g^{33} = \det\begin{pmatrix} 1 & 0 \\ 0 & 2x^1 \end{pmatrix}$$

$$= 2x^1,$$

$$g^{ij} = 0, \quad i, j \in \{1, 2, 3\}, \; i \neq j,$$

$$\frac{\partial g_{11}}{\Delta_1 x^1} = 0,$$

$$\frac{\partial g_{11}}{\Delta_2 x^2} = 0,$$

$$\frac{\partial g_{22}}{\Delta_1 x^1} = 2,$$

$$\frac{\partial g_{22}}{\Delta_2 x^2} = 0,$$

$$\frac{\partial g^{33}}{\Delta_1 x^1} = 0,$$

$$\frac{\partial g^{33}}{\Delta_2 x^2} = 2,$$

$$\frac{\partial g_{ij}}{\Delta_k x^k} = 0, \quad i, j, k \in \{1, 2, 3\}, \; i \neq j.$$

Hence

$$\Gamma_{11,1} = \frac{1}{2}\left(\frac{\partial g_{11}}{\Delta_1 x^1} + \frac{\partial g_{11}}{\Delta_1 x^1} - \frac{\partial g_{11}}{\Delta_1 x^1} \right)$$
$$= 0,$$

$$\Gamma_{11,2} = \frac{1}{2}\left(\frac{\partial g_{12}}{\Delta_1 x^1} + \frac{\partial g_{12}}{\Delta_1 x^1} - \frac{\partial g_{11}}{\Delta_2 x^2} \right)$$
$$= 0,$$

$$\Gamma_{12,1} = \frac{1}{2}\left(\frac{\partial g_{11}}{\Delta_2 x^2} + \frac{\partial g_{21}}{\Delta_1 x^1} - \frac{\partial g_{12}}{\Delta_1 x^1} \right)$$
$$= 0,$$

$$\Gamma_{12,2} = \frac{1}{2}\left(\frac{\partial g_{12}}{\Delta_2 x^2} + \frac{\partial g_{22}}{\Delta_1 x^1} - \frac{\partial g_{12}}{\Delta_2 x^2} \right)$$
$$= \frac{1}{2}\frac{\partial g_{22}}{\Delta_1 x^1}$$
$$= \frac{1}{2} \cdot 2$$
$$= 1,$$

$$\Gamma_{21,1} = \frac{1}{2}\left(\frac{\partial g_{21}}{\Delta_1 x^1} + \frac{\partial g_{11}}{\Delta_2 x^2} - \frac{\partial g_{21}}{\Delta_1 x^1} \right)$$
$$= 0,$$

$$\Gamma_{21,2} = \frac{1}{2}\left(\frac{\partial g_{22}}{\Delta_1 x^1} + \frac{\partial g_{12}}{\Delta_2 x^2} - \frac{\partial g_{21}}{\Delta_2 x^2} \right)$$
$$= \frac{1}{2}\frac{\partial g_{22}}{\Delta_1 x^1}$$

$$= \frac{1}{2} \cdot 2$$

$$= 1,$$

$$\Gamma_{22,1} = \frac{1}{2}\left(\frac{\partial g_{21}}{\Delta_2 x^2} + \frac{\partial g_{21}}{\Delta_2 x^2} - \frac{\partial g_{22}}{\Delta_1 x^1} \right)$$

$$= -\frac{1}{2} \frac{\partial g_{22}}{\Delta_1 x^1}$$

$$= -\frac{1}{2} \cdot 2$$

$$= -1,$$

$$\Gamma_{22,2} = \frac{1}{2}\left(\frac{\partial g_{22}}{\Delta_2 x^2} + \frac{\partial g_{22}}{\Delta_2 x^2} - \frac{\partial g_{22}}{\Delta_2 x^2} \right)$$

$$= 0,$$

$$\Gamma^1_{11} = g^{1a}\Gamma_{11,a}$$

$$= g^{11}\Gamma_{11,1} + g^{12}\Gamma_{11,2}$$

$$= 0,$$

$$\Gamma^2_{11} = g^{2a}\Gamma_{11,a}$$

$$= g^{21}\Gamma_{11,1} + g^{22}\Gamma_{11,2}$$

$$= 0,$$

$$\Gamma^1_{12} = g^{1a}\Gamma_{12,a}$$

$$= g^{11}\Gamma_{12,1} + g^{12}\Gamma_{12,2}$$

$$= 0,$$

$$\Gamma^2_{12} = g^{2a}\Gamma_{12,a}$$

$$= g^{21}\Gamma_{12,1} + g^{22}\Gamma_{12,2}$$

$$= 2x^2,$$

$$\Gamma^1_{21} = g^{1a}\Gamma - 21, a$$

$$= g^{11}\Gamma_{21,1} + g^{12}\Gamma_{21,2}$$

$$= 0,$$

$$\Gamma^2_{21} = g^{2a}\Gamma_{21,a}$$

$$= g^{21}\Gamma_{21,1} + g^{22}\Gamma_{21,2}$$

$$= 2x^2,$$

$$\Gamma^1_{22} = g^{1a}\Gamma_{22,a}$$

$$= g^{11}\Gamma_{22,1} + g^{12}\Gamma_{22,2}$$

$$= 1,$$

$$\Gamma^2_{22} = g^{2a}\Gamma_{22,a}$$

$$= g^{21}\Gamma_{22,1} + g^{22}\Gamma_{22,2}$$

$$= 0.$$

Therefore

$$
\begin{aligned}
R_{1212} ={}& \frac{1}{2}\left(\frac{\partial^2 g_{12}}{\Delta_1 x^1 \Delta_2 x^2} + \frac{\partial^2 g_{12}}{\Delta_1 x^1 \Delta_2 x^2} - \frac{\partial^2 g_{22}}{(\Delta_1 x^1)^2} - \frac{\partial^2 g_{11}}{(\Delta_2 x^2)^2} \right) \\
& + \Gamma_{21}^{\alpha}\Gamma - 12, \alpha - \Gamma_{22}^{\alpha}\Gamma_{11,\alpha} \\
& - \mu_1 \frac{\partial \Gamma_{22}^{\alpha}}{\Delta_1 x^1} \frac{\partial g_{21}}{\Delta_1 x^1} + \mu_2 \frac{\partial \Gamma_{21}^{\alpha}}{\Delta_2 x^2} \frac{\partial g_{\alpha 1}}{\Delta_2 x^2} \\
={}& \frac{1}{2}\left(\frac{\partial^2 g_{12}}{\Delta_1 x^1 \Delta_2 x^2} + \frac{\partial^2 g_{12}}{\Delta_1 x^1 \Delta_2 x^2} - \frac{\partial^2 g_{22}}{(\Delta_1 x^1)^2} - \frac{\partial^2 g_{11}}{(\Delta_2 x^2)^2} \right) \\
& + \Gamma_{21}^{1}\Gamma_{12,1} + \Gamma_{21}^{2}\Gamma_{12,2} - \Gamma_{22}^{1}\Gamma_{11,1} - \Gamma_{22}^{2}\Gamma_{11,2} \\
& - \mu_1 \frac{\partial \Gamma_{22}^{1}}{\Delta_1 x^1} \frac{\partial g_{11}}{\Delta_1 x^1} - \mu_1 \frac{\partial \Gamma_{22}^{2}}{\Delta_1 x^1} \frac{\partial g_{21}}{\Delta_1 x^1} \\
& + \mu_2 \frac{\partial \Gamma_{21}^{1}}{\Delta_2 x^2} \frac{\partial g_{11}}{\Delta_2 x^2} + \mu_2 \frac{\partial \Gamma - 21^2}{\Delta_2 x^2} \frac{\partial g_{21}}{\Delta_2 x^2} \\
={}& 2x^2.
\end{aligned}
$$

Consequently,

$$
\begin{aligned}
\kappa &= \frac{1}{g} R_{1212} \\
&= \frac{2x^2}{4x^1 x^2} \\
&= \frac{1}{2x^1}.
\end{aligned}
$$

Exercise 5.2. Find the Riemannian curvature for the metric

$$
\Delta s^2 = \left(\Delta_1 x^1\right)^2 + \left((x^1)^2 + 1\right)\left((\Delta_2 x^2)^2 + (\Delta_3 x^3)^2\right).
$$

Definition 5.9. If the Riemannian curvature κ is constant in a space V_N, then V_N is said to be a space of constant curvature.

Note that in an Euclidean space, we have that $\kappa = 0$.

5.5 The Einstein tensor

Definition 5.10. The covariant tensor

$$
\Gamma_{ij} = R_{ij} - \frac{1}{2}Rg_{ij}
$$

is called the Einstein tensor.

Definition 5.11. The mixed tensor

$$\Gamma_j^i R_j^i - \frac{1}{2}\delta_j^i R$$

is called the Einstein tensor.

In fact, we have

$$R_{ij} = R_{ij\alpha}^\alpha$$
$$= \frac{\partial \Gamma_{i\alpha}^\alpha}{\Delta_j x^j} - \frac{\partial \Gamma_{ij}^\alpha}{\Delta_\alpha x^\alpha} + \Gamma_{\gamma j}^\alpha \Gamma_{i\alpha}^\gamma - \Gamma_{\gamma\alpha}^\alpha \Gamma_{ij}^\gamma$$

and

$$R = R_i^i$$
$$= g^{\alpha i} R_{i\alpha}$$
$$= g^{\alpha i} R_{i\alpha\beta}^\beta$$
$$= g^{\alpha i}\left(\frac{\Gamma_{i\beta}^\beta}{\Delta_\alpha x^\alpha} - \frac{\partial \Gamma_{i\alpha}^\beta}{\Delta_\beta x^\beta} + \Gamma_{\gamma\alpha}^\beta \Gamma_{i\beta}^\gamma - \Gamma_{\gamma\beta}^\beta \Gamma_{i\alpha}^\gamma \right).$$

Thus

$$\Gamma_{ij} = \frac{\partial \Gamma_{i\alpha}^\alpha}{\Delta_j x^j} - \frac{\partial \Gamma_{ij}^\alpha}{\Delta_\alpha x^\alpha} + \Gamma_{\gamma j}^\alpha \Gamma_{i\alpha}^\gamma - \Gamma_{\gamma\alpha}^\alpha \Gamma_{ij}^\gamma$$
$$- \frac{1}{2} g^{\alpha i}\left(\frac{\Gamma_{i\beta}^\beta}{\Delta_\alpha x^\alpha} - \frac{\partial \Gamma_{i\alpha}^\beta}{\Delta_\beta x^\beta} + \Gamma_{\gamma\alpha}^\beta \Gamma_{i\beta}^\gamma - \Gamma_{\gamma\beta}^\beta \Gamma_{i\alpha}^\gamma \right) g_{ij}$$
$$= \frac{\partial \Gamma_{i\alpha}^\alpha}{\Delta_j x^j} - \frac{\partial \Gamma_{ij}^\alpha}{\Delta_\alpha x^\alpha} + \Gamma_{\gamma j}^\alpha \Gamma_{i\alpha}^\gamma - \Gamma_{\gamma\alpha}^\alpha \Gamma_{ij}^\gamma$$
$$- \frac{1}{2} \delta_j^\alpha \left(\frac{\Gamma_{i\beta}^\beta}{\Delta_\alpha x^\alpha} - \frac{\partial \Gamma_{i\alpha}^\beta}{\Delta_\beta x^\beta} + \Gamma_{\gamma\alpha}^\beta \Gamma_{i\beta}^\gamma - \Gamma_{\gamma\beta}^\beta \Gamma_{i\alpha}^\gamma \right)$$
$$= \frac{\partial \Gamma_{i\alpha}^\alpha}{\Delta_j x^j} - \frac{\partial \Gamma_{ij}^\alpha}{\Delta_\alpha x^\alpha} + \Gamma_{\gamma j}^\alpha \Gamma_{i\alpha}^\gamma - \Gamma_{\gamma\alpha}^\alpha \Gamma_{ij}^\gamma$$
$$- \frac{1}{2}\left(\frac{\Gamma_{i\beta}^\beta}{\Delta_\alpha x^\alpha} - \frac{\partial \Gamma_{i\alpha}^\beta}{\Delta_\beta x^\beta} + \Gamma_{\gamma\alpha}^\beta \Gamma_{i\beta}^\gamma - \Gamma_{\gamma\beta}^\beta \Gamma_{i\alpha}^\gamma \right).$$

Moreover, we have

$$R^i_j = g^{ia} R_{aj}$$

$$= g^{ia} R^{\beta}_{aj\beta}$$

$$= g^{ia} \left(\frac{\partial \Gamma^{\beta}_{a\beta}}{\Delta_j x^j} - \frac{\partial \Gamma^{\beta}_{aj}}{\Delta_\beta x^\beta} + \Gamma^{\beta}_{yj} \Gamma^{y}_{a\beta} - \Gamma^{\beta}_{y\beta} \Gamma^{y}_{aj} \right)$$

and

$$\Gamma^i_j = R^i_j - \frac{1}{2} \delta^i_j R$$

$$= g^{ia} \left(\frac{\partial \Gamma^{\beta}_{a\beta}}{\Delta_j x^j} - \frac{\partial \Gamma^{\beta}_{aj}}{\Delta_\beta x^\beta} + \Gamma^{\beta}_{yj} \Gamma^{y}_{a\beta} - \Gamma^{\beta}_{y\beta} \Gamma^{y}_{aj} \right)$$

$$- \frac{1}{2} \delta^i_j g^{ai} \left(\frac{\Gamma^{\beta}_{i\beta}}{\Delta_a x^a} - \frac{\partial \Gamma^{\beta}_{ia}}{\Delta_\beta x^\beta} + \Gamma^{\beta}_{ya} \Gamma^{y}_{i\beta} - \Gamma^{\beta}_{y\beta} \Gamma^{y}_{ia} \right)$$

$$= g^{ia} \left(\frac{\partial \Gamma^{\beta}_{a\beta}}{\Delta_j x^j} - \frac{\partial \Gamma^{\beta}_{aj}}{\Delta_\beta x^\beta} + \Gamma^{\beta}_{yj} \Gamma^{y}_{a\beta} - \Gamma^{\beta}_{y\beta} \Gamma^{y}_{aj} \right)$$

$$- \frac{1}{2} g^{ai} \left(\frac{\Gamma^{\beta}_{j\beta}}{\Delta_a x^a} - \frac{\partial \Gamma^{\beta}_{ja}}{\Delta_\beta x^\beta} + \Gamma^{\beta}_{ya} \Gamma^{y}_{j\beta} - \Gamma^{\beta}_{y\beta} \Gamma^{y}_{ja} \right).$$

Definition 5.12. A Riemannian space with identically zero curvature tensor is called a flat space.

Exercise 5.3. Determine if the following metric

$$\Delta s^2 = \left(\Delta_1 x^1 \right)^2 - \left(x^2 \right)^2 \left(\Delta_2 x^2 \right)^2$$

is flat or Euclidean.

Definition 5.13. The projective curvature tensor or Weyl tensor, denoted by W_{hijk}, is defined by

$$W_{hijk} = R_{hijk} + \frac{1}{1-N} \left(g_{kl} R_{hj} - g_{kh} R_{ij} \right),$$

where R_{hijk} is the Riemannian curvature tensor, and R_{ij} is the Ricci tensor.

Exercise 5.4. Find an expression of the Weyl tensor.

Let A_y be a covariant vector. We will find conditions under which

$$A_{y,v} = 0.$$

We have

$$\frac{\partial A_\gamma}{\Delta_v x^v} - A_\alpha \Gamma^\alpha_{\gamma v} = 0,$$

whereupon

$$\frac{\partial A_\gamma}{\Delta_v x^v} \Delta_v x^v = A_\alpha \Gamma^\alpha_{\gamma v} \Delta_v x^v \tag{5.2}$$

and

$$\int \frac{\partial A_\gamma}{\Delta_n u x^v} = \int A_\alpha \Gamma^\alpha_{\gamma v} \Delta_v x^v,$$

or

$$A_\gamma = \int \Delta A_\gamma$$
$$= \int A_\alpha \Gamma^\alpha_{\gamma v} \Delta_v x^v,$$

or

$$A_\gamma = \int \Delta A_\gamma$$
$$= \int A_\alpha \Gamma^\alpha_{\gamma v} \Delta_v x^v.$$

The last equation is integrable if

$$A_\alpha \Gamma^\alpha_{\gamma v}$$

is an exact differential. Thus we can write

$$A_\alpha \Gamma^\alpha_{\gamma v} \Delta_v x^v = \Delta B_\gamma$$
$$= \frac{\partial B_\gamma}{\Delta_v x^v} \Delta_v x^v,$$

or

$$\left(A_\alpha \Gamma^\alpha_{\gamma v} - \frac{\partial B_\gamma}{\Delta_v x^v} \right) \Delta_v x^v = 0,$$

or

$$A_\alpha \Gamma^\alpha_{\gamma v} = \frac{\partial B_\gamma}{\Delta_v x^v}.$$

Differentiating the last equation, we obtain

$$\frac{\partial A_\alpha}{\Delta_p x^p}\,\Gamma_{\gamma\nu}^{\alpha\sigma_p} + A_\alpha \frac{\partial \Gamma_{\gamma\nu}^\alpha}{\Delta_p x^p} = \frac{\partial^2 B\gamma}{\Delta_p x^p \Delta_\nu x^\nu}. \tag{5.3}$$

Now interchanging ν and p, we arrive at

$$\frac{\partial A_\alpha}{\Delta_\nu x^\nu}\,\Gamma_{\gamma p}^{\alpha\sigma_\nu} + A_\alpha \frac{\partial \Gamma_{\gamma p}^\alpha}{\Delta_\nu x^\nu} = \frac{\partial^2 B\gamma}{\Delta_\nu x^\nu \Delta_p x^p}. \tag{5.4}$$

Subtracting (5.4) from (5.3), we find

$$\begin{aligned}
0 &= \frac{\partial A_\alpha}{\Delta_p x^p}\Gamma_{\gamma\nu}^{\alpha\sigma_p} + A_\alpha \frac{\partial \Gamma_{\gamma\nu}^\alpha}{\Delta_p x^p} - \frac{\partial A_\alpha}{\Delta_\nu x^\nu}\Gamma_{\gamma p}^{\alpha\sigma_\nu} - A_\alpha \frac{\partial \Gamma_{\gamma p}^\alpha}{\Delta_\nu x^\nu}\\[4pt]
&= \Gamma_{ap}^c A_c \Gamma_{\gamma\nu}^{\alpha\sigma_p} + A_\alpha \frac{\partial \Gamma_{\gamma\nu}^\alpha}{\Delta_p x^p} - \Gamma_{a\nu}^c A_c \Gamma_{\gamma p}^{\alpha\sigma_\nu} - A_\alpha \frac{\partial \Gamma_{\gamma p}^\alpha}{\Delta_\nu x^\nu}\\[4pt]
&= \Gamma_{ap}^c A_c \Gamma_{\gamma\nu}^{\alpha\sigma_p} + A_c \frac{\partial \Gamma_{\gamma\nu}^c}{\Delta_p x^p} - \Gamma_{a\nu}^c A_c \Gamma_{\gamma p}^{\alpha\sigma_\nu} - A_c \frac{\partial \Gamma_{\gamma p}^c}{\Delta_\nu x^\nu}\\[4pt]
&= \Gamma_{ap}^c A_c \Gamma_{\gamma\nu}^a + A_c \frac{\partial \Gamma_{\gamma\nu}^c}{\Delta_p x^p} - \Gamma_{a\nu}^c A_c \Gamma_{\gamma p}^a - A_c \frac{\partial \Gamma_{\gamma p}^c}{\Delta_\nu x^\nu}\\[4pt]
&\quad + \mu_p \Gamma_{ap}^c A_c \frac{\partial \Gamma_{\gamma\nu}^a}{\Delta_p x^p} - \mu_\nu \Gamma_{a\nu}^c A_c \frac{\partial \Gamma_{\gamma p}^a}{\Delta\nu x^\nu}\\[4pt]
&= \left(\Gamma_{ap}^c \Gamma_{\gamma\nu}^a + \frac{\partial \Gamma_{\gamma\nu}^c}{\Delta_p x^p} - \Gamma_{a\nu}^c \Gamma_{\gamma p}^a - \frac{\partial \Gamma_{\gamma p}^c}{\Delta_\nu x^\nu} \right) A_c\\[4pt]
&\quad + \left(\mu_p \Gamma_{ap}^c \frac{\partial \Gamma_{\gamma\nu}^a}{\Delta_p x^p} - \mu_\nu \Gamma_{a\nu}^c \frac{\partial \Gamma_{\gamma p}^a}{\Delta\nu x^\nu} \right) A_c\\[4pt]
&= R_{c\gamma p\nu} A_c\\[4pt]
&\quad + \left(\mu_p\left(\Gamma_{ap}^c \frac{\partial \Gamma_{\gamma\nu}^a}{\Delta_p x^p} - \frac{\partial \Gamma_{\gamma\nu}^\beta}{\Delta_p x^p}\frac{\partial g_{bc}}{\Delta_p x^p} \right) - \mu_\nu\left(\Gamma_{a\nu}^c \frac{\partial \Gamma_{\gamma p}^a}{\Delta_\nu x^\nu} - \frac{\partial \Gamma_{\gamma p}^\beta}{\Delta_\nu x^\nu}\frac{\partial g_{\beta c}}{\Delta_\nu x^\nu} \right) \right) A_c\\[4pt]
&= \left(R_{c\gamma p\nu} + \left(\mu_p\left(\Gamma_{ap}^c \frac{\partial \Gamma_{\gamma\nu}^a}{\Delta_p x^p} - \frac{\partial \Gamma_{\gamma\nu}^\beta}{\Delta_p x^p}\frac{\partial g_{bc}}{\Delta_p x^p} \right) - \mu_\nu\left(\Gamma_{a\nu}^c \frac{\partial \Gamma_{\gamma p}^a}{\Delta_\nu x^\nu} - \frac{\partial \Gamma_{\gamma p}^\beta}{\Delta_\nu x^\nu}\frac{\partial g_{\beta c}}{\Delta_\nu x^\nu} \right) \right) \right) A_c.
\end{aligned}$$

Thus equation (5.2) is integrable if

$$R_{c\gamma p\nu} + \left(\mu_p\left(\Gamma_{ap}^c \frac{\partial \Gamma_{\gamma\nu}^a}{\Delta_p x^p} - \frac{\partial \Gamma_{\gamma\nu}^\beta}{\Delta_p x^p}\frac{\partial g_{bc}}{\Delta_p x^p} \right) - \mu_\nu\left(\Gamma_{a\nu}^c \frac{\partial \Gamma_{\gamma p}^a}{\Delta_\nu x^\nu} - \frac{\partial \Gamma_{\gamma p}^\beta}{\Delta_\nu x^\nu}\frac{\partial g_{\beta c}}{\Delta_\nu x^\nu} \right) \right) = 0.$$

Definition 5.14. The Einstein space is defined as a space that is homogeneous with regard to the Ricci tensor R_{ij}, i. e.,

$$R_{ij} = \lambda g_{ij}, \tag{5.5}$$

where λ is an invariant at all points.

Inner multiplication of g^{ij} shows that

$$g^{ij}R_{ij} = \lambda g^{ij}g_{ij}$$
$$= \lambda N,$$

whereupon

$$R = \lambda N,$$

and thus

$$\lambda = \frac{R}{N}.$$

Hence

$$R_{ij} = \lambda g_{ij}$$
$$= \frac{1}{N}Rg_{ij}.$$

5.6 Advanced practical problems

Problem 5.1. Compute R_{1321} for the metric

$$\Delta s^2 = \left(\Delta_1 x^1\right)^2 - x^1\left(\Delta_2 x^2\right)^2 + \left(\Delta_3 x^3\right)^2.$$

Problem 5.2. Compute R_{1222} for the metric

$$\Delta s^2 = x^1\left(\Delta_1 x^1\right)^2 + x^2\left(\Delta_2 x^2\right)^2.$$

Problem 5.3. Find

$$\Delta s^2 = \left(x^1\right)^2\left(\Delta_1 x^1\right)^2 + \left(\Delta_2 x^2\right)^2.$$

Problem 5.4. Find the Riemannian curvature

$$\Delta s^2 = \left(\Delta_1 x^1\right)^2 + 2x^1 x^2\left(\Delta_2 x^2\right)^2.$$

Problem 5.5. Find the Weyl tensor for the metric

$$\Delta s^2 = \left(\Delta_1 x^1 0^2 - x^1\left(\Delta_2 x^2\right)^2 + \left(x^3\right)^2\left(\Delta_3 x^3\right)^2.$$

Problem 5.6. Determine if the following metric is flat:

$$\Delta s^2 = \left(\Delta_1 x^1\right)^2 - \left(x^1\right)^2\left(\Delta_2 x^2\right)^2.$$

Problem 5.7. Find conditions under which

$$G_{lmn} = 0.$$

Problem 5.8. If Γ_i^j is the Einstein tensor, check if

$$\Gamma_{i,j}^j = 0.$$

Problem 5.9. If

$$R_{ij}g_{kl} - R_{il}g_{jk} + R_{jk}g_{il} - R_{kl}g_{ij} = 0,$$

check if the space is an Einstein space.

Problem 5.10. If

$$gR_{ij} = -g_{ij}R_{1212} \quad \text{and} \quad gR = -2R_{1212},$$

check if the space is an Einstein space.

6 Relativistic kinematics and dynamics

In this chapter, we present some applications of tensor calculus in relativistic dynamics and kinematics. Lorentz transformations are derived on arbitrary time scales. Velocity and acceleration vectors are defined and developed. Lagrange equations are derived. Conservation laws for the energy momentum vector and angular momentum tensor are obtained.

6.1 The principle of relativity

For the description of the motion of particles, it is necessary to have a coordinate system, which we attach with a clock, called a system of reference. The coordinate system is used to demonstrate the positions of particles in space, and the clock is used to measure the times at which the positions of particles in space are measured. There are systems of reference where the free motion of a particle is such that the particle has a constant velocity. By a free motion we mean the motion of a particle that is not acted by any external forces. Such systems of reference are called inertial systems. If two systems of reference are translated with constant velocity with respect to each other and one of them is an inertial system, then the other system of reference is also an inertial system. Now we are ready to write down the statement of the special principle of relativity: all laws of nature are equal in all inertial systems of reference. In other words, the equations that express the laws of nature are invariant with respect to the transformations of spatial coordinates and time from one inertial system of reference to another.

An event on nature is determined by four coordinates. These include the three space coordinates of the position where the event has taken place and the time coordinate when the event takes place. Each event is represented by a point, called the world point of the event with the coordinates

$$x^m = \left(x^0, x^1, x^2, x^3\right),\tag{6.1}$$

where the first coordinate is a time coordinate defined by

$$x^0 = ct, \quad t \in \mathbb{T},$$

and has the dimension of length. Here c is the speed of light. The other coordinates are the spatial Descartes coordinates

$$x^\alpha = \left(x^1, x^2, x^3\right) \in \mathbb{T}_1 \times \mathbb{T}_2 \times \mathbb{T}_3.\tag{6.2}$$

From now on, we will use Latin indices for the coordinates of the four-dimensional space time (6.1), and Greek indices for the usual three-dimensional spatial coordinates.

https://doi.org/10.1515/9783112228494-006

The set of all world points constitutes a four-dimensional manifold called the world. To each particle in such a world, there corresponds a line called the world line.

Now we consider two systems of reference denoted by K and K', which move with respect to each other along the common x-axis with some constant velocity V. The times of the considered systems are denoted by $t \in \mathbb{T}$ and $t' \in \mathbb{T}$, respectively. Let one event consist of a signal sent at time t_1 with constant velocity equal to the speed of light from the point with coordinates $(x_1, y_1, z_1) \in \mathbb{T}_1 \times \mathbb{T}_2 \times \mathbb{T}_3$ in the coordinate system K. Let the other event consist of the same signal being received at time t_2 at the point with coordinates (x_2, y_2, z_2) in the coordinate system K.

In the coordinate system K, the coordinates of these two events are related by the equation

$$c^2(t_2 - t_1)^2 - (x_2 - x_1)^2 - (y_2 - y_1)^2 - (z_2 - z_1)^2 = 0.$$

Because of the principle of invariance of the speed of light, in the coordinate system K' the coordinates of these two events are related by

$$c^2(t_2' - t_1')^2 - (x_2' - x_1')^2 - (y_2' - y_1')^2 - (z_2' - z_1')^2 = 0.$$

If (t_1, x_1, y_1, z_1) and (t_2, x_2, y_2, z_2) are the coordinates of any two events, then the quantity

$$s_{12} = \left(c^2(t_2 - t_1)^2 - (x_2 - x_1)^2 - (y_2 - y_1)^2 - (z_2 - z_1)^2 \right)^{\frac{1}{2}} \tag{6.3}$$

is called the interval between these two events. Analogously to (6.3), the square of the interval between two infinitesimally close events is determined by the metric

$$\begin{aligned}
\Delta s^2 &= c^2 \Delta t^2 - \Delta x^2 - \Delta y^2 - \Delta z^2 \\
&= g_{mn} \Delta x^m \Delta x^n, \quad m, n \in \{0, 1, 2, 3\},
\end{aligned} \tag{6.4}$$

and the contravariant space time coordinates in this four-dimensional metric space are given by

$$\begin{aligned}
x^m &= (x^0, x^1, x^2, x^3) \\
&= (ct, x, y, z).
\end{aligned}$$

The components of the covariant metric tensor g_{mn} are given by the matrix

$$(g_{mn}) = \begin{pmatrix} 1 & 0 & 0 & 0 \\ 0 & -1 & 0 & 0 \\ 0 & 0 & -1 & 0 \\ 0 & 0 & 0 & -1 \end{pmatrix}. \tag{6.5}$$

By (6.5) we conclude that the world is a four-dimensional pseudo-Euclidean metric space. We have

$$g = \det(g_{mn})$$
$$= -1.$$

The matrix of cofactors of the matrix (6.5) is

$$(G^{mn}) = \begin{pmatrix} -1 & 0 & 0 & 0 \\ 0 & 1 & 0 & 0 \\ 0 & 0 & 1 & 0 \\ 0 & 0 & 0 & 1 \end{pmatrix}.$$

Then

$$(g^{mn}) = \frac{(G^{mn})}{g}$$
$$= -\begin{pmatrix} -1 & 0 & 0 & 0 \\ 0 & 1 & 0 & 0 \\ 0 & 0 & 1 & 0 \\ 0 & 0 & 0 & 1 \end{pmatrix}$$
$$= \begin{pmatrix} 1 & 0 & 0 & 0 \\ 0 & -1 & 0 & 0 \\ 0 & 0 & -1 & 0 \\ 0 & 0 & 0 & -1 \end{pmatrix}. \tag{6.6}$$

By (6.5) and (6.6) we find

$$g^{mk} g_{kn} = \delta_n^m, \quad k, m, n \in \{0, 1, 2, 3\}.$$

The covariant space time coordinates in this four-dimensional metric space are given by

$$x_m = g_{mn} x^n$$
$$= (ct, -x, -y, -z) \tag{6.7}$$

and

$$x_m x^m = g_{mn} x^m x^n$$
$$= c^2 t^2 - x^2 - y^2 - z^2. \tag{6.8}$$

Using (6.7) and (6.8) and the principle of invariance of the speed of light, we conclude that an interval that is equal to zero in one inertial system of reference is also equal to zero in any other inertial system of reference. The quantities Δs and $\Delta s'$ are two infinitesimal quantities, and then we may write

$$\Delta s = a \Delta s', \tag{6.9}$$

where a is a parameter that may depend only on the absolute value of the relative velocity V of the two inertial systems of reference K and K'. As we may write (6.9), we may also write

$$\Delta s' = a\Delta s.$$

From this and (6.9) we find

$$a^2 = 1,$$

or

$$a = \pm 1.$$

From the special case of the identity transformation with

$$\Delta s = \Delta s'$$

we conclude that

$$a = 1.$$

6.2 Lorenz transformations

If we know the coordinates of a certain event

$$x^m = (ct', x', y', z')$$

in some inertial system of reference K', then we need expressions for the coordinates of that event

$$z^m = (ct, x, y, z)$$

in some other inertial system of reference K. As x^m is a contravariant vector, it is transformed according to the transformation law

$$
z^m = \frac{\partial z^m}{\Delta_n x^n} x^n \\
= \Lambda_n^m x^n.
$$

(6.10)

The last transformation is a linear transformation with the coefficients Λ_n^m independent of coordinates. The system Λ_n^m is a mixed second-order tensor in the pseudo-Euclidean metric space by metric (6.9), since it is determined with respect to the linear transformations (6.10).

To calculate the components of the tensor Λ_n^m, we introduce an imaginary-type coordinate system

$$\tau = ict, \quad i = \sqrt{-1},$$

and an imaginary metric

$$\Delta q = i\Delta s.$$

Thus

$$\Delta \tau = ic\Delta t,$$
$$\Delta \tau^2 = -c^2 \Delta t^2,$$

and the metric of the space becomes

$$\Delta q^2 = -\Delta s^2$$
$$= -(-\Delta \tau^2 - \Delta x^2 - \Delta y^2 - \Delta z^2)$$
$$= \Delta \tau^2 + \Delta x^2 + \Delta y^2 + \Delta z^2,$$

and we define the four-dimensional Descartes coordinates. The only suitable candidate for transformation (6.10) is a time scale rotation of the four-dimensional coordinate system given by

$$x = x' \cos_1(\psi, \psi_0) - \tau' \sin_1(\psi, \psi_0),$$
$$\tau = x' \sin_1(\psi, \psi_0) + \tau' \sin_1(\psi, \psi_0), \quad \psi, \psi_0 \in \mathbb{T}. \tag{6.11}$$

The coordinate system K' is moving with respect to the coordinate system K with constant velocity V. If we consider the motion of the origin of the coordinate system K', then we have $x' = 0$, and (6.12) takes the form

$$x = -\tau' \sin_1(\psi, \psi_0),$$
$$\tau = \tau' \sin_1(\psi, \psi_0). \tag{6.12}$$

Multiplying the first equation of (6.12) by $\cos_1(\psi, \psi_0)$ and the second equation of (6.12) by $\sin_1(\psi, \psi_0)$, we get

$$\cos_1(\psi, \psi_0) = -\tau' \sin_1(\psi, \psi_0) \cos_1(\psi, \psi_0),$$
$$\sin_1(\psi, \psi_0)\tau = \tau' \sin_1(\psi, \psi_0) \cos_1(\psi, \psi_0).$$

Adding the last two equations, we obtain

$$\cos_1(\psi, \psi_0)x + \sin_1(\psi, \psi_0)\tau = 0,$$

whereupon

$$-\frac{x}{\tau} = \tan_1(\psi, \psi_0)$$
$$= i\frac{V}{c}.$$

Because

$$\left(\cos_1(\psi, \psi_0)\right)^2 + \left(\sin_1(\psi, \psi_0)\right)^2 = e_\mu(\psi, \psi_0),$$

we get

$$\left(\cos_1(\psi, \psi_0)\right)^2 \left(1 + \left(\tan_1(\psi, \psi_0)\right)^2\right) = e_\mu(\psi, \psi_0),$$

or

$$\cos_1(\psi, \psi_0) = \sqrt{\frac{e_\mu(\psi, \psi_0)}{1 + (\tan_1(\psi, \psi_0))^2}},$$

and

$$\sin_1(\psi, \psi_0) = \tan_1(\psi, \psi_0)\cos_1(\psi, \psi_0)$$
$$= \tan_1(\psi, \psi_0)\sqrt{\frac{e_\mu(\psi, \psi_0)}{1 + (\tan_1(\psi, \psi_0))^2}}.$$

Thus

$$\cos_1(\psi, \psi_0) = \sqrt{\frac{e_\mu(\psi, \psi_0)}{1 - \frac{V^2}{c^2}}},$$
$$\sin_1(\psi, \psi_0) = i\frac{V}{c}\sqrt{\frac{e_\mu(\psi, \psi_0)}{1 - \frac{V^2}{c^2}}}.$$

Substituting this into (6.11), we find

$$x = x' \cos_1(\psi, \psi_0) - \tau' \sin_1(\psi, \psi_0)$$
$$= x'\sqrt{\frac{e_\mu(\psi, \psi_0)}{1 - \frac{V^2}{c^2}}} - \tau' i\frac{V}{c}\sqrt{\frac{e_\mu(\psi, \psi_0)}{1 - \frac{V^2}{c^2}}}$$
$$= \left(x' - i\tau'\frac{V}{c}\right)\sqrt{\frac{e_\mu(\psi, \psi_0)}{1 - \frac{V^2}{c^2}}}$$
$$= (x' + Vt')\sqrt{\frac{e_\mu(\psi, \psi_0)}{1 - \frac{V^2}{c^2}}},$$

$$\tau = x^1 \sin_1(\psi, \psi_0) + \tau' \cos_1(\psi, \psi_0)$$

$$= ix' \frac{V}{c} \sqrt{\frac{e_\mu(\psi, \psi_0)}{1 - \frac{V^2}{c^2}}} + \tau' \sqrt{\frac{e_\mu(\psi, \psi_0)}{1 - \frac{V^2}{c^2}}}$$

$$= ix' \frac{V}{c} \sqrt{\frac{e_\mu(\psi, \psi_0)}{1 - \frac{V^2}{c^2}}} + ict' \sqrt{\frac{e_\mu(\psi, \psi_0)}{1 - \frac{V^2}{c^2}}}$$

$$= ic\left(t' + \frac{V}{c^2}x'\right) \sqrt{\frac{e_\mu(\psi, \psi_0)}{1 - \frac{V^2}{c^2}}},$$

$$t = \frac{1}{ic}\tau$$

$$= \left(t' + \frac{V}{c^2}x'\right) \sqrt{\frac{e_\mu(\psi, \psi_0)}{1 - \frac{V^2}{c^2}}},$$

$$y = y',$$

$$z = z'.$$

Definition 6.1. The equations

$$x = (x' + Vt') \sqrt{\frac{e_\mu(\psi, \psi_0)}{1 - \frac{V^2}{c^2}}},$$

$$\tau = ic\left(t' + \frac{V}{c^2}x'\right) \sqrt{\frac{e_\mu(\psi, \psi_0)}{1 - \frac{V^2}{c^2}}},$$

$$t = \left(t' + \frac{V}{c^2}x'\right) \sqrt{\frac{e_\mu(\psi, \psi_0)}{1 - \frac{V^2}{c^2}}},$$

$$y = y',$$

$$z = z'$$

are called the Lorenz transformations.

In the special case, when $V \ll c$, the Lorentz transformations take the form

$$x = (x' + Vt') \sqrt{e_\mu(\psi, \psi_0)},$$

$$t = t',$$

$$y = y',$$

$$z = z'. \tag{6.13}$$

Definition 6.2. Transformations (6.13) are called the Galilei transformations.

Define

$$(\Lambda_n^m)(V) = \begin{pmatrix} \sqrt{\dfrac{e_\mu(\psi,\psi_0)}{1-\frac{V^2}{c^2}}} & \dfrac{V}{c}\sqrt{\dfrac{e_\mu(\psi,\psi_0)}{1-\frac{V^2}{c^2}}} & 0 & 0 \\[4mm] \dfrac{V}{c}\sqrt{\dfrac{e_\mu(\psi,\psi_0)}{1-\frac{V^2}{c^2}}} & \sqrt{\dfrac{e_\mu(\psi,\psi_0)}{1-\frac{V^2}{c^2}}} & 0 & 0 \\[4mm] 0 & 0 & 1 & 0 \\[2mm] 0 & 0 & 0 & 1 \end{pmatrix}.$$

We have

$$\det(\Lambda_n^m)(V) = \sqrt{\dfrac{e_\mu(\psi,\psi_0)}{1-\frac{V^2}{c^2}}}\,\det\begin{pmatrix} \sqrt{\dfrac{e_\mu(\psi,\psi_0)}{1-\frac{V^2}{c^2}}} & 0 & 0 \\[4mm] 0 & 1 & 0 \\[2mm] 0 & 0 & 1 \end{pmatrix}$$

$$-\,\dfrac{V}{c}\sqrt{\dfrac{e_\mu(\psi,\psi_0)}{1-\frac{V^2}{c^2}}}\,\det\begin{pmatrix} \dfrac{V}{c}\sqrt{\dfrac{e_\mu(\psi,\psi_0)}{1-\frac{V^2}{c^2}}} & 0 & 0 \\[4mm] 0 & 1 & 0 \\[2mm] 0 & 0 & 1 \end{pmatrix}$$

$$= \dfrac{e_\mu(\psi,\psi_0)}{1-\frac{V^2}{c^2}} - \dfrac{V^2}{c}\dfrac{e_\mu(\psi,\psi_0)}{1-\frac{V^2}{c^2}}$$

$$= \left(1-\dfrac{V^2}{c^2}\right)\dfrac{e_\mu(\psi,\psi_0)}{1-\frac{V^2}{c^2}}$$

$$= e_\mu(\psi,\psi_0),$$

and the cofactors of $(\Lambda_n^m)(V)$ are as follows:

$$(\Lambda_n^m)_{11}(V) = \det\begin{pmatrix} \sqrt{\dfrac{e_\mu(\psi,\psi_0)}{1-\frac{V^2}{c^2}}} & 0 & 0 \\[4mm] 0 & 1 & 0 \\[2mm] 0 & 0 & 1 \end{pmatrix}$$

$$= \sqrt{\dfrac{e_\mu(\psi,\psi_0)}{1-\frac{V^2}{c^2}}},$$

$$(\Lambda_n^m)_{12}(V) = (\Lambda_n^m)_{21}(V)$$

$$= -\det\begin{pmatrix} \dfrac{V}{c}\sqrt{\dfrac{e_\mu(\psi,\psi_0)}{1-\frac{V^2}{c^2}}} & 0 & 0 \\[4mm] 0 & 1 & 0 \\[2mm] 0 & 0 & 1 \end{pmatrix}$$

$$= -\frac{V}{c}\sqrt{\frac{e_\mu(\psi,\psi_0)}{1-\frac{V^2}{c^2}}},$$

$$(\Lambda_n^m)_{13}(V) = (\Lambda_n^m)_{31}(V)$$

$$= \det\begin{pmatrix} \frac{V}{c}\sqrt{\frac{e_\mu(\psi,\psi_0)}{1-\frac{V^2}{c^2}}} & \sqrt{\frac{e_\mu(\psi,\psi_0)}{1-\frac{V^2}{c^2}}} & 0 \\ 0 & 0 & 0 \\ 0 & 0 & 1 \end{pmatrix}$$

$$= 0,$$

$$(\Lambda_n^m)_{14}(V) = -\det\begin{pmatrix} \frac{V}{c}\sqrt{\frac{e_\mu(\psi,\psi_0)}{1-\frac{V^2}{c^2}}} & \sqrt{\frac{e_\mu(\psi,\psi_0)}{1-\frac{V^2}{c^2}}} & 0 \\ 0 & 0 & 1 \\ 0 & 0 & 0 \end{pmatrix}$$

$$= 0,$$

$$(\Lambda_n^m)_{22}(V) = \det\begin{pmatrix} \sqrt{\frac{e_\mu(\psi,\psi_0)}{1-\frac{V^2}{c^2}}} & 0 & 0 \\ 0 & 1 & 0 \\ 0 & 0 & 1 \end{pmatrix}$$

$$= \sqrt{\frac{e_\mu(\psi,\psi_0)}{1-\frac{V^2}{c^2}}},$$

$$(\Lambda_n^m)_{23}(V) = (\Lambda_n^m) - 32(V)$$

$$= -\det\begin{pmatrix} \frac{V}{c}\sqrt{\frac{e_\mu(\psi,\psi_0)}{1-\frac{V^2}{c^2}}} & \frac{V}{c}\sqrt{\frac{e_\mu(\psi,\psi_0)}{1-\frac{V^2}{c^2}}} & 0 \\ 0 & 0 & 0 \\ 0 & 0 & 1 \end{pmatrix}$$

$$= 0,$$

$$(\Lambda_n^m)_{24}(V) = (\Lambda_n^m)_{42}(V)$$

$$= \det\begin{pmatrix} \sqrt{\frac{e_\mu(\psi,\psi_0)}{1-\frac{V^2}{c^2}}} & \frac{V}{c}\sqrt{\frac{e_\mu(\psi,\psi_0)}{1-\frac{V^2}{c^2}}} & 0 \\ 0 & 0 & 1 \\ 0 & 0 & 0 \end{pmatrix}$$

$$= 0,$$

$$(\Lambda_n^m)_{33}(V) = \det\begin{pmatrix} \sqrt{\frac{e_\mu(\psi,\psi_0)}{1-\frac{V^2}{c^2}}} & \frac{V}{c}\sqrt{\frac{e_\mu(\psi,\psi_0)}{1-\frac{V^2}{c^2}}} & 0 \\ \frac{V}{c}\sqrt{\frac{e_\mu(\psi,\psi_0)}{1-\frac{V^2}{c^2}}} & \sqrt{\frac{e_\mu(\psi,\psi_0)}{1-\frac{V^2}{c^2}}} & 0 \\ 0 & 0 & 1 \end{pmatrix}$$

$$= \frac{e_\mu(\psi,\psi_0)}{1-\frac{V^2}{c^2}} - \frac{V^2}{c^2}\frac{e_\mu(\psi,\psi_0)}{1-\frac{V^2}{c^2}}$$

$$= \left(1 - \frac{V^2}{c^2}\right)\frac{e_\mu(\psi, \psi_0)}{1 - \frac{V^2}{c^2}}$$

$$= e_\mu(\psi, \psi_0),$$

$$(\Lambda_n^m)_{34}(V) = (\Lambda_n^m)_{43}(V)$$

$$= -\det\begin{pmatrix} \sqrt{\dfrac{e_\mu(\psi,\psi_0)}{1-\frac{V^2}{c^2}}} & \dfrac{V}{c}\sqrt{\dfrac{e_\mu(\psi,\psi_0)}{1-\frac{V^2}{c^2}}} & 0 \\[3ex] \dfrac{V}{c}\sqrt{\dfrac{e_\mu(\psi,\psi_0)}{1-\frac{V^2}{c^2}}} & \sqrt{\dfrac{e_\mu(\psi,\psi_0)}{1-\frac{V^2}{c^2}}} & 0 \\[3ex] 0 & 0 & 0 \end{pmatrix}$$

$$= 0,$$

$$(\Lambda_n^m)_{44}(V) = \det\begin{pmatrix} \sqrt{\dfrac{e_\mu(\psi,\psi_0)}{1-\frac{V^2}{c^2}}} & \dfrac{V}{c}\sqrt{\dfrac{e_\mu(\psi,\psi_0)}{1-\frac{V^2}{c^2}}} & 0 \\[3ex] \dfrac{V}{c}\sqrt{\dfrac{e_\mu(\psi,\psi_0)}{1-\frac{V^2}{c^2}}} & \sqrt{\dfrac{e_\mu(\psi,\psi_0)}{1-\frac{V^2}{c^2}}} & 0 \\[3ex] 0 & 0 & 1 \end{pmatrix}$$

$$= \frac{e_\mu(\psi, \psi_0)}{1 - \frac{V^2}{c^2}} - \frac{V^2}{c^2}\frac{e_\mu(\psi, \psi_0)}{1 - \frac{V^2}{c^2}}$$

$$= \left(1 - \frac{V^2}{c^2}\right)\frac{e_\mu(\psi, \psi_0)}{1 - \frac{V^2}{c^2}}$$

$$= e_\mu(\psi, \psi_0).$$

Consequently,

$$(\Lambda_n^m)^{-1}(V) = \frac{1}{e_\mu(\psi, \psi_0)}\begin{pmatrix} \sqrt{\dfrac{e_\mu(\psi,\psi_0)}{1-\frac{V^2}{c^2}}} & -\dfrac{V}{c}\sqrt{\dfrac{e_\mu(\psi,\psi_0)}{1-\frac{V^2}{c^2}}} & 0 & 0 \\[3ex] -\dfrac{V}{c}\sqrt{\dfrac{e_\mu(\psi,\psi_0)}{1-\frac{V^2}{c^2}}} & \sqrt{\dfrac{e_\mu(\psi,\psi_0)}{1-\frac{V^2}{c^2}}} & 0 & 0 \\[3ex] 0 & 0 & e_\mu(\psi, \psi_0) & 0 \\[2ex] 0 & 0 & 0 & e_\mu(\psi, \psi_0) \end{pmatrix}$$

$$= \begin{pmatrix} \dfrac{1}{\sqrt{e_\mu(\psi,\psi_0)(1-\frac{V^2}{c^2})}} & -\dfrac{\frac{V}{c}}{\sqrt{e_\mu(\psi,\psi_0)(1-\frac{V^2}{c^2})}} & 0 & 0 \\[3ex] -\dfrac{\frac{V}{c}}{\sqrt{e_\mu(\psi,\psi_0)(1-\frac{V^2}{c^2})}} & \dfrac{1}{\sqrt{e_\mu(\psi,\psi_0)(1-\frac{V^2}{c^2})}} & 0 & 0 \\[3ex] 0 & 0 & 1 & 0 \\[2ex] 0 & 0 & 0 & 1 \end{pmatrix}.$$

Observe that

$$\left(\Lambda_n^m\right)^{-1}(V) = \frac{1}{e_\mu(\psi, \psi_0)}\left(\Lambda_n^m\right)(-V).$$

Thus we may write

$$\frac{\partial x^m}{\Delta_n z^n} = \left(\Lambda_n^m\right)^{-1}(V)$$

$$= \frac{1}{e_\mu(\psi, \psi_0)}\left(\Lambda_n^m\right)(-V).$$

6.3 Velocity and acceleration vectors

In the special theory of relativity, the time differential Δt is not a scalar invariant, and the usual definition of the three-dimensional velocity

$$v^\alpha = \frac{\Delta x^\alpha}{\Delta t}, \quad \alpha \in \{1, 2, 3\},$$

is less useful, since it does not behave as a vector with respect to the transformations from one inertial system to another inertial system. Thus we introduce a four-velocity as a contravariant vector

$$u^m = \frac{\Delta x^m}{\Delta s}, \quad m \in \{0, 1, 2, 3\}.$$

On the other hand, by definition we have

$$\Delta s^2 = g_{mn}\Delta x^m \Delta x^n$$
$$= c^2 \Delta t^2 + g_{\alpha\beta}\Delta x^\alpha \Delta x^\beta$$
$$= c^2 \Delta t^2 - \Delta x^2 - \Delta y^2 - \Delta z^2,$$

and the square of the intensity of the three-dimensional velocity is given by

$$v^2 = \frac{\Delta x^2 + \Delta y^2 + \Delta z^2}{\Delta t^2}$$
$$= -g_{\alpha\beta}v^\alpha v^\beta.$$

From here we have

$$\Delta s^2 = c^2 \Delta t^2 - v^2 \Delta t^2$$
$$= c^2\left(1 - \frac{v^2}{c^2}\right)\Delta t^2$$
$$= c^2\left(1 + \frac{g_{\alpha\beta}v^\alpha v^\beta}{c^2}\right),$$

and then

$$u^m = \frac{\Delta x^m}{c\sqrt{1-\frac{v^2}{c^2}}\,\Delta t}, \qquad m \in \{0,1,2,3\}.$$

The temporal zeroth components of the four-velocity is then given by

$$u^0 = \frac{1}{\sqrt{1-\frac{v^2}{c^2}}},$$

and the three spatial components of the four-velocity are given by

$$u^a = \frac{v^a}{c\sqrt{1-\frac{v^2}{c^2}}}, \qquad a \in \{1,2,3\}.$$

Then

$$\begin{aligned}
u_m u^m &= g_{mn} u^m u^n \\
&= g_{mn} \frac{\Delta x^m \Delta x^n}{\Delta s^2} \\
&= \frac{\Delta s^2}{\Delta s^2} \\
&= 1.
\end{aligned}$$

Definition 6.3. The four-acceleration of a particle is defined as a contravariant vector

$$w^m = \frac{\Delta u^m}{\Delta s}, \qquad m \in \{0,1,2,3\}.$$

For the acceleration of a particle, we get

$$\begin{aligned}
w^m &= \frac{\Delta u^m}{\Delta s} \\
&= \frac{\Delta u^m}{c\sqrt{1-\frac{v^2}{c^2}}\,\Delta t}.
\end{aligned}$$

The temporal zeroth component of the four-acceleration is given by

$$\begin{aligned}
w^0 &= \frac{1}{c\sqrt{1-\frac{v^2}{c^2}}} \frac{\Delta u^0}{\Delta t} \\
&= \frac{1}{c\sqrt{1-\frac{v^2}{c^2}}} \frac{\Delta}{\Delta t}\left(\frac{1}{\sqrt{1-\frac{v^2}{c^2}}}\right),
\end{aligned}$$

and the three spatial components of the four-acceleration are given by

$$w^\alpha = \frac{1}{c\sqrt{1-\frac{v^2}{c^2}}}\frac{\Delta u^\alpha}{\Delta t}$$

$$= \frac{1}{c\sqrt{1-\frac{v^2}{c^2}}}\frac{\Delta}{\Delta t}\left(\frac{v^\alpha}{c\sqrt{1-\frac{v^2}{c^2}}}\right).$$

Since

$$g_{mn}u^m u^n = 1,$$

we get

$$0 = \frac{\Delta}{\Delta s}(g_{mn}u^m u^n)$$

$$= g_{mn}\frac{\Delta}{\Delta s}(u^m u^n)$$

$$= g_{mn}\frac{\Delta u^m}{\Delta s}u^n + g_{mn}u^{m\sigma}\frac{\Delta u^n}{\Delta s}$$

$$= g_{mn}\frac{\Delta u^m}{\Delta s}u^{n\sigma} + g_{mn}u^m\frac{\Delta u^n}{\Delta s}.$$

Thus

$$g_{mn}w^m u^n + g_{mn}u^{m\sigma}w^n = 0,$$

and

$$g_{mn}w^m u^{n\sigma} + g_{mn}u^m w^n = 0.$$

Therefore, in the general case of an arbitrary time scale, the vector of four-acceleration is not orthogonal to the vector of four-velocity.

6.4 Lagrange equations

Let us consider a free particle, i. e., a particle that is not under the influence of any forces. The equations describing the motion of this particle are obtained by using the variational principle. The action integral is given by

$$I = \int_{s_A}^{s_B} L\left(x^{n\sigma}, \frac{\Delta x^n}{\Delta s}\right)\Delta s$$

$$= \int_{s_A}^{s_B} L(x^{n\sigma}, u^n)\Delta s,$$

where u_n is the four-velocity of the particle. The equations of motion of the particle are obtained using the variational principle, i. e., the condition

$$\delta I = 0.$$

We have

$$\delta I = \delta\left(\int_{s_A}^{s_B} L(x^{n\sigma}, u^n)\Delta s\right)$$

$$= \int_{s_A}^{s_B} \delta(L(x^{n\sigma}, u^n))\Delta s$$

$$= \int_{s_A}^{s_B}\left(\frac{\partial L}{\partial x^{n\sigma}}(x^{n\sigma}, u^n)\delta(x^{n\sigma}) + \frac{\partial L}{\partial u^n}\delta u^n\right)\Delta s$$

$$= \int_{s_A}^{s_B}\left(\frac{\partial L}{\partial x^{n\sigma}}(x^{n\sigma}, u^n)\delta(x^{n\sigma}) + \frac{\partial L}{\partial(\frac{\Delta x^n}{\Delta s})}\delta\left(\frac{\Delta x^n}{\Delta s}\right)\right)\Delta s$$

$$= \int_{s_A}^{s_B}\left(\frac{\partial L}{\partial x^{n\sigma}}(x^{n\sigma}, u^n)\delta(x^{n\sigma}) + \frac{\partial L}{\partial(\frac{\Delta x^n}{\Delta s})}\frac{\Delta}{\Delta s}(\delta x^n)\right)\Delta s$$

$$= \int_{s_A}^{s_B}\left(\frac{\partial L}{\partial x^{n\sigma}}(x^{n\sigma}, u^n)(\delta x^n)^\sigma + \frac{\Delta}{\Delta s}\left(\frac{\partial L}{\partial(\frac{\Delta x^n}{\Delta s})}\delta x^n\right) - \frac{\Delta}{\Delta s}\left(\frac{\partial L}{\partial(\frac{\Delta x^n}{\Delta s})}\right)(\delta x^n)^\sigma\right)\Delta s$$

$$= \int_{s_A}^{s_B}\left(\frac{\partial L}{\partial x^{n\sigma}}(x^{n\sigma}, u^n)(\delta x^n)^\sigma - \frac{\Delta}{\Delta s}\left(\frac{\partial L}{\partial(\frac{\Delta x^n}{\Delta s})}\right)(\delta x^n)^\sigma\right)\Delta s$$

$$+ \int_{s_A}^{s_B}\frac{\Delta}{\Delta s}\left(\frac{\partial L}{\partial(\frac{\Delta x^n}{\Delta s})}\delta x^n\right)\Delta s$$

$$= \int_{s_A}^{s_B}\left(\frac{\partial L}{\partial x^{n\sigma}}(x^{n\sigma}, u^n) - \frac{\Delta}{\Delta s}\left(\frac{\partial L}{\partial(\frac{\Delta x^n}{\Delta s})}\right)\right)(\delta x^n)^\sigma \Delta s,$$

where we have used that

$$\int_{s_A}^{s_B}\frac{\Delta}{\Delta s}\left(\frac{\partial L}{\partial(\frac{\Delta x^n}{\Delta s})}\delta x^n\right)\Delta s = \frac{\partial L}{\partial(\frac{\Delta x^n}{\Delta s})}\delta x^n\Big|_{s=s_A}^{s_B}$$

$$= 0.$$

Thus the Lagrange equations are

$$\frac{\partial L}{\partial x^{n\sigma}} - \frac{\Delta}{\Delta s}\left(\frac{\partial L}{\partial\left(\frac{\Delta x^n}{\Delta s}\right)}\right) = 0, \quad n \in \{0,1,2,3\},$$

or

$$\frac{\partial L}{\partial x^{n\sigma}} - \frac{\Delta}{\Delta s}\left(\frac{\partial L}{\partial u^n}\right) = 0, \quad n \in \{0,1,2,3\}. \tag{6.14}$$

Next, we define the form of the Lagrangian function for a free particle. Equations (6.14) are equal in all inertial systems of reference, and the action integral is a scalar invariant with respect to the Lorentz transformations. Therefore the Lagrange function L itself is also a scalar invariant. Because of the homogeneity of the four-dimensional space time in the inertial frames of reference, the Lagrange function cannot be a function of space time coordinates. Then we have

$$L = L(u^n).$$

The only invariant that can be created using the four-velocity vector u^n is given by

$$u_n u^n = g_{kn} u^k u^n$$
$$= 1, \quad k, n \in \{0,1,2,3\}.$$

Then the action integral of a free particle is proportional to the arc length in the four-dimensional space time manifold, i. e.,

$$I = -mc \int_{s_A}^{s_B} \Delta s$$
$$= -mc \int_{s_A}^{s_B} \sqrt{g_{kn} \frac{\Delta x^k}{\Delta s} \frac{\Delta x^n}{\Delta s}} \, \Delta s.$$

Here m is a mass parameter of the free particle. Hence

$$L = -mc\sqrt{g_{kn} u^k u^n}.$$

Then

$$\frac{\partial L}{\partial u^n} = \frac{\partial}{\partial u^n}\left(-mc\sqrt{g_{kn} u^k u^n}\right)$$
$$= -\frac{1}{2}mc(g_{kj} u^k u^j)^{-\frac{1}{2}}(g_{kn} u^k + g_{nj} u^j).$$

Now using that

$$g_{kj} u^k u^j = 1,$$

we arrive at

$$\frac{\partial L}{\partial u^n} = -\frac{1}{2}mc(2g_{kn}u^k)$$
$$= -mcg_{kn}u^k$$
$$= -mcu_n$$

and

$$-\frac{\Delta}{\Delta s}\left(\frac{\partial L}{\partial u^n}\right) = mc\frac{\Delta u_n}{\Delta s}$$
$$= \frac{\Delta}{\Delta s}(mcu_n).$$

Therefore we obtain

$$mc\frac{\Delta u}{\Delta s} = \frac{\Delta}{\Delta s}(mcu_n)$$
$$= 0.$$

The equations of motion of a free particle are given in the form

$$w^n = 0, \quad n \in \{0,1,2,3\}.$$

Now we will investigate the nonrelativistic limit of the action integral

$$I = -mc\int_{s_A}^{s_B}\Delta s$$
$$= -mc^2\int_{t_A}^{t_B}\sqrt{1 - \frac{v^2}{c^2}}\,\Delta t$$
$$= \int_{t_A}^{t_B}L_T(x^{\alpha\sigma}, v^\alpha)\Delta t,$$

and the Lagrangian with respect to the nonscalar time variable t is given by

$$L_T(x^{\alpha\sigma}, v^\alpha) = -mc^2\sqrt{1 - \frac{v^2}{c^2}}$$
$$= -mc^2\sqrt{1 + \frac{g_{\alpha\beta}v^\alpha v^\beta}{c^2}}.$$

In the case of low velocities, $V \ll c$, we have

$$\sqrt{1 - \frac{v^2}{c^2}} \approx 1 - \frac{v^2}{2c^2},$$

and then

$$I \approx \int_{t_A}^{t_B} \left(-mc^2 + \frac{1}{2}mv^2 \right) \Delta t,$$

so that the Lagrangian can be written as

$$L(x^{a\sigma}, v^a) \approx \frac{1}{2}mv^2 - mc^2.$$

6.5 Energy momentum vector

In nonrelativistic mechanics, there are a number of constants of motion. These include the energy and momentum of a particle. We start with the action integral

$$I = \int_{t_A}^{t_B} L_T(x^{a\sigma}, v^a) \Delta t$$

and the Lagrange equations with respect to the nonscalar time variable t as the parameters are given by

$$\frac{\partial L_T}{\partial x^{a\sigma}} - \frac{\Delta}{\Delta t}\left(\frac{\partial L_T}{\partial v^a} \right) = 0, \quad \alpha \in \{1, 2, 3\},$$

whereupon

$$\frac{\partial L_T}{\partial x^{a\sigma}} = \frac{\Delta}{\Delta t}\left(\frac{\partial L_T}{\partial v^a} \right), \quad \alpha \in \{1, 2, 3\}.$$

Note that

$$\begin{aligned}
\frac{\Delta L_T}{\Delta t} &= \frac{\partial L_T}{\partial x^{a\sigma}}\frac{\Delta x^{a\sigma}}{\Delta t} + \frac{\partial L_T}{\partial v^a}{}^{\sigma}\frac{\Delta v^a}{\Delta t} \\
&= \frac{\partial L_T}{\partial x^{a\sigma}}v^a + \frac{\partial L_T}{\partial v^a}{}^{\sigma}\frac{\Delta v^a}{\Delta t} \\
&= \frac{\Delta}{\Delta t}\left(\frac{\partial L_T}{\partial v^a} \right)v^a + \frac{\partial L_T}{\partial v^a}{}^{\sigma}\frac{\Delta v^a}{\Delta t} \\
&= \frac{\Delta}{\Delta t}\left(v^a\frac{\partial L_T}{\partial v^a} \right),
\end{aligned}$$

and then

$$\frac{\Delta}{\Delta t}\left(v^\alpha \frac{\partial L_T}{\partial v^\alpha} - L_T\right) = 0.$$

Thus the quantity

$$\mathcal{E} = v^\alpha \frac{\partial L_T}{\partial v^\alpha} - L_T = \text{constant} \tag{6.15}$$

is a constant of motion.

Definition 6.4. Quantity (6.15) is called the total energy of a particle.

Definition 6.5. Define the momentum of a particle by

$$\vec{p} = \frac{\partial L_T}{\partial \vec{v}}.$$

We have

$$p_\alpha = -\frac{\partial L_T}{\partial v^\alpha}$$

and

$$p^\alpha = g^{\alpha\beta} p_\beta$$
$$= -g^{\alpha\beta} \frac{\partial L_T}{\partial v^\beta},$$

where

$$\{p^\alpha\} = \{p_x, p_y, p_z\}$$

and

$$\{p_\alpha\} = \{-p_x, -p_y, -p_z\}.$$

Then for the total energy, we find

$$\mathcal{E} = -p_\alpha v^\alpha - L_T$$
$$= -g_{\alpha\beta} p^\alpha v^\beta - L_T$$
$$= \vec{p} \cdot \vec{v} - L_T.$$

By the definition of the Lagrangian L_T we can compute

$$p_\alpha = -\frac{\partial L_T}{\partial v^\alpha}$$
$$= mc^2 \frac{\partial}{\partial v^\alpha}\left(1 + \frac{g_{\alpha\beta} v^\alpha v^\beta}{c^2}\right)^{\frac{1}{2}}$$

$$= \frac{1}{2} mc^2 \left(1 + \frac{g_{\alpha\beta} v^\alpha v^\beta}{c^2} \right)^{-\frac{1}{2}} \frac{1}{c^2} g_{\alpha\beta} \frac{\partial}{\partial v^\alpha} (v^\alpha v^\beta)$$

$$= \frac{1}{2} m \left(1 + \frac{g_{\alpha\beta} v^\alpha v^\beta}{c^2} \right)^{-\frac{1}{2}} 2 g_{\alpha\beta} v^\beta$$

$$= \frac{m v_\alpha}{\sqrt{1 + \frac{g_{\alpha\beta} v^\alpha v^\beta}{c^2}}}$$

$$= \frac{m v_\alpha}{\sqrt{1 - \frac{v^2}{c^2}}}.$$

Hence, for the total energy, we find

$$\mathcal{E} = -\frac{m v_\alpha v^\alpha}{\sqrt{1 - \frac{v^2}{c^2}}} + mc^2 \sqrt{1 - \frac{v^2}{c^2}}$$

$$= \frac{m v^2}{\sqrt{1 - \frac{v^2}{c^2}}} + mc^2 \sqrt{1 - \frac{v^2}{c^2}}$$

$$= m \frac{v^2 + c^2 (1 - \frac{v^2}{c^2})}{\sqrt{1 - \frac{v^2}{c^2}}}$$

$$= m \frac{v^2 + c^2 - v^2}{\sqrt{1 - \frac{v^2}{c^2}}}$$

$$= \frac{mc^2}{\sqrt{1 - \frac{v^2}{c^2}}}.$$

The particle at rest with $v = 0$ has the so-called rest energy

$$\mathcal{E}_0 = mc^2.$$

Definition 6.6. The kinetic energy of the particle is defined by

$$\mathcal{E}_K = \mathcal{E} - \mathcal{E}_0.$$

For the kinetic energy of a particle, we obtain

$$\mathcal{E}_K = \mathcal{E} - \mathcal{E}_0$$

$$= \frac{mc^2}{\sqrt{1 - \frac{v^2}{c^2}}} - mc^2$$

$$= mc^2 \left(\frac{1}{\sqrt{1 - \frac{v^2}{c^2}}} - 1 \right).$$

For the particles moving with low velocities $v \ll c$, we get

$$\frac{1}{\sqrt{1 - \frac{v^2}{c^2}}} \approx 1 + \frac{v^2}{2c^2},$$

and then

$$\mathscr{E}_K \approx mc^2 \left(1 + \frac{v^2}{2c^2} - 1 \right)$$

$$= mc^2 \left(\frac{v^2}{2c^2} \right)$$

$$= \frac{mv^2}{2},$$

$$p_a = mv_a,$$

and

$$p^a = mv^a.$$

Now we want to define the energy and momentum as constants of motion using the covariant Lagrangian

$$L = -mc\sqrt{g_{kn}u^k u^n}.$$

In analogy with

$$p_a = -\frac{\partial L_T}{\partial v^a},$$

we define the energy momentum four-vector p_n with the covariant and contravariant components

$$p_n = -\frac{\partial L}{\partial u^n}$$

$$= mc\frac{\partial}{\partial u^n}\left(\sqrt{g_{kn}u^k u^n} \right)$$

$$= \frac{mc}{2\sqrt{g_{kn}u^k u^n}} 2g_{kn}u^k$$

$$= mcu_n$$

and

$$p^n = mcu^n.$$

Because

$$\frac{\Delta u^n}{\Delta s} = 0,$$

we get

$$\frac{\Delta p^n}{\Delta s} = mc\frac{\Delta u^n}{\Delta s}$$
$$= 0.$$

Thus, for a free particle, the four components of the energy momentum four-vector are conserved. Since

$$u^0 = \frac{1}{\sqrt{1 - \frac{v^2}{c^2}}},$$

we obtain the temporal zeroth component of the energy momentum vector, which is proportional to the energy $\mathscr{E}$ of the particle, in the form

$$p^0 = mcu^0$$
$$= \frac{mc}{\sqrt{1 - \frac{v^2}{c^2}}}$$
$$= \frac{\mathscr{E}}{c}.$$

Now using that

$$u^\alpha = \frac{v^\alpha}{c\sqrt{1 - \frac{v^2}{c^2}}},$$

we obtain three spatial components of the energy momentum four-vector, which are equal to the components of the three-dimensional momentum of the particle p^α, i. e.,

$$p^\alpha = mcv^\alpha$$
$$= \frac{mv^\alpha}{\sqrt{1 - \frac{v^2}{c^2}}}.$$

From these results we see that the energy and momentum of a particle are not two independent quantities as in the nonrelativistic mechanics. In the relativistic mechanics, they are components of the same four-vector. Analogously to the equation

$$\frac{\partial L_T}{\partial v^\alpha}v^\alpha - L_T = \text{constant},$$

the constant of motion is identically equal to zero, i. e.,

$$\frac{\partial L}{\partial u^n} u^n - L = -p^n u_n - L$$
$$= 0.$$

The energy momentum vector as a four-vector transforms with respect to the transformations from one inertial system of reference to another, according to the transformation law

$$p^k = \Lambda^k_n p'^n,$$

where p^k are the components of the energy momentum tensor in the inertial system of reference K, and p'^k are the components of the energy momentum tensor in the inertial system of reference K' moving along the common x-axis with velocity V with respect to K. Using the explicit form of the tensor Λ^k_n, i. e.,

$$(\Lambda^n_k) = \begin{pmatrix} \dfrac{1}{\sqrt{1-\frac{v^2}{c^2}}} & \dfrac{\frac{v}{c}}{\sqrt{1-\frac{v^2}{c^2}}} & 0 & 0 \\[2ex] \dfrac{\frac{v}{c}}{\sqrt{1-\frac{v^2}{c^2}}} & \dfrac{1}{\sqrt{1-\frac{v^2}{c^2}}} & 0 & 0 \\[2ex] 0 & 0 & 1 & 0 \\[1ex] 0 & 0 & 0 & 1 \end{pmatrix},$$

we get

$$\mathcal{E} = \frac{1}{\sqrt{1-\frac{v^2}{c^2}}}(\mathcal{E}' + V'p_x)$$

$$p_x = \frac{1}{\sqrt{1-\frac{v^2}{c^2}}}\left(p'_x + \frac{V}{c^2}\mathcal{E}'\right),$$

$$p_y = p'_y,$$
$$p_z = p'_z.$$

Now using that

$$u_m u^m = 1,$$

we obtain

$$p^n p_n = m^2 c^2 u^n u_n$$
$$= m^2 c^2.$$

Thus we can write

$$\frac{\mathcal{E}^2}{c^2} + p^\alpha p_\alpha = m^2 c^2$$

and

$$p^\alpha p_\alpha = g_{\alpha\beta} p^\alpha p^\beta$$
$$= -\vec{p} \cdot \vec{p}$$
$$= -p^2.$$

Therefore

$$\frac{\mathcal{E}^2}{c^2} - p^2 = m^2 c^2,$$

$$\frac{\mathcal{E}^2}{c^2} = p^2 + m^2 c^2,$$

and thus

$$\mathcal{E}^2 = c^2(p^2 + m^2 c^2),$$
$$\mathcal{E} = c\sqrt{p^2 + m^2 c^2}. \tag{6.16}$$

Definition 6.7. Function (6.16) is called the Hamiltonian of the particle and is denoted by H, i. e.,

$$H = c\sqrt{p^2 + m^2 c^2}.$$

The conservation law of the momentum of a particle is a consequence of the homogeneity of space, and the conservation law of the energy of a particle is a consequence of the homogeneity of time. Because of the homogeneity of space-time, the mechanical properties of a free particle remain unchanged after the translation of a particle from a point with coordinates x^n to a point with coordinates $x^n + \lambda^n$, where λ^n is an infinitesimally small constant four-vector. Thus the Lagrange function must be invariant with respect to this translation, and its variation must be zero. Therefore

$$\delta L = \frac{\partial L}{\partial x^{k\sigma}} \delta x^{k\sigma}$$
$$= \frac{\partial L}{\partial x^{k\sigma}} \lambda^{k\sigma}$$
$$= 0.$$

Hence, using the equation of motion

$$\frac{\partial L}{\partial x^{k\sigma}} - \frac{\Delta}{\Delta s}\left(\frac{\partial L}{\partial u^n}\right) = 0,$$

we obtain

$$\delta L = -\frac{\Delta}{\Delta s}\left(\frac{\partial L}{\partial u^n}\right)\lambda^{k\sigma}$$
$$= 0.$$

Since λ^n is a nonzero infinitesimally small four-vector, the last expression gives

$$-\frac{\Delta}{\Delta s}\left(\frac{\partial L}{\partial u^n}\right) = 0,$$

from which we find

$$0 = -\frac{\Delta}{\Delta s}\left(\frac{\partial L}{\partial u^n}\right)$$
$$= mc\frac{\Delta u_n}{\Delta s}$$
$$= \frac{\Delta p_n}{\Delta s}.$$

Then

$$\frac{\Delta p^n}{\Delta s} = \frac{1}{c\sqrt{1-\frac{v^2}{c^2}}}\frac{\Delta p^n}{\Delta t}$$
$$= 0,$$

and

$$p^n = \text{constant}.$$

Thus we obtain the three-dimensional momentum conservation law

$$\frac{\Delta\vec{p}}{\Delta t} = 0,$$

that is,

$$\vec{p} = \text{constant},$$

and the energy conservation law

$$\frac{\Delta\mathscr{E}}{\Delta t} = 0,$$

that is,

$$\mathscr{E} = \text{constant}.$$

6.6 Angular momentum tensor

Definition 6.8. In the relativistic mechanics the angular momentum tensor is defined as follows:

$$M_{nk} = x_n p_k - x_k p_n, \quad k, n \in \{0, 1, 2, 3\}.$$

Only the spatial components of the angular momentum tensor with $k, n \in \{1, 2, 3\}$ have a physical meaning and coincide with the usual definition of the angular momentum in nonrelativistic mechanics. In nonrelativistic mechanics, it is customary to form an axial three-dimensional angular momentum vector

$$M^\nu = \frac{1}{2} e^{\nu\alpha\beta} M_{\alpha\beta}$$
$$= e^{\nu\alpha\beta} x_\alpha p_\beta, \quad \alpha, \beta \in \{1, 2, 3\},$$

or in the classical vector motion,

$$\vec{M} = \det \begin{pmatrix} \vec{e}_1 & \vec{e}_2 & \vec{e}_3 \\ x_1 & x_2 & x_3 \\ p_1 & p_2 & p_3 \end{pmatrix}$$
$$= \vec{r} \cdot \vec{p},$$

where $\vec{e}_1$, $\vec{e}_2$, and $\vec{e}_3$ are the unit spatial coordinate vectors. The conservation law of the angular momentum tensor is a consequence of the isotropic nature of the four-dimensional space-time. The three-dimensional angular momentum conservation law is a consequence of the isotropic nature of the three-dimensional space. Because of the isotropic nature of the four-dimensional space-time, the mechanical properties of a free particle remain unchanged after rotations in the four-dimensional space-time. Now we consider a special case of a time scale rotation for some angle $\theta \in \mathscr{R}$ about the 3-axis in the three-dimensional space. Then the rotation between the coordinates z^m in the rotated inertial system of reference K' and coordinates x^k in the original inertial system of reference K is given by

$$z^n = \Omega_j^n x^j, \quad j, n \in \{0, 1, 2, 3\}.$$

This can be rewritten in the form

$$\begin{pmatrix} z^0 \\ z^1 \\ z^2 \\ z^3 \end{pmatrix} = \begin{pmatrix} 1 & 0 & 0 & 0 \\ 0 & \cos_\theta(t, t_0) & \sin_\theta(t, t_0) & 0 \\ 0 & -\sin_\theta(t, t_0) & \cos_\theta(t, t_0) & 0 \\ 0 & 0 & 0 & 1 \end{pmatrix} \begin{pmatrix} x^0 \\ x^1 \\ x^2 \\ x^3 \end{pmatrix}, \tag{6.17}$$

where $t, t_0 \in \mathbb{T}$. If we consider the rotation of some infinitesimal angle $\delta\theta \approx 0$, then we have

$$\cos_{\delta\theta}(t, t_0) \approx 1$$

and

$$\sin_{\delta\theta}(t, t_0) \approx \delta\theta.$$

Substituting this into (6.17), we find

$$
\begin{pmatrix} z^0 \\ z^1 \\ z^2 \\ z^3 \end{pmatrix}
=
\begin{pmatrix}
1 & 0 & 0 & 0 \\
0 & 1 & \delta\theta & 0 \\
0 & -\delta\theta & 1 & 0 \\
0 & 0 & 0 & 1
\end{pmatrix}
\begin{pmatrix} x^0 \\ x^1 \\ x^2 \\ x^3 \end{pmatrix}
$$

or

$$
\begin{aligned}
z^n &= x^n + \delta\Omega^n_j x^j \\
&= x^n + \delta\Omega^{nk} g_{kj} x^j \\
&= x^n + \delta\Omega^{nk} x_k,
\end{aligned}
$$

where $\delta\Omega^n_j$ is a mixed tensor defined by the antisymmetric matrix

$$
(\delta\Omega^n_j) =
\begin{pmatrix}
0 & 0 & 0 & 0 \\
0 & 0 & \delta\theta & 0 \\
0 & -\delta\theta & 0 & 0 \\
0 & 0 & 0 & 0
\end{pmatrix}.
$$

The covariant coordinates x_k of the four-dimensional space-time in matrix form are

$$
\begin{aligned}
(x_k) &= (g_{jk})(x^j) \\
&=
\begin{pmatrix}
1 & 0 & 0 & 0 \\
0 & -1 & 0 & 0 \\
0 & 0 & -1 & 0 \\
0 & 0 & 0 & -1
\end{pmatrix}
\begin{pmatrix} x^0 \\ x^1 \\ x^2 \\ x^3 \end{pmatrix} \\
&=
\begin{pmatrix} x^0 \\ -x^1 \\ -x^2 \\ -x^3 \end{pmatrix}.
\end{aligned}
$$

We can also compute the components of the antisymmetric contravariant tensor $\delta\Omega^{nk}$ in the matrix form:

$$(\delta\Omega^{nk}) = (g^{jk})(\delta\Omega^n_j)$$

$$= \begin{pmatrix} 1 & 0 & 0 & 0 \\ 0 & -1 & 0 & 0 \\ 0 & 0 & -1 & 0 \\ 0 & 0 & 0 & -1 \end{pmatrix} \begin{pmatrix} 0 & 0 & 0 & 0 \\ 0 & 0 & \delta\theta & 0 \\ 0 & -\delta\theta & 0 & 0 \\ 0 & 0 & 0 & 0 \end{pmatrix}$$

$$= \begin{pmatrix} 0 & 0 & 0 & 0 \\ 0 & 0 & -\delta\theta & 0 \\ 0 & \delta\theta & 0 & 0 \\ 0 & 0 & 0 & 0 \end{pmatrix}.$$

Thus we always have

$$\delta\Omega^{nk} = -\delta\Omega^{kn}, \quad k, n \in \{0, 1, 2, 3\},$$

and the most general infinitesimal rotation of the coordinates is given by the transformation relations

$$z^n = x^n + \delta x^n,$$
$$\delta x^n = \delta\Omega^{nk} x_k.$$

The most general infinitesimal rotation of a four-velocity vector is given by the transformation relations

$$u'^n = u^n + \delta u^n,$$
$$\delta u^n = \delta\Omega^{nk} u_k.$$

The Lagrange function $L(x^{n\sigma}, u^n)$ of a particle must be invariant with respect to this infinitesimal rotation, i. e.,

$$\delta L = 0.$$

Now we will compute the variation of the Lagrangian L with respect to the infinitesimal rotation:

$$\delta L = \frac{\partial L}{\partial x^{n\sigma}} \delta x^{n\sigma} + \frac{\partial L}{\partial u^n} \delta u^n$$
$$= 0,$$

or

$$\delta L = \frac{\partial L}{\partial x^{n\sigma}} \delta\Omega^{nk} x^\sigma_k + \frac{\partial L}{\partial u^n} \delta\Omega^{nk} u_k$$
$$= 0.$$

Substituting the equation of motion into the last equation, we obtain

$$\delta L = \delta\Omega^{nk}\left(\frac{\Delta}{\Delta s}\left(\frac{\partial L}{\partial u^n}\right)x_k^\sigma + \frac{\partial L}{\partial u^n}u_k\right)$$
$$= 0.$$

Using the definition of the energy momentum four-vector and the definition of the four-velocity, we find

$$\delta L = -\delta\Omega^{nk}\left(\frac{\Delta p_n}{\Delta s}x_k^\sigma + p_n\frac{\Delta x_k}{\Delta s}\right)$$
$$= -\frac{\Delta}{\Delta s}(\delta\Omega^{nk}p_n x_k)$$
$$= 0.$$

Using the antisymmetry of the tensor $\delta\Omega^{nk}$, the last equation becomes

$$\delta L = -\frac{1}{2}\delta\Omega^{nk}\frac{\Delta}{\Delta s}(p_n x_k - p_k x_n)$$
$$= \frac{1}{2}\delta\Omega^{nk}\frac{\Delta M_{nk}}{\Delta s}$$
$$= 0.$$

From here, as a direct consequence of the isotropic nature of space-time, we obtain the conservation law of the angular momentum tensor,

$$\frac{\Delta M_{nk}}{\Delta s} = 0,$$

whereupon

$$\frac{\Delta M_{nk}}{\Delta t} = 0,$$

that is,

$$M_{nk} = \text{constant.}$$

The spatial part of this conservation law for $n, k \in \{1, 2, 3\}$, i. e.,

$$M^\nu = e^{\nu\alpha\beta}x_\alpha p_\beta$$
$$= \text{constant,}$$

is a useful conservation law of the three-dimensional angular momentum vector.

Index

https://doi.org/10.1515/9783112228494-007